An Introduction to Optimization Techniques

T0331105

An Introduction to Optimization Techniques

Vikrant Sharma, Vinod Kumar Jain and
Atul Kumar

CRC Press
Taylor & Francis Group
Boca Raton London New York

CRC Press is an imprint of the
Taylor & Francis Group, an **informa** business

First edition published 2021
by CRC Press
6000 Broken Sound Parkway NW, Suite 300, Boca Raton, FL 33487-2742

and by CRC Press
2 Park Square, Milton Park, Abingdon, Oxon, OX14 4RN

CRC Press is an imprint of Taylor & Francis Group, LLC

Library of Congress Cataloging-in-Publication Data

ISBN: 978-0-367-49324-0 (hbk)
ISBN: 978-0-367-49326-4 (pbk)
ISBN: 978-1-003-04576-2 (ebk)

Typeset in Palatino
by SPi Global, India

Contents

Preface

Humankind has been looking for the best since the beginning of life. Several improvements have been made from Archimedes to the present day, distributed through multiple disciplines. Optimization is the process that provides the optimum solution to a problem. This book introduces the basic ideas and techniques of optimization. A substantial portion of the book is devoted to applications. This book is designed to give the reader a sense of the challenge of analysing a given situation and formulating a model for it while explaining the assumptions and inner structure of the methods discussed as fully as possible. The main aim of this book is to present the key contemporary optimization techniques. It has been written for undergraduates and graduates studying engineering, management, physical, or social sciences. The use of the mathematical argument is treated as a means instead of an end; hence, it is held at the minimum level required to justify the methods presented. All optimization concepts in the book are explained in a simple way that helps readers to easily understand the subject.

Mapping with Bloom's Taxonomy

The main feature of this book is the mapping of chapters with the revised Bloom Taxonomy Level (BT level). Bloom's Taxonomy was created to promote a higher level of cognitive thought, such as analysis, evaluation, and creation, rather than simply to remember facts. Bloom's Taxonomy has six learning stages from a lower level through higher-level thinking.

> *BT level 1. Remembering:* Recall information or data
>
> *BT level 2. Understanding:* Understand the meaning of written, oral, and graphic information
>
> *BT level 3. Applying:* Applying a concept in a new situation
>
> *BT level 4. Analysing:* Breaking down the concepts into the constituent part and drawing inferences
>
> *BT level 5. Evaluating:* Making judgements based on standards and a set of criteria
>
> *BT level 6. Creating:* Create a new thing or model based on prior learning

All chapters' learning outcomes, solved examples, questions, and practice problems are mapped with the aforementioned Bloom's Taxonomy levels.

Pedagogical Features

This book includes a variety of valuable pedagogical features to help readers understand and retain the content.

1. Each chapter begins with the learning outcomes (LO) section, which highlights the critical points of each chapter.
2. LOs are presented in a box within that specific section of each chapter.

3. Solved examples are presented in each section after the theoretical discussion to clarify the concept of that section.

4. At the end of each chapter, a summary reinforces key ideas and helps readers recall the concepts discussed.

5. Questions are provided at the end of each chapter to test the understanding of the concept discussed in the chapter.

6. At the end of each chapter, the practice questions test readers' knowledge of the application of the concept discussed in the chapter.

7. Competency checks using the six levels of Bloom's Taxonomy help readers assess their understanding of the material.

Excel Used to Solve Optimization Techniques

Another feature of this book is the use of a Microsoft Excel spreadsheet to solve optimization techniques. A separate chapter (Chapter 11) is dedicated to the application of spreadsheets to solve different optimization techniques. This chapter is intended to help readers use Excel and provide guidance on how to use it to carry out quantitative work. Examples for the spreadsheet application are included in the text, and corresponding easy-to-use Excel files are provided separately. Many software products are available to resolve optimization techniques, but we rely on Excel because it is used worldwide, and a spreadsheet approach allows us to solve many problems quite easily.

Acknowledgements

We would like to thank our colleagues at Mody University of Science and Technology, Lakshmangarh, and our family members for their encouragement and moral support in bringing this book to fruition. We do appreciate CRC Press for its diligent effort in rendering this book elegant in a prompt manner.

Vikrant Sharma
Vinod Kumar Jain
Atul Kumar

Authors

Vikrant Sharma is an assistant professor in the Department of Mechanical Engineering, Mody University of Science and Technology, Lakshmangarh, Rajasthan, India. He graduated from the University of Pune in 2004 with a degree in production engineering. He earned a master's degree in manufacturing system engineering from the Malaviya National Institute of Technology Jaipur in 2007. He has 13 years of research and teaching experience and ten publications in international journals of repute. He attended a two-month course on operation research held at National Programme on Technology Enhanced Learning and conducted by IIT Madras in April 2015. He also won first prize (100,000 INR) for his book CNC Machines and Automation from All India Council for Technical Education in the all-India-level competition under the scheme "TAKNIKI PATHYAPUSTAK PURASKAR YOJANA – 2014" (Technical Text Book Prize distribution Scheme of 2014). He is a lifetime member of the Institution of Engineers (India) and Institution of Engineering and Technology (IET, United Kingdom).

His research interests include industrial engineering, optimization techniques, production and operations management, plant layout, multicriteria decision making, and product design.

Dr. Vinod Kumar Jain earned an MBA, M Tech (CS), and PhD in computer science from Devi Ahilya University, Indore. He has been involved in teaching, training, research, and administration for the past 24 years, which includes his 14 years of experience as director/dean/dean of academics in various institutions/universities of repute in India. He has rich experiences in autonomy implementation, Accreditation Board for Engineering and Technology, United States; IET accreditation, United Kingdom; National Board of Accreditation, India; and National Assessment and Accreditation Council, India. He is the 2016 recipient of the Rashtriya Shiksha Gaurav Puruskar awarded by the Centre for Education Growth and Research, New Delhi; the 2018 academic leader of the year awarded by Integrated Chambers of Commerce and Industry, New Delhi; and an accredited management teacher accredited by All India Management Association (AIMA), New Delhi.

Dr. Jain has more than 220 publications to his credit, including books, monographs, research papers, and popular articles. Dr. Jain has organized many conferences/ seminars/ quality improvement programs and delivered more than 50 keynote/expert lectures. He is actively associated with professional and social organizations, such as the Institute of Electrical and Electronics Engineers, Computer Society of India, Indian Society for Training and Development, Indian Society for Technical Education, Institute of Electronics and Telecommunication Engineering, Institution of Engineers (IETE; India), AIMA, Association of Indian Management Schools, Global Conference on Flexible Systems Management, Quality Circle Forum of India, and Bharat Vikas Parishad. He is a PhD supervisor at Mody University Science and Technology, Lakshmangarh, and has guided 12 PhD scholars. He is a fellow of IET, United Kingdom; IETE, New Delhi; and Institution of Engineers, India.

Dr. Jain recently published a book titled *The Stances of E-Government: Policies, Processes and Technologies* published by CRC Press, Taylor and Francis Group, Chapman & Hall Publication, London, New York.

 **Dr. Atul Kumar**, Associate Professor and Head of Mechanical Engineering Department; holds a PhD from Suresh Gyan Vihar University, Jaipur, M. Tech. from (Kurukshetra University) and B.Tech from JNTU Hyderabad. He has been involved in teaching, training, research, and administration for the past 24 years in leading organizations out of which 21 years' experience in teaching (Mechanical Engineering). A seasoned professional with rich experience in various institutions/universities of repute in India. He has rich experiences in ABET (Accreditation Board for Engineering and Technology), United States; NAAC (National Assessment and Accreditation Council), India, IET, UK accreditations. There are more than 50 publications to his credit, including four books, Monographs, Research Papers in international journals and conferences, popular articles, etc. Dr. Atul has organized many conferences / seminars /QIPs and delivered many expert lectures. His research area is machining of aluminum metal matrix composite and Multi criteria decision making, particularly Taguchi optimization and ANOVA analysis. He has bagged best book award worth Rs. 1 Lac from AICTE, New Delhi for the book titled *CNC Machine & Automation* in the year 2016. He has also received the best paper award from many SCOPUS indexed international journals. He has also completed 3 months duration course in Operation Research offered by NPTEL and conducted by IIT Chennai in 2015. He is actively associated with professional and social organizations such as ISTE (Indian Society for Technical Education), IEI (Institution of Engineers (India), Solar Energy Society of India (SESI), International Association of Engineers (IAENG), Reviewer body of International Journal of Emerging Technology and Innovative Research. (JETIR), Professional Member of IFERP, Editorial board member/reviewer of IJTES (International Journal of Technology and Engineering Sciences), Member of Association for Machines and Mechanisms (AMM) Chennai. At present, he is guiding as a PhD supervisor at Mody University Lakshmangarh besides completing one DST sponsored and number of Industry sponsored research projects.

1

Introduction to Optimization Techniques

Learning Outcomes: After studying this chapter, the reader should be able to

LO 1. Understand optimization methods. (BT level 2)

LO 2. Explain the need for optimization techniques. (BT level 2)

LO 3. Understand the history of optimization techniques. (BT level 2)

LO 4. Understand different terms in optimization. (BT level 2)

LO 5. Understand and classify different optimization techniques. (BT level 2 and 5)

LO 6. List advanced optimization techniques. (BT level 1)

LO 7. Understand optimization by design of experiment. (BT level 2)

LO 8. Know the various applications of optimization techniques. (BT level 1 and 2)

LO 9. Know the limitations of optimization techniques. (BT level 1)

1.1 Introduction

Optimization is the technique of finding the optimal solution/design/output under given inputs and constraints. Optimization's primary objective is not to achieve an ideal solution but the optimal or most efficient solution and thereby acquire the best response under the given set of constraints. System optimization requires various constraints related to logistics, supply chain management, preventive and predictive maintenance, and economic and administrative operations related to the system under study. Thus the optimization aims at minimizing input (such as effort, cost, surface roughness) and maximizing output (such as yield, profit, output, efficiency) under a given set of constraints.

LO 1. Understand optimization methods

An optimization model can be given as follows:
Objective function: Maximize (or minimize) $f(x)$ subjected to the constraints

$$g_j(x) > 0,$$

$$j = 1, 2, \ldots h_j(x), \ldots 0,$$

$$j = m + J, m + 2 \ldots p,$$

where x = vectors decision variables, $g(x)$ = inequality constraints, and $h(x)$ = equality constraints.

> **Definition:** Ascertain a set of design variables that reduce (or boost) the response while at the same time fulfilling all the constraints.

> **Examples:** When you purchase a vehicle, you search for one with high speed (objective) and a cost that you can manage (constraint). Similar is searching for the best school (variable) for your ward (objective) while at the same time looking for one with affordable school fees (constraint), etc.

- How to make a matrix/formulation solution perfect?
- What should be the characteristics?
- What should be the conditions?

1.2 Need for Optimization

- Describe a variable.
- Quantify responses with respect to variables.
- Find the optimum objective within the field of study.
- Require fewer experiments to achieve optimum solutions at a lower cost.
- Trace and rectify the "problem" in a considerably easier way.

LO 2. Explain the need for optimization techniques

1.3 Historical Perspective

Optimization existence can be traced back to Newton, Lagrange, and Cauchy. Developing differential optimization approaches was feasible thanks to contributions from Newton and Leibnitz.

LO 3. Understand the history of optimization techniques

- Isaac Newton (1642–1727): He devised optimization differential calculus techniques.
- Joseph-Louis Lagrange (1736–1813): He contributed significantly to solving optimization problems with given restrictions. Augustin-Louis Cauchy (1789–1857): He proposed the steepest descent approach for unconstrained optimization.
- George Bernard Dantzig (1914–2005): He developed linear and simplex programming techniques (1947).

- Albert William Tucker (1905–1995): He established essential and adequate criteria required for non-linear programming problems.
- Bernoulli Euler, Lagrange, and Weierstate implemented simple calculus variants. First, Lagrange researched the definition of restricted optimization proposed by Cauchy.
- Notwithstanding all of these earlier works, relatively little advancement was made until later decades, when computer processors allowed optimization procedures to be applied and, thanks to machine speed, stimulated study in this field.
- Major advances in computational techniques for unconstrained computation were made in Europe. These include the introduction of Dantzig's simplex approach in 1947, Bellman's Theory of Optimality in 1957, etc.
- In 1939 and 1951, two separate groups found the Karush-Kuhn-Tucker condition: a necessary condition for restricted optimization. Between the 1940s and the 1970s, the traditional optimization methods were quickly established.
- In 1960, Zoutendijk Rosen's work on non-linear programming was quite important. In 1961, Chames and Cooper created multi-objective programming methods, also called goal programming.

1.4 Optimization Terms/Parameters

1. **Objective function:** Objective function describes one or more quantities to minimize or maximize. Optimization challenges can include a single objective feature or multiple objective functions. Typically, separate targets

   LO 4. Understand different terms in optimization

   aren't compatible. Variables that maximize one target can be far from ideal for others. Multi-objective problems may be reformulated as single-objective problems either by formulating a weighted combination of different priorities or by treating those targets as constraints.

2. **Constraint:** A set of constraints requires the unknown to presume certain values while excluding others. There are conditions for rendering a proposal feasible. If the design variables, restrictions, objectives, and interactions are selected, then the problem of optimization can be defined.

3. **Decision variables/factors and their levels:** It is an assigned variable, such as concentration or temperature.

4. **Quantitative factor:** An assigned numerical value (e.g., percent of the concentration of a solution).

5. **Qualitative factor:** A numerical value that cannot be assigned to a parameter, such as colour, grade of a material, or attitude.

6. **Levels:** Factor levels' ranges are the values ranging from a minimum value to a maximum value assigned to a particular variable.

- Parameters (data)
- Constraints

7. **Effect and interactions:** The response change is caused by variation in levels. It gives the relationship between different factors and levels. It provides the overall effect of two or more variables (e.g., the mixed effect of depth of cut and revolutions per minute on the surface roughness of the machined workpiece).
8. **Performance objective:** The result of an experiment result. It is the assessment effect (e.g., surface roughness of a machined workpiece or disintegration time).

1.5 Types of Optimization

Problems with optimization can be classified in many ways, as described next:

LO 5. Understand and classify different optimization technique

1.5.1 Based on Constraints

(i) **Constrained:** Limitations on the process or a system because of functional restrictions
(ii) **Unconstrained:** No restrictions but almost non-existent in pharmaceuticals

1.5.2 Based on the Physical Configuration

Based on the physical configuration, design problems are known as optimal control and arbitrary control.

Optimal control problems: An optimal control problem consists of many steps in which each stage is derived from the preceding point. Two variables describe this problem: control and state variables. Control variables define the process and track how one stage transitions. State variables describe device the state at every level. The issue is to find a collection of control variables to minimize total objective function according to a number of control and state variables.

1.5.3 Based on the Nature of the Variables

(i) **Deterministic programming:** A deterministic system will always produce the same output for the same input. All design variables are deterministic in this type of problem.
(ii) **Stochastic (probabilistic) programming:** In this optimization, any or all variables are represented probabilistically.

1.5.4 Based on the Number of Objective Functions

In this context, objective functions may be categorized as single-objective and multi-objective programming problems.

(i) **Single-objective programming problems:** Problems with one objective.
(ii) **Multi-objective programming problems:** Problems with more than one objective.

1.5.5 Geometric Programming Problem

A geometric programming problem is a category of non-linear optimization problem defined by unique purpose and constrained functions. Decision factors will be non-zero and non-negative.

1.5.6 Integer Programming Problem

When two or more of a problem's design variables are limited to only integer values, the problem is classified as integer programming.

1.5.7 Real-Value Programming Problem

If all design variables can only take real value, then the optimization approach is known as a real programming problem.

1.5.8 Non-linear Programming (NLP)

NLP problems take the general shape of linear programming problems, except that the objective function and/or the constraint involve non-linear terms. Typically, seeking an exact solution to NLP problems is quite challenging. The different algorithms usually find an estimated answer within a permissible optimum error. In a few NLP problems, there is no accurate way to calculate the global maximum. Existing algorithms sometimes stop after hitting a local maximum. The following are various types of NP problems:

1. Unconstrained optimization
2. Linear-constrained design
3. Quadratic programming
4. Convex programming

1.6 Advanced Optimization Techniques

Most of the real-world optimization challenges include problems such as discrete, numerous overlapping targets, nonlinearity, and discontinuity. Newer and sophisticated optimization

LO 6. To list advanced optimization techniques.

approaches are emerging day by day to solve these real-life challenges, and some are mentioned as follows:

1. Simulated annealing
2. Evolutionary optimization algorithms
3. Genetic algorithms
4. Estimation of the distribution algorithm
5. Particle swarm optimization
6. Ant colony optimization
7. Virtual bee algorithms

1.7 Optimization by Design of Experiments (DOE) (Experimental Designs)

DOE is a computational technique for rapidly maximizing system output with known input variables. It begins with a testing plan for all known variables that are suspected of influencing the efficiency of the system.

LO 7 Understand optimization by design of experiment

1.7.1 Types of Experimental Design

1. **Completely randomized design:** This involves the experimental randomized runs of various factor-level combinations.
2. **Randomized block design:** This is one of the most commonly applied methods. Blocking helps to minimize experimental error by minimizing the input of established variance factors within experimental units. This is achieved by combining the experimental units into blocks to minimize variance within each block. Because even variance within a region is part of experimental error, blocking becomes more successful where the experimental area has a stable pattern of variability.

3. **Factorial design:** This involves measuring all variables (n) at specific levels(x), including associations of the total number of tests as Xn. See Figure 1.1.
   a) **Fractional factorial design:** It is a fraction (1/xp) of a complete or maximum factorial design, where 'p' is the degree of fractionation and the total number of experiments needed is Xn. See Figure 1.2.
   b) **Plackett-Burman design:** R. L. Plackett and J. P. Burman developed the Plackett-Burman design in 1946. The design requires the simultaneous assessment of multiple factors to distinguish the critical few from the trivial ones. This is an important design method for screening a large number of factors.
4. **Response surface design:** The innovative DOE techniques that are used to better understand and maximize responses are a response surface design. It is used mostly

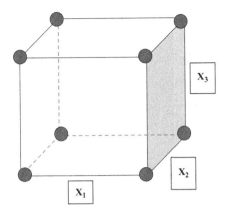

FIGURE 1.1
Factorial Designs.

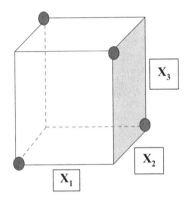

FIGURE 1.2
Fractional Factorial Designs.

to refine models after important factors are determined by means of screening or factorial designs.

a) **Central composite design:** Central composite design: Box and Wilson (JRSS, B 1951) produced a single common second-order response surface architecture. The concept has three parts (see Figure 1.3):

   (i) factorial points (⬭)

   (ii) centre points (⬤)

   (iii) axial points (◉)

b) **Box-Behnken design:** These are designs that are composed of 2p-design variations (see Figure 1.4).

   Properties:

   (i) *efficient* (few runs)

   (ii) nearly rotatable

   (iii) no hypercube corner points

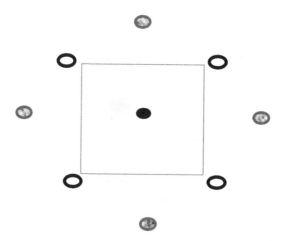

FIGURE 1.3
Response Surface Designs.

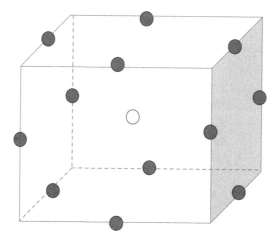

FIGURE 1.4
Box-Behnken Design.

1.8 Applications of Optimization Techniques

Different optimization methods are used for varied optimization problems.

- Ensures that a business receives more income and, in turn, more profit
- Boosts cycle efficiency
- Reduces variability in production quality
- Increases the production period

LO 8. Know the various applications of optimization techniques

- Reduce overall cost
- Evaluates and compares substitutes
- Provides analogue circuit design, very-large-scale integration designs, etc.
- Optimizes electrical power systems (turbines, motors) and networks
- Optimizes mobile networks
- Offers the shortest routes for multi-distributor sites
- Places company options
- Provides inventory management and regulations
- Uses quantitative and systematic analysis methods in economic and feasibility studies (research and development)
- **Distribution:** Throughout the following delivery environments, quantitative methods may be used in vehicle routing, warehouse location, etc.
- **Sales and marketing:** Check consumer analysis and demand forecasting efficacy
- **Financial management:** Provide planning, revenue modelling, economic and financial research, etc.

1.9 Limitations of Optimization Techniques

Despite numerous useful functions and the wide range of applicability mentioned earlier, optimization techniques suffer from a range of limitations since these strategies are focused on empirical results and analysis, which have their own limitations.

LO 9. Know the limitations of optimization techniques

1. Ideal utilization of the available resources is impractical, as accurate and successful judgment is not feasible because of numerous constraints.
2. The quantitative approach requires the statistical representation of specific data in the form of calculations, formulas, etc., which prove difficult and hard to comprehend and implement.
3. Quantitative approaches are costly since they typically need the expertise of a variety of professional specialists and costly machinery.
4. Most of the decisions are based on empirical facts which do not have any relevance to present cases.
5. Optimization strategies cannot be implemented universally since the market and development process systems differ from location to location.
6. Optimization strategies do not address qualitative factors (i.e., skill, traits, mentality, and decision-making capacity of a person).
7. Optimization strategies cannot override human judgment since they act as a tool for evaluating and understanding a question. Human thinking makes the ultimate call.

1.10 Optimization Method in Engineering and Management Applications

Several techniques have been developed in this field and several more have been added to the list. Nonetheless, the following are some common techniques and approaches that we consider and address in depth in this book.

1. Linear programming
2. Allocation model (i.e., transportation model and assignment model)
3. Network models
4. Scheduling and sequencing
5. Replacement model
6. Game theory
7. Queueing theory
8. Dynamic programming
9. Integer programming
10. Goal programming
11. Decision making

SUMMARY

- Optimization means maximizing or minimizing one or more functions with possible constraints.
- Classical optimization methods are helpful to seek the optimal solution.
- These are all empirical techniques that use differential calculus to find the optimal solution.
- For functional implementations, traditional approaches have restricted reach because others include objective processes that are not constant and/or differentiable. Nevertheless, the analysis of these traditional optimization strategies is a foundation for improving much of the computational techniques that have developed into modern strategies more suited for the realistic problems of today.

Questions

1. Define optimization techniques. Explain the nature of optimization techniques. *(BT level 2)*
2. Explain the importance of optimization techniques in the field of business and marketing management. *(BT level 2)*

3. Explain the uses of optimization techniques in engineering. *(BT level 2)*

4. Explain the drawbacks of optimization techniques. *(BT level 2)*

5. Optimization techniques are the eyes of an administration. Explain. *(BT level 2)*

6. Optimization techniques provide a tool for experimental research. Discuss. *(BT level 5)*

7. Write a note on the application of various optimization techniques in different fields of business decision making. *(BT level 2)*

8. Explain the significance of optimization techniques in the field of transport. *(BT level 2)*

9. Explain the nature of optimization techniques. *(BT level 2)*

10. Optimization methods are the foundation of decision making. Justify this statement. *(BT level 5)*

11. Explain the limitations of optimization techniques. *(BT level 2)*

12. Discuss the necessity and usefulness of optimization techniques. *(BT level 2)*

2

Linear Programming

Learning Outcomes: After studying this chapter, the reader should be able to

LO 1. *Understand the linear programming problem (LPP). (BT level 2)*

LO 2. *Name examples of the linear programming problem. (BT level 1)*

LO 3. *Mathematically draft an applied word problem and set up a linear program. (BT level 6)*

LO 4. *Understand general form and terms used in LPP. (BT level 2)*

LO 5. *Understand and apply a graphical method for solving an LPP with two variables. (BT level 2 and 3)*

LO 6. *Understand and apply a simple method for solving an LPP. (BT level 2 and 3)*

LO 7. *Understand and apply the Big M method for solving an LPP. (BT level 2 and 3)*

LO 8. *Understand and apply the two-phase method for solving an LPP. (BT level 2 and 3)*

LO 9. *Formulate the dual of an LPP. (BT level 6)*

LO 10. *Understand sensitivity analysis. (BT level 2)*

LO 11. *Write advantages of LP model. (BT level 1)*

LO 12. *Write limitations of LP model. (BT level 1)*

2.1 Introduction

Linear programming (LP): LP is an effective mathematical modelling methodology that seeks to assign "scarce or minimal" means to a variety of limited resources, such as labour, tangible assets, energy, and warehouse space, to many competitive tasks – for example, a commodity, operation, work, new equipment, or programs – on the basis of a specified optimum criterion. Programming is just another term for "planning" and refers to the method of drawing up a specific action plan from a variety of alternatives. A statistical methodology is used in resource management problems, such as production planning, to achieve an optimal solution. The modelling of physical phenomena by linear inequalities is known as LP. It has three main components:

1. *Decision variables:* These are the unknowns to be determined, and these variables are always greater than or equal to zero

2. *Objective function:* It is denoted by Z. It is a linear function of decision variables to be optimized. Each problem is about maximizing or minimizing a certain amount, usually profit or expense.

3. *Constraints:* These are linear equality/inequalities. They limit our capacity to achieve our objective.

When a problem is solved by LP, a number of linear inequalities will check on the system, and the inputs will be optimized (i.e., maximized [or minimized]).

LO 1. Understand the linear programming problem (LPP)

Definition of Linear Programming Problem (LPP)

▶ It is a quantitative technique that uses limited resources efficiently and effectively to achieve organizational objectives

2.2 Examples of LPP

1. *Development of production plan*

   LO 2. Name examples of the LPP

   - Meet the expected demands of a firm's production
   - Reduce the overall costs of production, as well as inventory.

2. *Determination of a distribution network that eliminates overall transportation costs from various warehouses to specific marketplaces.*

3. *A problem of production scheduling*

A supplier knows that a certain number of products of a given commodity will be supplied for the next n months each month. Either it can be generated on a daily basis, subject to a limit of one month each, or in overtime. The cost of making an item during overtime is higher than routine production. The produced item that has not been sold at the end of the month is associated with a storage cost. The problem is to establish a production schedule that reduces to a minimum the sum of production and storage costs.

4. *Optimization of flow problems*

Flows like data flow in an Internet or telephone network, traffic flow, or a fluid flow in various channels, etc., pose challenges to the assessment of the flow network efficiency with maximum flow capacity.

2.3 Formulation of LPP

Formulation of LPP is the logical description of a problem scenario in a mathematical form in terms of articulated decision variables, an objective function, and a set of constraints.

LO 3. To be able to mathematically draft an applied word problem and set up a linear program

The steps in the creation of a linear program are as follows:

1. Identify the objective and define decision variables.
2. Define and articulate the collection of constraints on the decision variables in the form of linear equations/inequations.
3. Express in the decision variables the objective function in the form of a linear equation.
4. Graphically or mathematically optimize the objective function.

2.4 General Form of LPP

Maximizeorminimize $Z = C_1 x_1 + C_2 x_2 + \ldots + C_n x_n$

subject to

LO 4. Understand general form and terms used in LPP

$$a_{11} x_1 + a_{12} x_2 + \ldots + a_{1n} x_n \left(\leq , = , \geq \right) b_1$$

Similarly

$$a_{21} x_1 + a_{22} x_2 + \ldots + a_{2n} x_n \left(\leq , = , \geq \right) b_2$$

$$a_{31} x_1 + a_{32} x_2 + \ldots + a_{3n} x_n \left(\leq , = , \geq \right) b_3$$

$$a_{41} x_1 + a_{42} x_2 + \ldots + a_{41n} x_n \left(\leq , = , \geq \right) b_4$$

$$\vdots$$

$$A_{n1} x_1 + a_{n2} x_2 + \ldots + a_{nn} x_n \left(\leq , = , \geq \right) b_n$$

$$x_i \geq 0, i = 1, 2, 3, \ldots, n$$

2.4.1 Important Concepts of LP

▶ *Objective function:* The problem is determined and transformed into an appropriate objective function. Typically, the aim in all of these situations is either to increase income or reduce expenses or time.

▶ *Constraints:* In actual life, organizations usually have limited resources, such as raw material, time, or labour, with which they have to do a job efficiently. These are called constraints.

▶ *Non-negativity constraint:* Negative values of actual objects are impractical, such as generating negative numbers of seats or tables, so the non-negativity factor must be used as a constraint.

▶ *Feasible solution:* A set of variables is termed as a feasible solution to LPP that optimizes an objective and satisfies its constraints and non-negativity constraints.

▶ *Unbounded solution:* If the objective function can be increased or reduced forever, then the solution is considered unbounded.

2.4.2 Variables Used in LP

▶ *Slack variable:* The non-negative variable which is added to the left-hand side (LHS) of the constraint (when the inequalities are of the type ($<$ =)) to transform into an equation.

▶ *Surplus variable:* The non-negative variable which is subtracted from the LHS of a constraint (when the inequalities are of the type ($>$ =) or (=)) to transform into an equation.

▶ *Artificial variable:* Artificial variables are imaginary and not physical.

2.5 Basic Assumptions of the LPP Model

▶ *Proportionality:* We presume proportionality occurs in the objective and constraints and constancy between output increase and utilization of input resources.

▶ *Additivity:* Objective function and constraint equations must be additive. We presume that additivity in all activities equals the sum of the specific tasks.

▶ *Divisibility:* Decision parameters may hold any fractional value and are, therefore, continuous in existence as opposed to the integer value.

▶ *Certainty:* It is presumed that all parameter values are determined with consistency on the basis that certainty prevails and that the number of tasks and constraints will not alter throughout the study process.

Examples based on LP formulation

EXAMPLE 1:

A manufacturer produces three types of products (P1, P2, P3) on three different machines (M1, M2, M3). The times required to manufacture a unit of each of the three products and the daily capacity of the three machines are given in Table 2.1.

TABLE 2.1

Data for Example 1

Machine	Time per unit (minutes)			Machine capacity (minute per day)
	P1	P2	P3	
M1	4	6	4	800
M2	8	–	6	900
M3	4	10	–	750

P1 takes six minutes to process, P2 takes six minutes to process, and P3 takes four minutes to process.

It is necessary to determine the daily number of units for each product to be produced. Profit per unit is Rs. 5, Rs. 4, and Rs. 7 for P1, P2, and P3, respectively. Formulate a mathematical model that maximizes daily profit. *(BT level 6)*

Solution:

Decision variable:

Let x_1 = number of unites of P1 produced

x_2 = number of unites of P2 produced

x_3 = number of unites of P3 produced

Objective function : Maximize profit $Z = 4x_1 + 3x_2 + 6x_3$

Constraints:

The total capacity in machine M1, $4x_1 + 6x_2 + 4x_3 \leq 800$

Each unit of P1 takes four minutes to process in M2.
So, x_1 no P1 product takes $4x$ minutes to process in M2.

$$M_2 \rightarrow 8x_1 + 6x_3 \leq 900$$

$$M_3 \rightarrow 4x_1 + 10x_2 \leq 750$$

So the LPP model of this problem is

$$\text{Maximize } Z = 4x_1 + 3x_2 + 6x_3$$

subject to

$$4x_1 + 6x_2 + 4x_3 \leq 800 \text{ (Machine } M1 \text{ capacity)}$$

$$8x_1 + 6x_3 \leq 900 \qquad \text{(Machine } M2 \text{ capacity)}$$

$$4x_1 + 10x_2 \leq 750 \qquad \text{(Machine } M3 \text{ capacity)}$$

$$x_1, x_2, x_3 \geq 0$$

EXAMPLE 2:

A furniture manufacturer makes three models of chairs. Each chair is first made in the carpentry shop and is next sent to the finishing shop where it is polished and painted. The number of work hours of labour required in both shops is given in Table 2.2.

The profit from the sale of chair 1 is Rs. 120, chair 2 is Rs. 200, and chair 3 is Rs. 180. If raw materials and equipment are sufficiently supplied and all the manufactured chairs are sold, the manufacturer wishes to decide the optimum product mix (i.e., the quantities that will maximize profit from each type of chair). Make the mathematical model of this problem. *(BT level 6)*

TABLE 2.2

Data for Example 2

	Chair 1 (hrs)	Chair 2 (hrs)	Chair 3 (hrs)	Available (hrs) for one month
Carpentry shop	5	10	8	1,000
Finishing and painting shop	2	2	4	700

Solution:

Decision variables:

Let x_1 = number of unites of chair 1 produced
x_2 = number of unites of chair 2 produced
x_3 = number of unites of chair 3 produced

Objective function:

$$\text{Maximize } Z = 120x_1 + 200x_2 + 180x_3$$

Constraints:

$$5x_1 + 10x_2 + 8x_3 \leq 1,000 \, (\text{carpentry shop capacity})$$

$$2x_1 + 2x_2 + 4x_3 \leq 700 \, (\text{painting and finishing shop capacity})$$

$$x_1, x_2, x_3 \geq 0$$

EXAMPLE 3:

A manufacturer wishes to determine how advertisements can be placed in three selected monthly magazines X, Y, and Z. The goal is to advertise in a manner that maximizes overall exposure to key buyers. There are known percentages of readers of each magazine. The number of ads placed in any particular magazine is multiplied by the number of principal buyers. Table 2.3 shows the number of readers, principal buyers, and advertising costs. The estimated amount for advertising is at most Rs. 3 lakhs. Marketing agreed that X magazine would have no more than ten ads, Y magazine should have at least five ads, and Z magazine should have at least eight ads. Formulate this problem as an LP model. *(BT level 6)*

Solution:

TABLE 2.3

Data for Example 3

	Magazine X	Magazine Y	Magazine Z
Readers	0.9 lakh	0.7 lakh	0.5 lakh
Principal buyers	25%	20%	15%
Cost per ads in Rs.	40,000	30,000	25,000

Decision variables:

Let x_1 = number of ads in magazine X
x_2 = number of ads in magazine Y
x_3 = number of ads in magazine Z

Objective function:

$$\text{Maximize (total exposure)} \, Z = (25\% \text{ of } 90,000)x_1 + (20\% \text{ of } 70,000)x_2$$
$$+ (15\% \text{ of } 50,000)x_3$$

$$Z = 22,500x_1 + 14,000x_2 + 7,500x_3$$

Constraints:

$$40,000x_1 + 30,000x_2 + 25,000x_3 \leq 300,000 \, (\text{Maximum budget})$$

$$x_1 \leq 10 \ \left(\text{Advertisement in } X \text{ magazine}\right)$$

$$x_2 \geq 5 \quad \left(\text{Advertisement in } Y \text{ magazine}\right)$$

$$x_3 \geq 8 \qquad \left(\text{Advertisement in } Z \text{ magazine}\right)$$

$$x_1, x_2, x_3 \geq 0$$

EXAMPLE 4:

A person needs to select which constituents of a diet meet his regular protein, fat, and carbohydrate requirements at a minimal cost. The choice consists of three different food styles. Table 2.4 shows the yield per unit of these foods.

TABLE 2.4

Data for Example 4

	Food 1	Food 2	Food 3	Minimum requirement
Proteins	2	3	7	750
Fats	1	1	5	150
Carbohydrates	5	3	6	650
Cost per unit (Rs.)	40	35	70	

Formulate the LP model for the problem. *(BT level 6)*
Solution:
Decision variables:

        Let x_1 = units of Food 1
             x_2 = units of Food 2
             x_3 = units of Food 3

Objective function:

$$\text{Minimize } Z\left(\text{total cost}\right) = 40x_1 + 35x_2 + 70x_3$$

Constraints:

$$2x_1 + 3x_2 + 7x_3 \geq 750 \ \left(\text{Proteins / day}\right)$$

$$x_1 + x_2 + 5x_3 \geq 150 \ \left(\text{Fats / day}\right)$$

$$5x_1 + 3x_2 + 6x_3 \geq 650 \ \left(\text{Carbohydrates / day}\right)$$

$$x_1, x_2, x_3 \geq 0$$

Example 5: Resident doctors work five consecutive days in a multi-disciplinary hospital and have two consecutive days off. Their five-day work can start on any weekday. The following minimum number of doctors are required in the hospital (Table 2.5).

TABLE 2.5

Data for Example 5

Day	Sunday	Monday	Tuesday	Wednesday	Thursday	Friday	Saturday
Number of doctors required	20	30	30	25	30	25	25

On the same day, not more than 30 doctors can begin their five working days. Formulate the LP model to minimize the number of doctors employed by the hospital. (*BT level 6*)

Solution:

Decision variables:

Let x_1 = number of doctors beginning their work on Sunday

x_2 = number of doctors beginning their work on Monday

x_3 = number of doctors beginning their work on Tuesday

x_4 = number of doctors beginning their work on Wednesday

x_5 = number of doctors beginning their work on Thursday

x_6 = number of doctors beginning their work on Friday

x_7 = number of doctors beginning their work on Saturday

Objective function:

$$\text{Minimize } Z \text{ (total number of doctors)} = x_1 + x_2 + x_3 + x_4 + x_5 + x_6 + x_7$$

Constraints:

$$x_1 + x_4 + x_5 + x_6 + x_7 \geq 20 \quad \text{(Sunday's requirement is fulfilled by the number}$$
$$\text{of doctors beginning their work on Sunday and}$$
$$\text{previously for consecutive days – i.e., Saturday, Friday,}$$
$$\text{Thursday, Wednesday and so on)}$$

$$x_1 + x_2 + x_5 + x_6 + x_7 \geq 30$$

$$x_1 + x_2 + x_3 + x_6 + x_7 \geq 30$$

$$x_1 + x_2 + x_3 + x_4 + x_7 \geq 25$$

$$x_1 + x_2 + x_3 + x_4 + x_5 \geq 30$$

$$x_2 + x_3 + x_4 + x_5 + x_6 \geq 25$$

$$x_3 + x_4 + x_5 + x_6 + x_7 \geq 25$$

$$x_j \leq 30$$

$$x_j \geq 0 \left(j = 1,2,3\ldots\ldots,7 \right)$$

2.6 Solutions to LPP

The solution to an LPP (also known as a viable solution) is basically the degree to which decision variables satisfy all the constraints. There are two methods to seeking an optimal solution:

1. *Graphical technique/process*
2. *Simplex technique/process*

2.6.1 Graphical Method

The graphical approach is appropriate for solving the LPP involving two decision variables x_1, and x_2. The graphical method includes two main steps to solve an LP. The assessment of the solution region determines the feasible solution. Remember that the collection of values of variables $x_1, x_2, x_3 \ldots x_n$, which fulfils both constraints and non-negative parameters, is considered the feasible LP solution.

LO 5. Understand and apply the graphical method for solving an LPP with two variables

2.6.2 Procedure of Graphical Method

1. Transform every inequality into an equation.
2. Plot every equation graphically.
3. Highlight the "feasible region" and common feasible region.
4. Assess the coordinates of the corner points.
5. Put the coordinates of all corner points into the objective function and find out one which provides the optimal value.

2.6.3 Types of Graphical Methods to Solve LPP

The graphical approach is confined to LP models comprising just two decision variables. Graphical methods provide visualization of the solution rendered. Graphical methods may be divided into two categories:

1. *Corner-point evaluation method*
2. *Isoprofit (cost) line method*

1. *Method of Corner-Point Solution*

1. Plot all restrictions for a feasible area and locate the corner positions of the feasible region.
2. Calculate the gain at every corner point.
3. Select the best value corner point for the previously identified objective function. It is the optimal option.

2. *Method of Isoprofit Line*

1. Calculate all constraints to assess the feasible region.
2. To locate the slope, pick a particular line of benefit.
3. Transfer the objective function line to maximize revenue (or reduce expenses) while maintaining slope. The last point indicates the optimal solution.
4. At this last point, locate the value of decision variables and calculate profit (or cost).

2.7 Types of Solutions to LPP

▶ *Optimal solution:* The solution that maximizes or minimizes (optimizes) the target function.
▶ *Unbounded solution:* The solution occurs when the value of the decision variables may be increased indefinitely without violating any of the constraints.
▶ *Inconsistent solution:* This solution implies that the problem is not solved, and this is so when there is no common feasible region.

Examples based on the LP graphical solution

EXAMPLE 1:

Apply the graphical method to solve the following LPP. *(BT level 3)*

$$Max\ Z = 12x_1 + 16x_2$$

subject to

$$10x_1 + 20x_2 \leq 120$$

$$8x_1 + 8x_2 \leq 80$$

$$x_1, x_2 \geq 0$$

Solution: First convert the inequalities into equalities and compute the coordinated x_1 and x_2, as noted next.

$$10x_1 + 20x_2 = 120$$

We get $x_2 = 6$ when $x_1 = 0$ and $x_1 = 12$ when $x_2 = 0$.

$$8x_1 + 8x_2 = 80$$

We get $x_2 = 10$ when $x_1 = 0$ and $x_1 = 10$ when $x_2 = 0$.
Now plot the constraints as shown in Figure 2.1 and the obtained feasible region.

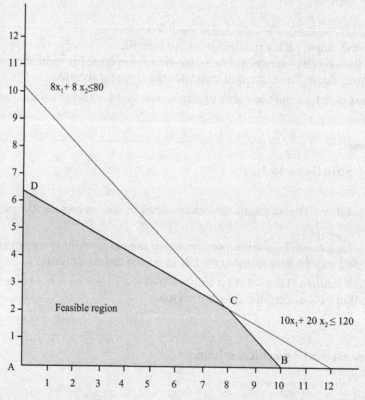

FIGURE 2.1
Feasible Region for Example 1.

The feasible region is ABCD. The objective function value at each of the corner points is determined by substituting its coordinates in the objective function, as indicated next.

The objective function at corner point A = $12 \times 0 + 16 \times 0 = 0$.
The objective function at corner point B = $12 \times 10 + 16 \times 0 = 120$.
The objective function at corner point C = $12 \times 8 + 16 \times 2 = 128$.
The objective function at corner point D = $12 \times 0 + 16 \times 6 = 96$.

Since the problem is the maximization type, the optimal solution is

$$x_1 = 8 \, and \, x_2 = 2 \, with \, objective \, function \, Z = 128.$$

EXAMPLE 2:

Apply the graphical method to solve the following LPP. *(BT level 3)*

$$Min \, Z = 4x_1 + 6x_2$$

subject to

$$2x_1 + 2x_2 \geq 12$$

$$12x_1 + 2x_2 \geq 24$$

$$x_1, x_2 \geq 0$$

Solution: First convert the inequalities into equalities and compute the coordinated x_1 and x_2.

$$2x_1 + 2x_2 = 12$$

We get $x_2 = 6$ when $x_1 = 0$ and $x_1 = 6$ when $x_2 = 0$.

$$12x_1 + 2x_2 = 24$$

We get $x_2 = 12$ when $x_1 = 0$ and $x_1 = 2$ when $x_2 = 0$.
Now plot the constraints as shown in Figure 2.2 and obtain the feasible region.

The feasible region is ABC. The objective function value at each of the corner points is determine by substituting its coordinates into the objective function, as indicated next.

The objective function at corner point A = $4 \times 0 + 6 \times 12 = 72$.
The objective function at corner point B = $4 \times 1.2 + 6 \times 4.8 = 33.6$.
The objective function at corner point C = $4 \times 6 + 6 \times 0 = 24$.

Since the problem is of the minimization type, the optimal solution is

$$x_1 = 6 \, and \, x_2 = 0 \, with \, objective \, function \, Z = 24.$$

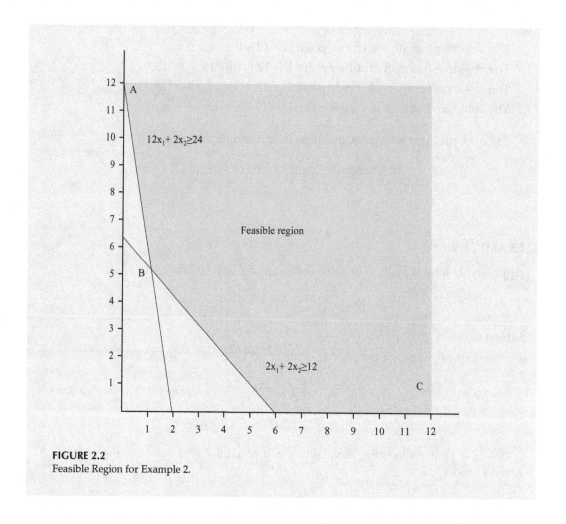

FIGURE 2.2
Feasible Region for Example 2.

2.8 Simplex Method

Using the simplex technique, we can solve LPP when there are more than two decision variables. The simplex method is not used to search for possible solutions. This just describes a limited and special set of feasible alternatives, the set of vertex points of the convex feasible space describing the ideal or optimized solution.

Two cases generally arise.

LO 6. Understand and apply the simplex method for solving an LPP

► Maximization
► Minimization

2.8.1 Some Important Terms of the Simplex Technique

Basic feasible solution (BFS): This solution often satisfies the constraints on non-negotiability. There are two types of BFS:

1. *Degenerate BFS:* This type occurs if single or multiple basic variables attain zero value.
2. *Non-degenerate BFS:* All basic variables are non-zero.

2.8.2 Simplex Method Steps

1. Formulate the LPP.
2. Introduce slack/auxiliary variables.
3. Find the initial BFS by setting a zero value to the decision variables.
4. Create a table and add all coefficients.
5. Using the most negative coefficient, Zj determines the key column (the related vector is an input variable from the next iterating table).
6. Find the ratio by dividing the values of the X_B column by the positive values of the key column. The corresponding row of the minimum ratio is called pivot row and corresponding variables leave the solution.
7. The element at the intersection of row and column is termed the pivot element.
8. Construct the next table by removing the left variable and inserting the input variable.
9. Translate the pivot element into unity by performing the row.
10. Repeat the procedure until all the values are $cj - zj \leq 0$.

Examples based on the LP simplex method

EXAMPLE 1:

Apply the simplex method to solve the following LPP. (*BT level 3*)
$$Max\ Z = 12x_1 + 16x_2$$

subject to
$$10x_1 + 20x_2 \leq 120$$

$$8x_1 + 8x_2 \leq 80$$

$$x_1, x_2 \geq 0$$

Solution: First convert the problem to the canonical form by adding slack variables S_1 and S_2.

After introducing the slack variables, the objective function and constraints are modified as follows.
$$Max\ Z = 12x_1 + 16x_2 + 0S_1 + 0S_2$$

subject to

$$10x_1 + 20x_2 + S_1 = 120$$

$$8x_1 + 8x_2 + S_2 = 80$$

$$x_1, x_2, S_1, S_2 \geq 0$$

The initial simplex table of the previous problem is shown in Table 2.6.

TABLE 2.6

Initial Simplex Table for Example 1

Iteration-1	C_j	12	16	0	0		
Basic variable	C_B	x_1	x_2	S_1	S_2	X_B	Min Ratio
S_1	0	10	(20)	1	0	120	$120/20 = 6 \rightarrow$
S_2	0	8	8	0	1	80	$80/8 = 10$
$Z = 0$	Z_j	0	0	0	0		
	$C_j - Z_j$	12	16↑	0	0		

The first row in Table 2.6 includes the coefficient of variables from the objective function. The basic variable has a unit coefficient in one of the constraints and zero in the remaining constraints. Here the simplex method starts with the basic variables S_1 and S_2. The value of Z_j is determined as

$$Zj = \sum_{i=1}^{2} C_B a_{ij},$$

where a_{ij} is the coefficient for the ith row and jth column.

When the values of $C_j - Z_j$ are less than or equal to zero, then optimality is reached. In Table 2.6 all values of $C_j - Z_j$ are either greater than or equal to zero, and the solution can be improved further. Select the variable with maximum $C_j - Z_j$ value as the entering variable in the next iteration. Here it is x_2 (with a value of 16), and the corresponding column is the key column.

Now, for each row, determine the ratio between the X_B column and the key column. The minimum ratio value is the leaving variable (break tie randomly). The corresponding row is known as the key row. Here the leaving variable is S_1. The key element is the intersection of the key column and key row. Here the key element is 20.

The second iteration of the problem is shown in Table 2.7.

The following row operations are performed to determine the values of the third and fourth rows of Table 2.7.

TABLE 2.7

Iteration 2 Simplex Table for Example 1

Iteration-2	C_j	12	16	0	0		
Basic variable	C_B	x_1	x_2	S_1	S_2	X_B	*Min Ratio*
x_2	16	1/2	1	1/20	0	6	6/ (1/2) =12
S_2	0	(4)	0	−2/5	1	32	32/4 = 8→
$Z = 96$	Z_j	8	16	4/5	0		
	$C_j − Z_j$	4↑	0	−4/5	0		

TABLE 2.8

Iteration 3 Simplex Table for Example 1

Iteration-3	C_j	12	16	0	0	
Basic variable	C_B	x_1	x_2	S_1	S_2	X_B
x_2	16	0	1	1/10	−1/8	2
x_1	12	1	0	−1/10	1/4	8
$Z = 128$	Z_j	12	16	2/5	1	
	$C_j − Z_j$	0	0	−2/5	−1	

Third row new (in Table 2.7) = Third row old (from Table 2.6) ÷ 20

Fourth row new (in Table 2.7) = Fourth row old (from Table 2.6) – 8 × Third row new (in Table 2.7)

In Table 2.7, the maximum positive $C_j − Z_j$ is 4, so the entering variable is x_1, and the minimum ratio is 8, so the leaving variable is S_2. The key element is 4.

The third iteration of the problem is shown in Table 2.8

The following row operations are performed to determine the values of the third and fourth rows of Table 2.8.

Fourth row new (in Table 2.8) = Fourth row old (from Table 2.7) ÷ 4

Third row new (in Table 2.8) = Third row old (from Table 2.7) – ½ × Fourth row new (in Table 2.8)

In Table 2.8, all the values of $C_j - Z_j$ are either negative or zero. Hence optimality is reached.

The optimal solution is

$$x_1 = 8 \text{ and } x_2 = 2 \text{ with objective function } Z = 128.$$

Example 2: Apply the simplex method to solve the following LPP. *(BT level 3)*

$$\max Z = 4x_1 + 3x_2 + 6x_3$$

subject to

$$2x_1 + 4x_2 \leq 450$$

$$4x_1 + 3x_3 \leq 500$$

$$2x_1 + 3x_2 + 2x_3 \leq 450$$

$$x_1, x_2, x_3 \geq 0$$

Solution: First convert the problem to the canonical form by adding slack variables S_1, S_2, and S_3.

After introducing slack variables, the objective function and constraints are modified as follows.

$$\max Z = 4x_1 + 3x_2 + 6x_3 + 0S_1 + 0S_2 + 0S_3$$

subject to

$$2x_1 + 4x_2 + S_1 = 450$$

$$4x_1 + 3x_3 + S_2 = 500$$

$$2x_1 + 3x_2 + 2x_3 + S_3 = 450$$

$$x_1, x_2, x_3, S_1, S_2, S_3 \geq 0$$

The initial simplex table of the previous problem is shown in Table 2.9.

TABLE 2.9

Initial Simplex Table for Example 2

Iteration-1	Cj	4	3	6	0	0	0		
Basic variable	C_B	x_1	x_2	x_3	S_1	S_2	S_3	X_B	Min Ratio
S_1	0	2	4	0	1	0	0	450	–
S_2	0	4	0	(3)	0	1	0	500	$500/3 \rightarrow$
S_3	0	2	3	2	0	0	1	450	$450/2 = 225$
$Z = 0$	Zj	0	0	0	0	0	0		
	$Cj - Zj$	4	3	6↑	0	0	0		

Select the variable with maximum $C_j - Z_j$ value as the entering variable in the next iteration. Here it is x_3 (with a value of 6) and the corresponding column is the key column. Now, for each row, determine the ratio between the X_B column and the key column. The minimum ratio value is the leaving variable. The corresponding row is known as the key row. Here the leaving variable is S_2. The key element is the intersection of the key column and key row. Here the key element is 3.

The second iteration of the problem is shown in Table 2.10.

The following row operations are performed to determine the values of the third, fourth, and fifth rows of Table 2.10.

Fourth row new (in Table 2.10) = Fourth row old (in Table 2.9) ÷ 3

Third row new (in Table 2.10) = Third row old (in Table 2.9)

Fifth row new (in Table 2.10) = Fifth row old (in Table 2.9) – 2 Fourth row new (in Table 2.10)

Select the variable with the maximum $C_j - Z_j$ value as the entering variable in the next iteration. Here it is x_3 (with a value of 3), and the corresponding column is the key column. Now, for each row, determine the ratio between X_B column and the key column. The minimum ratio value is the leaving variable. The corresponding row is known as the key row. Here the leaving variable is S_3. The key element is the intersection of the key column and key row. Here the key element is 3.

The third iteration of the problem is shown in Table 2.11.

TABLE 2.10

Iteration 2 Simplex Table for Example 2

Iteration-2	Cj	4	3	6	0	0	0		
Basic variable	C_B	x_1	x_2	x_3	S_1	S_2	S_3	X_B	*Min Ratio*
S_1	0	2	4	0	1	0	0	450	4,504
x_3	6	4/3	0	1	0	1/3	0	500/3	–
S_3	0	–2/3	(3)	0	0	–2/3	1	350/3	$350/(3/3) = 350/9 \rightarrow$
$Z = 1,000$	Zj	8	0	6	0	2	0		
	$Cj - Zj$	–4	3↑	0	0	–2	0		

TABLE 2.11

Iteration 3 Simplex Table for Example 2

Iteration-3	Cj	4	3	6	0	0	0	
Basic variable	C_B	x_1	x_2	x_3	S_1	S_2	S_3	X_B
S_1	0	26/9	0	0	1	8/9	–4/3	2650/9
x_3	6	4/3	0	1	0	1/3	0	500/3
x_2	3	–2/9	1	0	0	–2/9	1/3	350/9
$Z = 3350/3$	Zj	22/3	3	6	0	4/3	1	
	$Cj - Zj$	–10/3	0	0	0	–4/3	1	

The following row operations are performed to determine the values of the third, fourth, and fifth rows of Table 2.11.

Fifth row new (in Table 2.11) = Fifth row new (in Table 2.10) $\div$ 3

Third row new (in Table 2.11) = Third row old (in Table 2.10) – 4 Fifth row new (in Table 2.11

Fourth row new (in Table 2.11) = Fourth row old (in Table 2.10)

In Table 2.11, all values of $C_j - Z_j$ are either negative or zero. Hence, optimality is reached. The optimal solution is

$$x_1 = 0, x_2 = 350 / 9 \text{ and } x_3 = 500 / 3 \text{ with objective function } Z = 3350 / 3$$

2.9 Big M Method/Method of Penalty

The Big M method of solving an LPP involves artificial variables. In the last section, LPPs with less than or equal ($\leq$) type constraints were addressed. This property, along with the fact that all the constraints are on the right side, offers us a ready start solution that includes all the slack variables.

LO 7: Understand and apply the Big M method for solving an LPP

Nevertheless, several LLPs are available that cannot be solved by pure slack variables. To achieve this, we add artificial variables that play the function of slack variables. Nonetheless, because artificial variables in the original model have no physical sense, arrangements must be made to render them null and void at an optimum iteration stage. In other terms, we use them to start and give up the solution until their work is over. A rational way to accomplish this purpose is to penalize the objective function of the artificial variable. In this method a very high penalty (big M) is assigned to the artificial variables in the objective function.

Example based on the LP big M method

EXAMPLE 1:

Apply the Big M method to solve the following LLP. (BT level 3)

$$\min Z = 4x_1 + 3x_2$$

subject to

$$2x_1 + 2x_2 \geq 12$$

$$14x_1 + 2x_2 \geq 28$$

$$x_1, x_2 \geq 0$$

Solution: First convert the problem to canonical form by adding surplus variables S_1, S_2, and artificial variables A_1 and A_2.

After introducing surplus variables and artificial variables, the objective function and constraints are modified as follows.

$$\min Z = 4x_1 + 3x_2 + 0S_1 + 0S_2 + MA_1 + MA_2$$

subject to

$$2x_1 + 2x_2 - S_1 + A_1 = 12$$

$$14x_1 + 2x_2 - S_2 + A_2 = 28$$

$$x_1, x_2, S_1, S_2, A_1, A_2 \geq 0$$

The initial simplex table of the previous problem is shown in Table 2.12.

TABLE 2.12

Initial Table for Example 1

Iteration-1	Cj	4	3	0	0	M	M		
Basic variable	C_B	x_1	x_2	S_1	S_2	A_1	A_2	X_B	Min Ratio
A_1	M	2	2	−1	0	1	0	12	12/2 = 6
A_2	M	(14)	2	0	−1	0	1	28	28/14 = 2→
Z = 40M	Zj	16M	4M	−M	−M	M	M		
	Cj − Zj	4−16M↑	3−4M	M	M	0	0		

In Table 2.12, all the values of $C_j - Z_j$ are either greater than or equal to zero, and the solution can be improved further. Select the variable with the maximum $C_j - Z_j$ value as the entering variable in the next iteration. Here it is x_1, and the corresponding column is the key column.

Now, for each row, determine the ratio between the X_B column and the key column. The minimum ratio value is the leaving variable. The corresponding row is known as the key row. Here the leaving variable is A_2. The key element is the intersection of the key column and key row. Here the key element is 14.

The second iteration of the problem is shown in Table 2.13.

The following row operations are performed to determine the values of the third and fourth rows of Table 2.13.

Fourth row new (in Table 2.13) = Fourth row old (from Table 2.12) ÷ 14

Third row new (in Table 2.13) = Third row old (from Table 2.12) − 2 × Fourth row new
(in Table 2.12)

TABLE 2.13

Iteration 2 for Example 1

Iteration-2	C_j	4	3	0	0	M	M		
Basic variable	C_B	x_1	x_2	S_1	S_2	A_1	A_2	X_B	Min Ratio
A_1	M	0	(12/7)	-1	1/7	1	$-1/7$	8	$8/(12/7) = 14/3 \to$
x_1	4	1	1/7	0	$-1/14$	0	1/14	2	$2/(1/7) = 14$
$Z = 8M + 8$	Z_j	4	$12M/7 + 4/7$	$-M$	$M/7 - 2/7$	M	$-M/7 + 2/7$		
	$C_j - Z_j$	0	$-12M/7 + 17/7 \uparrow$	M	$-M/7 + 2/7$	0	$8M/7 - 2/7$		

TABLE 2.14

Third Iteration for Example 1

Iteration-3	C_j	4	3	0	0	M	M		
Basic variable	C_B	x_1	x_2	S_1	S_2	A_1	A_2	X_B	Min Ratio
x_2	3	0	1	$-7/12$	1/12	7/12	$-1/12$	14/3	
x_1	4	1	0	1/12	$-1/12$	$-1/12$	1/12	4/3	
$Z = 58/3$	Z_j	4	3	$-17/12$	$-1/12$	17/12	1/12		
	$C_j - Z_j$	0	0	17/12	1/12	$M - 17/12$	$M - 1/12$		

In Table 2.13, the optimum solution is not reached. The entering variable is x_2, and the minimum ratio is for A_2. The key element is (12/7).

The third iteration of the problem is shown in Table 2.14

The following row operations are performed to determine the values of the third and fourth rows of Table 2.14.

Third row new (in Table 2.14) = Third row old (in Table 2.13) × 7/12

Fourth row new (in Table 2.14) = Fourth row old (in Table 2.13) – 1/7 × Third row new (in Table 2.14)

In Table 2.14, all the values of $C_j - Z_j$ are either positive or zero. Hence optimality is reached.

The optimal solution is

$$x_1 = 4/3 \text{ and } x_2 = 14/3 \text{ with objective function } Z = 58/3.$$

2.10 Two-Phase Method

The simplex two-phase method is another way to solve an LPP that includes certain artificial variables. In two phases, the solution can be obtained. In the first phase of this approach, the number of artificial variables is reduced to achieve a BFS to the original LPP. The second phase optimizes the original objective function from the BFS obtained at the end of phase 1.

LO 8. Understand and apply the two-phase method for solving an LPP

Example based on the LP two-phase method

EXAMPLE 1:

Apply the two-phase method to solve the following LPP. *(BT level 3)*

$$\max Z = 4x_1 + 7x_2$$

subject to

$$2x_1 + x_2 \geq 2$$

$$x_1 + 3x_2 \geq 3$$

$$x_1 + x_2 \leq 4$$

$$x_1, x_2, x_3 \geq 0$$

Solution:

Phase 1:

The problem is converted to canonical form by adding slack, surplus, and artificial variables.

1. As constraint 1 is of the type "$\geq$," we should subtract the surplus variable S_1 and add the artificial variable A_1.

2. As constraint 2 is of the type "$\geq$," we should subtract the surplus variable S_2 and add the artificial variable A_2.

3. As constraint 3 is of the type "$\leq$," we should add the slack variable S_3.

After introducing the slack variable, surplus variables, and artificial variables, the objective function and constraints are modified as follows.

$$\max Z = -A_1 - A_2$$

subject to

$$2x_1 + x_2 - S_1 + A_1 = 2$$

$$x_1 + 3x_2 - S_2 + A_2 = 3$$

$$x_1 + x_2 + S_3 = 4$$

$$x_1, x_2, S_1, S_2, S_3, A_1, A_2 \geq 0$$

The initial simplex table of the previous problem for phase 1 is shown in Table 2.15.

TABLE 2.15

Initial Simplex Table (Phase 1) for Example 1

Iteration-1	Cj	0	0	0	0	0	−1	−1		
Basic variable	C_B	x_1	x_2	S_1	S_2	S_3	A_1	A_2	X_B	Min Ratio
A_1	−1	2	1	−1	0	0	1	0	2	2/1 = 2
A_2	−1	1	(3)	0	−1	0	0	1	3	3/3 = 1→
S_3	0	1	1	0	0	1	0	0	4	4/1 = 4
$Z = -5$	Zj	−3	−4	1	1	0	−1	−1		
	$Cj - Zj$	3	4↑	1	1	0	0	0		

Maximum $Cj - Zj$ is 4. So the entering variable is x_2. The minimum ratio is 1. So the leaving basis variable is A_2. The key element is 3.

The second iteration of the problem for phase 1 is shown in Table 2.16.

The following row operations are performed to determine the values of the third, fourth, and fifth rows of Table 2.16.

TABLE 2.16

Iteration 2 (Phase 1) for Example 1

Iteration-2	C_j	0	0	0	0	0	−1		
Basic variable	C_B	x_1	x_2	S_1	S_2	S_3	A_1	X_B	Min Ratio
A_1	−1	(5/3)	0	−1	1/3	0	1	1	1/ (5/3) = 3/5→
x_2	0	1/3	1	0	−1/3	0	0	1	1/ (1/3) = 3
S_3	0	2/3	0	0	1/3	1	0	3	3/ (2/3) = 9/2
$Z = -1$	Z_j	−5/3	0	1	−1/3	0	−1		
	$C_j - Z_j$	5/3↑	0	1	1/3	0	0		

TABLE 2.17

Iteration 3 (Phase 1) for Example 1

Iteration-3	C_j	0	0	0	0	0	
Basic variable	C_B	x_1	x_2	S_1	S_2	S_3	X_B
x_1	0	1	0	−3/5	1/5	0	3/5
x_2	0	0	1	1/5	−2/5	0	4/5
S_3	0	0	0	2/5	1/5	1	13/5
$Z = 0$	Z_j	0	0	0	0	0	
	$C_j - Z_j$	0	0	0	0	0	

Fourth row new (in Table 2.16) = Fourth row old (in Table 2.15) ÷ 3

Third row new (in Table 2.16) = Third row old (in Table 2.15) − Fourth row new (in Table 2.16)

Fifth row new (in Table 2.16) = Fifth row old (in Table 2.15) − Fourth row new (in Table 2.16)

In Table 2.16, the optimum solution is not reached. The entering variable is x_1, and the minimum ratio is for A_1. The key element is (5/3).

The third iteration of the problem for phase 1 is shown in Table 2.17.

The following row operations are performed to determine the values of the third, fourth, and fifth rows of Table 2.17.

Third row new (in Table 2.17) = Third row old (in Table 2.16) × 3/5

Fourth row new (in Table 2.17) = Fourth row old (in Table 2.16) − 1/3 Third row new (in Table 2.17)

Fifth row new (in Table 2.17) = Fifth row old (in Table 2.16) − 2/3 Third row new (in Table 2.17)

In Table 2.17, all values of $C_j - Z_j$ are zero. Hence optimality is reached.

The optimal solution is

TABLE 2.18

Initial Simplex Table (Phase 2) for Example 1

Iteration-1	C_j	4	7	0	0	0		
Basic variable	C_B	x_1	x_2	S_1	S_2	S_3	X_B	Min Ratio
x_1	4	1	0	−3/5	(1/5)	0	3/5	(3/5)/(1/5) = 3→
x_2	7	0	1	1/5	−2/5	0	4/5	−
S_3	0	0	0	2/5	1/5	1	13/5	(13/5)/(1/5) = 13
$Z = 8$	Z_j	4	7	−1	−2	0		
	$C_j - Z_j$	0	0	1	2↑	0		

TABLE 2.19

Iteration 2 (Phase 2) for Example 1

Iteration-2	C_j	4	7	0	0	0		
Basic variable	C_B	x_1	x_2	S_1	S_2	S_3	X_B	Min Ratio
S_2	0	5	0	−3	1	0	3	−
x_2	7	2	1	−1	0	0	2	−
S_3	0	−1	0	(1)	0	1	2	2/1 = 2→
$Z = 14$	Z_j	14	7	−7	0	0		
	$C_j - Z_j$	−10	0	7↑	0	0		

$$x_1 = 3/5 \text{ and } x_2 = 4/5 \text{ with objective function } Z = 0.$$

Phase 2:

The initial BFS for this phase is the one obtained at the end of phase I. Here the artificial variables are eliminated, and the objective function becomes

$$\max Z = 4x_1 + 7x_2 + 0S_1 + 0S_2 + 0S_3.$$

The initial simplex table of the previous problem for phase 2 is shown in Table 2.18.

The maximum $C_j - Z_j$ is 2. So the entering variable is S_2. The minimum ratio is 3. So the leaving basis variable is x_1. The key element is $1/5$.

The second iteration of the problem for phase 2 is shown in Table 2.19.

The following row operations are performed to determine the values of the third, fourth, and fifths row of Table 2.19.

Third row new (in Table 2.19) = Third row old (in Table 2.18) × 5

Fourth row new (in Table 2.19) = Fourth row old (in Table 2.18) + 2/5 Third row new (in Table 2.19)

Fifth row new (in Table 2.19) = Fifth row old (in Table 2.18) − 1/5 Third row new (in Table 2.19)

TABLE 2.20

Iteration 3 (Phase 2) for Example 1

Iteration-3	Cj	4	7	0	0	0	
Basic variable	C_B	x_1	x_2	S_1	S_2	S_3	X_B
S_2	0	2	0	0	1	3	9
x_2	7	1	1	0	0	1	4
S_1	0	−1	0	1	0	1	2
Z = 28	Zj	7	7	0	0	7	
	$Cj - Zj$	−3	0	0	0	−7	

In Table 2.19, optimum solution is not reached. The entering variable is S_1, and the minimum ratio is for S_3. The key element is (1).

The third iteration of the problem for phase 2 is shown in Table 2.20.

The following row operations are performed to determine the values of the third, fourth, and fifth rows of Table 2.20.

Fifth row new (in Table 2.20) = Fifth row old (in Table 2.19)

Third row new (in Table 2.20) = Third row old (in Table 2.19) × 5 Fifth row new (in Table 2.20)

Fourth row new (in Table 2.20) = Fourth row old (in Table 2.20) + Fifth row new (in Table 2.20)

In Table 2.20, all values of $C_j - Z_j$ are either negative or zero. Hence optimality is reached. The optimal solution is

$$x_1 = 0 \text{ and } x_2 = 4 \text{ with objective function } Z = 28.$$

2.11 Duality in LPP

LO 9. Formulate the dual of an LPP

LPPs are related to a dual problem. In other words, every minimization problem is linked to the maximization problem and vice versa. The first LPP is known as the primal. The problem derived from the primal is known as its dual. There are common optimized solutions for primary and secondary problems.

The conversion from primal to dual is carried out for many reasons. The dual form of the problem is often easy to solve. The dual problem variables also provide information that is useful for strategic management.

Procedure

1. Transform the objective function to minimization in the dual and vice versa. Write the equation such that it acknowledges right-hand side (RHS) constraints.
2. The number of main variables is the number of dual constraints and vice versa.
3. The coefficient in the primal objective function is the dual RHS constraints and vice versa.
4. Consider transposing the body matrix of the primary problems in shaping the dual constraints.

Example based on duality in LPP

EXAMPLE 1:

Determine the dual of the following LPP (primal). *(BT level 3)*

$$\text{Min } Z = 4x_1 + 3x_2 + 2x_3$$

subject to

$$3x_1 + 4x_2 + 2x_3 \geq 25$$

$$5x_1 + 7x_2 + 4x_3 \geq 35$$

$$x_1 + x_2 + 2x_3 \geq 40$$

$$2x_1 + 2x_2 + 3x_3 \geq 50$$

$$x_1, x_2, x_3 \geq 0$$

Solution: As the primal contains four constraints, the dual contains four variables. Also, the objective of the primal is minimization, and the objective in dual will be maximization.

Let $y_1, y_2, y_3,$ and y_4 be variables in the dual.
The dual of the given problem is the following.

$$\max W = 25y_1 + 35y_2 + 40y_3 + 50y_4$$

subject to

$$3y_1 + 5y_2 + y_3 + 2y_4 \leq 4$$

$$4y_1 + 7y_2 + y_3 + 2y_4 \leq 3$$

$$2y_1 + 4y_2 + 2y_3 + 3y_4 \leq 2$$

$$y_1, y_2, y_3, y_4 \geq 0$$

EXAMPLE 2:

Determine the dual of the following LPP (primal). *(BT level 3)*

$$\text{Min } Z = 4x_1 + 7x_2$$

subject to

$$2x_1 + 4x_2 \geq 50$$

$$5x_1 + 3x_2 \leq 30$$

$$2x_1 + 3x_2 \geq 60$$

$$x_1, x_2, \geq 0$$

Solution: First, we write the given problem symmetrically as follows.

$$\min Z = 4x_1 + 7x_2$$

subject to

$$2x_1 + 4x_2 \geq 50$$

$$-5x_1 - 3x_2 \geq -30$$

$$2x_1 + 3x_2 \geq 60$$

$$x_1, x_2, \geq 0$$

The dual of the given problem is as follows.

$$\max W = 50y_1 - 30y_2 + 60y_3$$

subject to

$$2y_1 - 5y_2 + 2y_3 \leq 4$$

$$4y_1 - 3y_2 + 3y_3 \leq 7$$

$$y_1, y_2, y_3 \geq 0$$

EXAMPLE 3:

Determine the dual of the following LPP (primal). *(BT level 3)*

$$\max Z = 8x_1 + 8x_2 + 20x_3$$

subject to

$$3x_1 + 5x_2 + 7x_3 \leq 20$$

$$3x_1 + 7x_2 + 6x_3 \leq 35$$

$$2x_1 + 3x_2 + x_3 \leq 30$$

$$x_1, x_2, x_3 \geq 0$$

Solution: As the primal contains three constraints, the dual contains three variables. Also, the objective of the primal is maximization, and the objective in the dual will be minimization.

Let y_1, y_2, and y_3 be variables in the dual.

The dual of the given problem is as follows.

$$\min W = 20y_1 + 35y_2 + 30y_3$$

subject to

$$3y_1 + 3y_2 + 2y_3 \geq 8$$

$$5y_1 + 7y_2 + 3y_3 \geq 8$$

$$7y_1 + 6y_2 + y_3 \geq 20$$

$$y_1, y_2, y_3 \geq 0$$

2.12 Sensitivity Analysis

In the physical world, the scenario continues to shift as prices for raw materials fluctuate, availability of machines tends to decrease, and profit increases for one product and decreases for another product. Decision makers need to determine how

LO 10. Understand sensitivity analysis

these changes affect the optimum solution. The examination of a response can be used to reassure and define the responses to these changes. Sensitivity analysis deals with the sum of individual shifts in the objective function and constraints. This explores how changes in the coefficient of an LPP influence the optimal solution. Sensitivity analysis can be performed as follows:

- Changing the right-hand coefficient of constraints,
- Modifying objective function coefficients,
- Adding new variables, and
- Adding new constraints.

2.13 Advantages of the LP Model

1. It enables decision makers to use their productive resource effectively.

LO 11. Write the advantages of the LP model

2. The decision-making process is more rational and less subjective.

3. Bottlenecks can occur in a production process (e.g., some machines may be in great demand in a factory whereas others may be idle for a certain time). These bottlenecks can be checked with the help of LP.

4. Linear modelling illustrates how administrators can best leverage their efficient variables by sharing these elements.

5. A well-structured LPP can be achieved through the more efficient use of scarce resources, such as labour and machines, leading to improved decision quality.

6. In the basic calculations and constraints, the planner can better understand the problem and its optimal solution with a comprehensive view of relationships.

7. Using sensitivity analysis, the LPP can be a quickly obtained, updated, and personalized solution.

8. Simple methodology lets consumers measure shadow costs. Using shadow prices offers details relevant to decisions requiring the acquisition of additional resources to meet certain aims, such as maximizing commitment.

2.14 Limitations of the LP Model

1. LP applies when constraints and objectives are linear – that is to say, where they can be represented as straight-line equations. In real-life cases, this approach cannot be used without linear constraints or objective functions.

LO 12. Write the limitations of the LP model

2. Factors like uncertainty and time are not addressed.

3. Model parameters are believed to be constant, but actually, they are not constant in real-life situations.

4. LP addresses one objective, whereas real-life scenarios can have several competing agendas.

5. There is no guarantee that when we solve an LPP we will get an integer value. A non-integer value is useless in several instances of real life.

CHAPTER SUMMARY

- LP is used to find an optimal solution to problems with a number of constraints.
- The goal is to determine the optimal (maximum or minimum) value for a linear function of several variables (objective) subject to the conditions under which the variables are non-negative and satisfy a set of linear constraints.
- An LPP with two variables can be solved graphically.
- The Slack and surplus variables are used to convert inequalities into an equation.
- For more than two variables, the simplex method is useful.
- An LPP problem might have an optimum solution, multiple solution, unbounded solution, and infeasible solution.
- Where artificial variables are involved, the Big M method is used.
- For every LPP, there is another problem associated with it known as dual.
- Sensitivity analysis is used to answer how the solution of an LPP changes when the coefficient of constraints or objective function will change.

Questions

1. What are the key elements of LP? *(BT level 2)*
2. What is the objective role of the LP model? What is the limitation? *(BT level 2)*
3. List and explain in brief the steps involved in graphically solving an LPP. *(BT level 2)*
4. List the various types of constraints in formulating an LP model. *(BT level 1)*
5. Define the feasible region. *(BT level 1)*
6. What does an unbounded solution mean? *(BT level 2)*
7. What is the concept of duality? *(BT level 2)*
8. Explain the simplex process protocol to determine the optimal solution. *(BT level 2)*
9. What is a standard, non-basic variable? *(BT level 2)*
10. Explain an LPP's unbounded solution. *(BT level 2)*
11. Differentiate between primary and dual problems. *(BT level 5)*
12. Why is the simplex technique better than the graphical method? *(BT level 4)*
13. Discuss the sensitivity analysis role in LP. *(BT level 2)*

3

Transportation Problems and Assignment Problem

Learning Outcomes: After studying this chapter, the reader should be able to

LO 1. Understand the transportation problem and introduce the various concepts behind it. (BT level 2)

LO 2. Formulate and solve the transportation problems. (BT level 6 and 3)

LO 3. Understand and apply the basic feasible solution method. (BT level 2 and 3)

LO 4. Understand and apply the optimal solution method. (BT level 2 and 3)

LO 5. Understand additional issues in transportation modelling. (BT level 2)

LO 6. Understand assignment problems and the use of assignment models in industry and business. (BT level 2 and 3)

LO 7. Formulate an assignment problem. (BT level 6)

LO 8. Understand the use of the Hungarian method. (BT level 3)

LO 9. Understand the solving of the maximization assignment problem. (BT level 2)

LO 10. Understand and solve unbalanced assignment problems. (BT level 3)

3.1 Introduction

LO 1. Understand the transportation problem and introduce the various concepts behind it

Designing supply chain networks and locating factories, warehouses, and distribution centres are strategic decisions that will affect costs. Consequently, companies spend a significant amount of time and resources on considering and evaluating alternatives to choose the one that is most suitable for them. Specifically, transportation models can help businesses minimize total transportation and distribution costs to demand destinations from various sources of supply by considering the location of facilities or the sources of supply and by deciding how many units of a particular product should be transported from each source of supply to satisfy the existing demand for a company's products.

The transportation model is a special case of linear programming, with the objective of minimizing the costs required for transporting the goods from the different sources of supply (suppliers) to the various demand destinations (customers). The model is often classified as a linear programming problem (LPP) because the relationship between the variable's transportation costs and the number of units shipped is assumed to be linear. There are important decision rules that must be applied to the transportation problem. The transportation model requires the following assumptions:

a. The capacity at each supply location or origin is limited.

b. The demand requirements at each destination are known.

c. The items shipped are the same (homogeneous) regardless of their origin or destination.

d. The shipping cost on a per-unit basis remains the same regardless of the number of units shipped.

e. The mode of transportation being used does not change between each origin and destination, and there is only one route used.

3.2 Mathematical Form of the Transportation Problem

LO 2. Formulate and solve the transportation problems

Let $a_i \rightarrow$ no. of supply units available at ith source/supply point

$b_j \rightarrow$ no. of demand units required at jth destination

$C_{ij} \rightarrow$ unit transportation cost to transporting unit forints

$rp_j \rightarrow$ always ≥ 0 be the no. of units shift from jth destination.
Then the LPP has the following mathematical form.

$$\text{Min } Z = \sum_{i=0}^{m}\sum_{j=0}^{n} C_{ij}X_{ij}$$

subject to

$$\sum_{j=1}^{n} X_{ij} = a_i \quad i = 1, 2 \ldots m$$

$$\sum_{i=1}^{m} X_{ij} = b_j \quad j = 1, 2 \ldots n$$

$$X_{ij} \geq 0$$

The transportation problem is balanced if the total supply is equal to the total requirement; otherwise, it would be called an unbalanced transportation problem. The unbalanced transportation problem can be solved considering following two cases.

Case I

If $\sum a_i > \sum b_j$, then we add one column to the transportation table with zero unit cost in all cells.

Case II

$\sum ai < \sum bj$, then we add one row to the transportation table with zero unit cost in all cells.

3.3 Solution of a Transportation Problem

LO 3. Understand and apply the basic feasible solution method

The transportation problem solution consists of two steps:

1. Find an initial feasible basic solution.
2. Achieve an optimal solution by successively enhancing the original feasible simple solution (obtained in step 1) before no more reduction in transport costs is feasible.

3.3.1 Methods to Find an Initial Basic Feasible Solution (BFS)

Here we describe some simple methods to obtain the initial BFS.

1. **North-west corner rule**
   In this rule, we have the following steps:

   Step 1. Start with the cell (1, 1) at the north-west corner (i.e., the top-left corner) and allocate as much as possible. Thus $x_{11} = \min(a_1, b_1)$.

   Step 2. (i) If $b_1 < a_1$, then $x_{11} = b_1$, and there is still some quantity left in row 1.So move to the right-hand cell (1, 2) and make a second allocation in the amount of $x_{12} = \min(a_1, x_{11} - b_2)$ in that cell (1, 2).

   (ii) If $b_1 > a_1$, then $x_{11} = a_1$, and there is still some requirement remaining in column 1, then move vertically down to the cell (2, 1) and render the second allocation in the amount of $x_{21} = \min(a_1, b_1 - x_{11})$ in that cell.

   (iii) If $b_1 = a_1$, then $x_{12} = 0$, or $x_{21} = 0$.

   Begin from the transportation table's current north-west corner and allocate as much as feasible.

   Step 3. Repeat steps 1 and 2 until all the available quantity is exhausted or the entire requirement is met.

2. **Lowest-cost entry (matrix minima) method**
   In this method, we have the following steps:

   Step 1. Carefully analyse the cost matrix to identify the lowest cost. Let it be c_{ij}. Allocate x_{ij} as much as possible in the cell (i, j). $x_{ij} = \min(a_i, b_j)$.

   Step 2. (i) If $x_{ij} = a_i$, then the capacity or the ith root is fully exhausted. In this case, cross out the ith row of the transportation table and decrease b_j by a_1. Go over to step 3.

(ii) If $x_{ij} = b_j$, then the requirement of the jth destination is completely met; cross out the jth column of the transportation table and decrease a_i by b_j. Now go to step 3.

(iii) If $x_{ij} = a_i = b_j$, then either cross out the ith row or jth column but not both. Now go over to step 3.

Step 3. Repeat steps 1 and 2 for the reduced transport table until all available is depleted or all requirements are met.

3.3.2 Unit Cost Penalty Method (Vogel's Approximation Method)

Here we have the following basic process:

Step 1. Identify the smallest costs for each row of the transportation table. Find the difference for each row. Write these differences in parentheses alongside the transportation table against the respective rows. Write similar variations or differences for each section below the column. They are called "penalties."

Step 2. Now pick the row or column with the highest penalties. Use any predetermined tie-breaking option if a tie arises. Allocate the maximum possible amount to the lowest-cost cell in that row or column. Let the biggest penalty pertain to the ith row and let it be the smallest cost in the ith row. Allocate $xij = \min (ai, bj)$ in cell (i, j). Then, as usual, we cross out the ith row or the jth column and build a reduced matrix.

Step 3. Now calculate the reduced transportation table row and column penalties and repeat step 2. We continue this cycle until all the available quantity is exhausted or the requirement is fulfilled.

Examples based on the initial BFS

EXAMPLE 1:

Using the north-west corner method, determine an initial feasible basic solution to the following transport problem (Table 3.1). *(BT level 3)*

Solution:
First, we construct an empty three by four matrix complete with row and column requirements (Table 3.2).

TABLE 3.1

Data for Example 1

	I	II	III	IV	Supply
A	13	11	15	20	2
B	17	14	12	13	6
C	18	18	15	12	7
Demand	3	3	4	5	15

TABLE 3.2

North-West Solution

	I	II	III	IV	b_j
A	2(13)				2
B	1(17)	3(14)	2(12)		6
C			2(15)	5(12)	7
a_i	3	3	4	5	

To Start with cell (1, 1) (top-left corner) and allocate the maximum. Therefore, $X_{11} = 2$ because minimum $a1 = 2$ and $b1 = 3$ are 2. There is no amount left at source A, so we go down to the cell (2, 1) and allocate it as much as possible. Since column 1 still needs an amount 1 and the amount 6 is available in row 2 at source B, we allocate the maximum possible amount 1 to the cell (i.e., $X_{21} = 1$). Column 1 allocation 3 is complete. Now we move to the right of the cell (2,1) (*i.e.* to the cell (2, 2)) and allocate as much as possible. Since the quantity 5 is already accessible in row 2, and the quantity 3 is accessible in column 2, then we can assign the maximum quantity of 3 in cell (2, 2) (i.e., $X_{22} = 3$). Since the allocation of 3 in column 2 is complete, we move to the cell (2, 3) (i.e., $X_{23} = 2$). No amount is available in row 2, so we go down to the cell (3, 3) and allocate as much as possible. Since the amount 2 will still be required in column 3, and the amount 7 is available in row 3, the amount 2 is in the cell (3, 3) (i.e., $X_{33} = 2$). Thus column 3 allocation 4 is complete. Now the amount 5 is still available in row 3, and the amount 5 is required in column 5 so that the amount 5 is allocated in the cell (3, 4) (i.e., $X_{34} = 5$), which completes the allocation.

The initial feasible solution is

$$X_{11} = 2, x_{21} = 1, x_{22} = 3, x_{23} = 2, x_{33} = 2, x_{34} = 5.$$

EXAMPLE 2:

Using the "lowest-cost entry method," find the initial BFS to the following transportation problem (Table 3.3). *(BT level 3)*

TABLE 3.3

Data for Example 2

		Destinations				
		A	B	C	D	Supply
	I	1	5	3	3	34
	II	3	3	1	2	15
Origins	III	0	2	2	3	12
	IV	2	7	2	4	19
	Demand	21	25	17	17	80

TABLE 3.4

Lowest Cost Solution

		A	B	C	D	Supply
	I	(1)	(5)	(3)	(3)	34
	II	(3)	(3)	(1)	(2)	15
Origins	III	12(0)	(2)	(2)	(3)	12
	IV	(2)	(7)	(2)	(4)	19
		21	25	17	17	

Solution:

First, we construct an empty four by four matrix complete with costs, row, and column requirements (Table 3.4).

Examining the cost matrix carefully, we observe that the minimum cost 0 is in cell (3, 1), and so we allocate the maximum possible amount 12 to this cell. This exhausts the availability at origin III, and still, nine more units are required at A. Leaving the third row, we have the matrix shown in Table 3.5.

Now the lowest cost 1 is at the cells (1, 1) and (2, 3). We can allocate the maximum possible amount 9 at cell (1, 1) and 15 at cell (2, 3).

Allocating these amounts, we observe that requirements of A are completed, and origin II is also exhausted. Also, two units more are required at destination C, and 25 units are still available at source I.

Currently, leaving the first row and the second column, we have the following cost matrix (Table 3.6).

The lowest cost in this cost matrix is 2 in cell (2, 2), so we allocate a maximum of 2 in this cell (2, 2). This also fulfils demand at destination C. Eliminating row C, we are left with the following matrix (Table 3.7).

TABLE 3.5

Lowest Cost Solution

		A	B	C	D	
	I	9 (1)	(5)	(3)	(3)	34
	II	(3)	(3)	15 (1)	(2)	15
Origins	IV	(2)	(7)	(2)	(4)	19
		9	25	17	17	

TABLE 3.6

Lowest Cost Solution

		B	C	D	
	I	(5)	(3)	(3)	25
Origins	IV	(7)	2 (2)	(4)	19-2
		25	2–2 17		

TABLE 3.7

Lowest Cost Solution

		B	D	
	I	8 (5)	17 (3)	25
Origins	IV	17(7)	(4)	17
		25	17	

In this table, the lowest cost is 3 in cell (1, 2). Allocating the maximum possible amount 17 in cell (1, 2), we complete the allocations as shown in Table 3.7.

Finally, all the allocations are shown in Table 3.8.

Hence the initial BFS of the problem is

$$X_{11} = 9, x_{12} = 8, x_{14} = 17, x_{23} = 15, x_{31} = 12, x_{42} = 17, x_{43} = 2.$$

TABLE 3.8

Lowest Cost Solution

		Destination				
		A	B	C	D	Supply
	I	9 (1)	8 (5)	(3)	17 (3)	34
	II	(3)	(3)	15(1)	(2)	15
Origins	III	12 (0)	(2)	(2)	(3)	12
	IV	(2)	17 (7)	(2)	(4)	19
	Demand	21	25	17	17	

EXAMPLE 3:

Find the initial BFS of the following transportation problem (Table 3.9) using Vogel's approximation method. *(BT level 3)*

TABLE 3.9

Data for Example 3

		To				
		I	II	III	IV	Available
	A	5	1	3	3	34
	B	3	3	5	4	15
From	C	6	4	4	3	12
	D	4	1	4	2	19
	Requirement	21	25	17	17	80

TABLE 3.10

Vogel's Solution

		I	II	III	IV		Penalties
	A	(5)	25 (1)	(3)	(3)	34-25	(2)←
From	B	(3)	(3)	(5)	(4)	15	(1)
	C	(6)	(4)	(4)	(3)	12	(1)
	D	(4)	(1)	(4)	(2)	19	(1)
		21	25-25	17	17		
Penalties		(1)	(2)	(1)	(1)		

Solution:

First, we write the cost and requirement matrix and compute the penalties (difference of the smallest two terms in each row and each column; Table 3.10).

Since the maximum penalty 2 is associated with row 1 and column 1, we choose one of them to say row 1 and allocate the maximum possible amount 25 in the cell $(1, 2)$ with minimum cost (i.e., $X_{12} = 25$). Thus the requirements of II are completed.

Leaving the second column, we have a matrix with the left requirements and availabilities Table 3.11).

Since the maximum penalty is 2 in row 1, we allocate the maximum possible amount 9 in the cell $(1, 2)$, which has minimum cost. Thus row 1 (i.e., A) is exhausted, so leaving this row, we have the following matrix with the remaining requirements and availabilities, etc. (Table 3.12).

The highest penalty is 2 in row 3; therefore, we assign a maximum feasible amount 17 in the lowest-cost cell $(3, 3)$. This completes column IV requirements. Leaving column IV, the following matrix shows the needed requirements and availability. The maximum penalty is now in rows 1 and 2. We allocate the maximum possible amount 15 in the cell $(1, 1)$ by selecting row 1 and thus exhausting row B (Table 3.13).

Leaving row B, we get the following cost matrix in which the maximum penalty is 2 in row 1 and column 1. Selecting row 1 and allocating the maximum possible amount 8 at $(1, 2)$, we complete the remaining allocations (Table 3.14).

Thus the final table showing all allocations is as follows (Table 3.15).

TABLE 3.11

Vogel's Solution

		I	III	IV		Penalties
	A	(5)	9 (3)	(3)	9-9	(2) ←
From	B	(3)	(5)	(4)	15	(1)
	C	(6)	(4)	(3)	12	(1)
	D	(4)	(4)	(2)	19	(2)
		21	17-9	17		
Penalties		(1)	(1)	(1)		

TABLE 3.12

Vogel's Solution

		I	III	IV		Penalties
	B	(3)	(5)	(4)	15	(1)
From	C	(6)	(4)	(3)	12	(1)
	D	(4)	(4)	17 (2)	19-17	(2)
		21	8	17-17		←
Penalties		(1)	(1)	(1)		

TABLE 3.13

Vogel's Solution

		I	III		Penalties
	B	15(3)	(5)	15-15	(2) ←
From	C	(6)	(4)	12	(2)
	D	(4)	(1)	2	(0)
		21-15	8		
Penalties		(1)	(1)		

TABLE 3.14

Vogel's Solution

		I	III		Penalties
From	C	4 (6)	8 (4)	12	(2) ←
	D	2 (4)	(1)	2	(0).
		6	8		
Penalties		(1)	(1)		

TABLE 3.15

Vogel's Final Solution

	I	II	III	IV	
A	(5)	25 (1)	9 (3)	(3)	34
B	15 (3)	(3)	(5)	(4)	15
C	4 (6)	(4)	8 (4)	(3)	12
D	2 (4)	(1)	(4)	17 (2)	19
	21	25	17	17	

Hence the initial BFS of the problem is

$$X_{12} = 25, \ X_{13} = 9, \ X_{21} = 15, \ X_{31} = 4, \ X_{33} = 8, \ X_{41} = 2, \ X_{44} = 17.$$

3.3.3 Methods to Find an Optimal Solution

1. **Stepping-stone method**

   In this rule, we have the following steps:

   Step 1. Find the initial BFS of the transportation problem.

   LO 4. Understand and apply the optimal solution method

   Step 2. Check the number of occupied cells. If there are less than $m + n - 1$, there exists degeneracy. Introduce a very small positive assignment of e (= 0) in suitable independent positions so that the number of occupied cells is exactly equal to $m + n - 1$.

   Step 3. Compute the improvement index for each of the unoccupied cells. This is done by calculating the opportunity cost of an unoccupied cell. This means that if we shift one unit from a cell containing positive shipment to the unoccupied cell, that will be the net cost. If all the unoccupied cells have a positive improvement index, then the given solution is an optimum solution.

   Step 4. If there are several unoccupied cells with negative improvement indices, then we select the cell having the largest negative improvement index and shift the maximum possible units to that cell without violating the supply and demand constraints. After this, go to step 3.

2. **U–V method or MODI method**

   In the U–V method, cell evaluations of all the unoccupied cells are calculated simultaneously, and only one closed path for the most negative cell is traced. So it is a more time-saving method than the stepping-stone method. This method consists of the following steps:

   Step 1. Find the initial BFS of the transportation problem.

   Step 2. Check the number of occupied cells. If there are less than $m + n - 1$, there exists degeneracy. Introduce a very small positive assignment of e (= 0) in suitable independent positions so that the number of occupied cells is exactly equal to $m + n - 1$.

   Step 3. Assign $U_1, U_2, \ldots .. U_m$ to m rows and $V_1, V_2, \ldots \ldots V_n$ to n columns, respectively.

   Step 4. For each occupied cell in the current solution, solve the system of equations $U_i + V_j = C_{ij}$, staring initially with some $U_i = 0$ or $V_j = 0$ and entering successively the values of U_i and V_j in the transportation table margins.

   Step 5. Compute the cell evaluations $d_{ij} = C_{ij} - (U_i + V_j)$ for each unoccupied cell (i, j) and enter them into the corresponding cell (i, j).

   Step 6. If all $d_{ij} \leq 0$, then the current BFS is an optimum one. If at least one $d_{ij} > 0$, select the unoccupied cell, allowing the largest positive net evaluation to enter the basis.

   Step 7. Form a new BFS by giving maximum allocation to the cell for which d_{ij} is most negative by making an occupied cell empty. For that, draw a closed path consisting of horizontal and vertical lines beginning and ending at the cell for which d_{ij} is most negative and having its other corners at some allocated cells. Along this closed loop, indicate $+\theta$ and $-\theta$ alternative at the corners. Choose the minimum of θ from the cells having $-\theta$. Add this minimum of θ to the cells with $+\theta$ and subtract this minimum of θ from the allocation to the cells with $-\theta$.

   Step 8. Return to step 3 and repeat the process until an optimum BFS has been obtained.

Examples based on an optimal solution

EXAMPLE 1:

Determine the optimum distribution of the following transportation problem (Table 3.16) using the stepping-stone method. *(BT level 3)*

Solution:
Following, Vogel's approximation method, an initial BFS is obtained in Table 3.17).

Iteration 1 of the Optimality Test

1. Creating a closed loop for unoccupied cells, we get Table 3.18.

Since all net cost change ≥ 0, the final optimal solution is arrived at in Table 3.19. The minimum total transportation cost = 17 × 280 + 13 × 120 + 15 × 80 + 9 × 240 + 17 × 180 = 12,740.

TABLE 3.16

Data for Example 1

Factory	Warehouse			Supply
	D1	D2	D3	
S1	17	21	13	400
S2	15	9	19	320
S3	27	25	17	180
Demand	360	240	300	900

TABLE 3.17

BFS Using Vogel's Method

	D1	D2	D3	Supply
S1	17 (280)	21	13 (120)	400
S2	15 (80)	9 (240)	19	320
S3	27	25	17 (180)	180
Demand	360	240	300	

TABLE 3.18

Iteration 1

Unoccupied cell	Closed path	Net cost change
S1D2	S1D2→S1D1→S2D1→S2D2	21 − 17 + 15 − 9 = 10
S2D3	S2D3→S2D1→S1D1→S1D3	19 − 15 + 17 − 13 = 8
S3D1	S3D1→S3D3→S1D3→S1D1	27 − 17 + 13 − 17 = 6
S3D2	S3D2→S3D3→S1D3→S1D1→S2D1→S2D2	25 − 17 + 13 − 17 + 15 − 9 = 10

TABLE 3.19

Iteration 2

	D1	D2	D3	Supply
S1	17 (280)	21	13 (120)	400
S2	15 (80)	9 (240)	19	320
S3	27	25	17 (180)	180
Demand	360	240	300	

EXAMPLE 2:

Determine the optimum solution of the following transportation problem (Table 3.20) using the stepping-stone method. *(BT level 3)*

Solution:

Following the north-west corner method, an initial BFS is obtained in Table 3.21.

The minimum total transportation cost = $22 \times 200 + 26 \times 50 + 36 \times 175 + 28 \times 125 + 26 \times 150 + 20 \times 250 = 24,400$.

Here the number of allocated cells is $6 = m + n - 1 = 3 + 4 - 1 = 6$.

This solution is non-degenerate.

TABLE 3.20

Data for Example 2

Factory	Warehouse				Supply
	D1	D2	D3	D4	
S1	22	26	34	28	250
S2	32	36	28	20	300
S3	42	48	26	20	400
Demand	200	225	275	250	

TABLE 3.21

BFS Using the North-West Method

	D1	D2	D3	D4	Supply
S1	22 (200)	26 (50)	34	28	250
S2	32	36 (175)	28 (125)	20	300
S3	42	48	26 (150)	20 (250)	400
Demand	200	225	275	250	

Iteration 1 of the Optimality Test

1. Creating a closed loop for unoccupied cells, we get Table 3.22.
2. Select the unoccupied cell having the highest negative net cost change (i.e., cell S2D4 = −2) and draw a closed path from S2D4.

   The closed path is S2D4→S2D3→S3D3→S3D4.

   The closed path and plus/minus allocation for the current unoccupied cell S2D4 is shown in Table 3.23.
3. Minimum allocated value among all negative positions (−) on the closed path = 125.

   Subtract 125 from all (−) and add it to all (+), as shown in Table 3.24.
4. Repeat steps 1 to 3 until an optimal solution is obtained.

TABLE 3.22

Iteration 1

Unoccupied cell	Closed path	Net cost change
S1D3	S1D3→S1D2→S2D2→S2D3	34 − 26 + 36 − 28 = 16
S1D4	S1D4→S1D2→S2D2→S2D3→S3D3→S3D4	28 − 26 + 36 − 28 + 26 − 20 = 16
S2D1	S2D1→S2D2→S1D2→S1D1	32 − 36 + 26 − 22 = 0
S2D4	S2D4→S2D3→S3D3→S3D4	20 − 28 + 26 − 20 = −2
S3D1	S3D1→S3D3→S2D3→S2D2→S1D2→S1D1	42 − 26 + 28 − 36 + 26 − 22 = 12
S3D2	S3D2→S3D3→S2D3→S2D2	48 − 26 + 28 − 36 = 14

TABLE 3.23

Iteration 1

	D1	D2	D3	D4	Supply
S1	22 (200)	26 (50)	34 [16]	28 [16]	250
S2	32 [0]	36 (175)	28 (125) (−)	20 [−2] (+)	300
S3	42 [12]	48 [14]	26 (150) (+)	20 (250) (−)	400
Demand	200	225	275	250	

TABLE 3.24

Iteration 1

	D1	D2	D3	D4	Supply
S1	22 (200)	26 (50)	34	28	250
S2	32	36 (175)	28	20 (125)	300
S3	42	48	26 (275)	20 (125)	400
Demand	200	225	275	250	

Iteration 2 of the Optimality Test

1. Creating a closed loop for unoccupied cells, we get Table 3.25.

Since all net cost change ≥ 0, the final optimal solution is arrived at in Table 3.26.

The minimum total transportation cost = $22 \times 200 + 26 \times 50 + 36 \times 175 + 20 \times 125 + 26 \times 275 + 20 \times 125 = 24{,}150$.

Note that the alternate solution is available with the unoccupied cell S2D1 = 0, but with the same optimal value.

TABLE 3.25

Iteration 2

Unoccupied cell	Closed path	Net cost change
S1D3	S1D3→S1D2→S2D2→S2D4→S3D4→S3D3	$34 - 26 + 36 - 20 + 20 - 26 = 18$
S1D4	S1D4→S1D2→S2D2→S2D4	$28 - 26 + 36 - 20 = 18$
S2D1	S2D1→S2D2→S1D2→S1D1	$32 - 36 + 26 - 22 = 0$
S2D3	S2D3→S2D4→S3D4→S3D3	$28 - 20 + 20 - 26 = 2$
S3D1	S3D1→S3D4→S2D4→S2D2→S1D2→S1D1	$42 - 20 + 20 - 36 + 26 - 22 = 10$
S3D2	S3D2→S3D4→S2D4→S2D2	$48 - 20 + 20 - 36 = 12$

TABLE 3.26

Iteration 2

	D1	D2	D3	D4	Supply
S1	22 (200)	26 (50)	34	28	250
S2	32	36 (175)	28	20 (125)	300
S3	42	48	26 (275)	20 (125)	400
Demand	200	225	275	250	

EXAMPLE 3:

Determine the optimum solution of the following transportation problem (Table 3.27) using the U–V method. *(BT level 3)*

Solution:

Following, the north-west corner method, an initial BFS is obtained in Table 3.28.

The minimum total transportation cost = $22 \times 200 + 26 \times 50 + 36 \times 175 + 28 \times 125 + 26 \times 150 + 20 \times 250 = 24400$

Here the number of allocated cells is $6 = m + n - 1 = 3 + 4 - 1 = 6$.

This solution is non-degenerate.

TABLE 3.27

Data for Example 3

Factory	Warehouse				Supply
	D1	D2	D3	D4	
S1	22	26	34	28	250
S2	32	36	28	20	300
S3	42	48	26	20	400
Demand	200	225	275	250	

TABLE 3.28

BFS Using the North-West Method

	D1	D2	D3	D4	Supply
S1	22 (200)	26 (50)	34	28	250
S2	32	36 (175)	28 (125)	20	300
S3	42	48	26 (150)	20 (250)	400
Demand	200	225	275	250	

Iteration 1 of the Optimality Test

1. Find u_i and v_j for all occupied cells (i,j), where $c_{ij} = u_i + v_j$ (Table 3.29).
   1. Substituting, $u_1 = 0$, we get the following:
   2. $c_{11} = u_1 + v_1 \Rightarrow v_1 = c_{11} - u_1 \Rightarrow v_1 = 22 - 0 \Rightarrow v_1 = 22$
   3. $c_{12} = u_1 + v_2 \Rightarrow v_2 = c_{12} - u_1 \Rightarrow v_2 = 26 - 0 \Rightarrow v_2 = 26$
   4. $c_{22} = u_2 + v_2 \Rightarrow u_2 = c_{22} - v_2 \Rightarrow u_2 = 36 - 26 \Rightarrow u_2 = 10$
   5. $c_{23} = u_2 + v_3 \Rightarrow v_3 = c_{23} - u_2 \Rightarrow v_3 = 28 - 10 \Rightarrow v_3 = 18$
   6. $c_{33} = u_3 + v_3 \Rightarrow u_3 = c_{33} - v_3 \Rightarrow u_3 = 26 - 18 \Rightarrow u_3 = 8$
   7. $c_{34} = u_3 + v_4 \Rightarrow v_4 = c_{34} - u_3 \Rightarrow v_4 = 20 - 8 \Rightarrow v_4 = 12$

TABLE 3.29

Iteration 1

	D1	D2	D3	D4	Supply	u_i
S1	22 (200)	26 (50)	34	28	250	$u_1 = 0$
S2	32	36 (175)	28 (125)	20	300	$u_2 = 10$
S3	42	48	26 (150)	20 (250)	400	$u_3 = 8$
Demand	200	225	275	250		
v_j	$v_1 = 22$	$v_2 = 26$	$v_3 = 18$	$v_4 = 12$		

2. Find d_{ij} for all unoccupied cells (i,j), where $d_{ij} = c_{ij} - (u_i + v_j)$ (Table 3.30).
   1. $d_{13} = c_{13} - (u_1 + v_3) = 34 - (0 + 18) = 16$
   2. $d_{14} = c_{14} - (u_1 + v_4) = 28 - (0 + 12) = 16$

3. $d21=c21-(u2+v1)=32-(10+22)=0$
4. $d24=c24-(u2+v4)=20-(10+12)=-2$
5. $d31=c31-(u3+v1)=42-(8+22)=12$
6. $d32=c32-(u3+v2)=48-(8+26)=14$

TABLE 3.30

Iteration 1

	D1	D2	D3	D4	Supply	u_i
S1	22 (200)	26 (50)	34 [16]	28 [16]	250	u1 = 0
S2	32 [0]	36 (175)	28 (125)	20 [–2]	300	u2 = 10
S3	42 [12]	48 [14]	26 (150)	20 (250)	400	u3 = 8
Demand	200	225	275	250		
v_j	v1 = 22	v2 = 26	v3 = 18	v4 = 12		

3. Now choose the minimum negative value from all *dij* (opportunity cost) = d24 = [–2] and draw a closed path from S2D4.

    The closed path is S2D4→S2D3→S3D3→S3D4.

    Closed path and plus/minus sign allocation (Table 3.31).

TABLE 3.31

Iteration 1

	D1	D2	D3	D4	Supply	u_i
S1	22 (200)	26 (50)	34 [16]	28 [16]	250	u1 = 0
S2	32 [0]	36 (175)	28 (125) (–)	20 [–2] (+)	300	u2 = 10
S3	42 [12]	48 [14]	26 (150) (+)	20 (250) (–)	400	u3 = 8
Demand	200	225	275	250		
v_j	v1 = 22	v2 = 26	v3 = 18	v4 = 12		

4. Minimum allocated value among all negative positions (–) on closed path = 125. Subtract 125 from all (–) and add it to all (+) (Table 3.32).

TABLE 3.32

Iteration 1

	D1	D2	D3	D4	Supply
S1	22 (200)	26 (50)	34	28	250
S2	32	36 (175)	28	20 (125)	300
S3	42	48	26 (275)	20 (125)	400
Demand	200	225	275	250	

5. Repeat steps 1 to 4 until an optimal solution is obtained.

Iteration 2 of the Optimality Test

1. Find u_i and v_j for all occupied cells(i,j), where $c_{ij} = u_i + v_j$ (Table 3.33).
   1. Substituting $u_1 = 0$, we get the following:
   2. $c_{11}=u_1+v_1 \Rightarrow v_1=c_{11}-u_1 \Rightarrow v_1=22-0 \Rightarrow v_1=22$
   3. $c_{12}=u_1+v_2 \Rightarrow v_2=c_{12}-u_1 \Rightarrow v_2=26-0 \Rightarrow v_2=26$
   4. $c_{22}=u_2+v_2 \Rightarrow u_2=c_{22}-v_2 \Rightarrow u_2=36-26 \Rightarrow u_2=10$
   5. $c_{24}=u_2+v_4 \Rightarrow v_4=c_{24}-u_2 \Rightarrow v_4=20-10 \Rightarrow v_4=10$
   6. $c_{34}=u_3+v_4 \Rightarrow u_3=c_{34}-v_4 \Rightarrow u_3=20-10 \Rightarrow u_3=10$
   7. $c_{33}=u_3+v_3 \Rightarrow v_3=c_{33}-u_3 \Rightarrow v_3=26-10 \Rightarrow v_3=16$

TABLE 3.33

Iteration 2

	D1	D2	D3	D4	Supply	u_i
S1	22 **(200)**	26 **(50)**	34	28	250	$u_1 = 0$
S2	32	36 **(175)**	28	20 **(125)**	300	$u_2 = 10$
S3	42	48	26 **(275)**	20 **(125)**	400	$u_3 = 10$
Demand	200	225	275	250		
v_j	$v_1 = 22$	$v_2 = 26$	$v_3 = 16$	$v_4 = 10$		

2. Find d_{ij} for all unoccupied cells(i,j), where $d_{ij} = c_{ij} - (u_i + v_j)$ (Table 3.34).
   1. $d_{13}=c_{13}-(u_1+v_3)=34-(0+16)=18$
   2. $d_{14}=c_{14}-(u_1+v_4)=28-(0+10)=18$
   3. $d_{21}=c_{21}-(u_2+v_1)=32-(10+22)=0$
   4. $d_{23}=c_{23}-(u_2+v_3)=28-(10+16)=2$
   5. $d_{31}=c_{31}-(u_3+v_1)=42-(10+22)=10$
   6. $d_{32}=c_{32}-(u_3+v_2)=48-(10+26)=12$

TABLE 3.34

Iteration 2

	D1	D2	D3	D4	Supply	u_i
S1	22 **(200)**	26 **(50)**	34 [18]	28 [18]	250	$u_1 = 0$
S2	32 [0]	36 **(175)**	28 [2]	20 **(125)**	300	$u_2 = 10$
S3	42 [10]	48 [12]	26 **(275)**	20 **(125)**	400	$u_3 = 10$
Demand	200	225	275	250		
v_j	$v_1 = 22$	$v_2 = 26$	$v_3 = 16$	$v_4 = 10$		

Since all $dij \geq 0$, the final optimal solution is arrived at in Table 3.35.

The minimum total transportation cost = $22 \times 200 + 26 \times 50 + 36 \times 175 + 20 \times 125 + 26 \times 275 + 20 \times 125 = 24,150$.

Note that the alternate solution is available with unoccupied cell S2D1:d21 = [0] but with the same optimal value.

TABLE 3.35

Iteration 2

	D1	D2	D3	D4	Supply
S1	22 (200)	26 (50)	34	28	250
S2	32	36 (175)	28	20 (125)	300
S3	42	48	26 (275)	20 (125)	400
Demand	200	225	275	250	

3.4 Degeneracy in Transport Problems

We remember that a BFS is a m-origin and n-destination transportation problem which should include at most $(m + n - 1)$ independent non-zero allocations. If this number is equal to $(m + n - 1)$, the BFS is known to be non-degenerate; otherwise, it is stated to be degenerate. Thus degeneration occurs when the number of independent individual allocations is less than $(m + n - 1)$.

LO 5. Understand additional issues in transportation modelling

A transportation problem will degenerate in two ways:

(i) BFS can degenerate from the outset (i.e., from the beginning).
(ii) It may degenerate at some intermediate point when the selection of one entry cell empties two or more preoccupied cells simultaneously. In these situations, to overcome degeneration, we assign an extremely small amount (close to zero) to one or more empty matrix cells (usually the lowest-cost cells if possible) such that the total number of occupied (allocated) cells is $(m + n - 1)$.

Very limited quantities denoted by (delta) or (epsilon) satisfy the following conditions:

1. $\Delta < x_{ij}$ for $x_{ii} > 0$
2. $x_{ij} + \Delta = x_{ij} = x_{ij} - \Delta, x_{ij} > 0$
3. $\Delta + 0 = \Delta$.
4. If the solution contains more than one Δ then,
   i. If Δ, Δ' are in the same row, $\Delta < \Delta'$ when Δ is to the left of Δ'.
   ii. If Δ, Δ' are in the same column, $\Delta < \Delta'$ when Δ is above Δ'.

These rules indicate that even after implementation, the original solution of the problem is not modified. Applying the optimality check is simply a technique. Because there is no physical significance, it is ultimately omitted. The following examples explain the procedure.

Examples based on degeneracy

EXAMPLE 1:

A manufacturer wants to ship eight loads of his product, as shown in Table 3.36. The matrix gives the mileage from origin 0 to destination D. Shipping costs are Rs.10 per load per mile. What shipping schedule should be used? (*BT level 3*)

Since the total number of allocations is four, which is less than $m + n - 1 = 5$, this is a degenerate solution. The attempt to assign ui and vj values to Table 3.36 won't be successful.

To solve this degeneration, we allocate very small amounts to some suitable cell. The cell (1, 2) receives five allocations at independent positions (Table 3.37)

Test the optimal solution: To check the best or optimized solution, we have Table 3.38.

Since $d_{22} = -25 < 0$, the solution under test is not optimal. Now we shall allocate to this cell (2, 2) as much as possible. Thus we take Δ from cell (1, 2) to cell (2, 2) and form the new table to check the solution for optimality (Table 3.39).

Because $dij > 0$ for all empty cells, the solution under check is optimal.

The solution to the problem is

$$X_{11} = 1, x_{21} = 3, x_{32} = 2, x_{33} = 2.$$

And the minimum mileage $= 50.1 + 90.3 + 200.2 + 50.2 = 820$ (i.e., minimum cost = Rs. 8200).

TABLE 3.36

Data for Example 1

	D_1	D_2	D_3	$a_i\downarrow$
0_1	(50) 1	(30)	(220)	1
0_2	(90) 3	(45)	(170)	3
0_3	(250)	(200) 2	(50) 2	4
$b_j\rightarrow$	4	2	2	

TABLE 3.37

Step 1

(50) 1	(30) Δ	(220)	$1 + \Delta = 1$
(90) 3	(45)	(170)	3
(250)	(200) 2	(50) 2	4
4	$2 + \Delta = 2$	2	

TABLE 3.38

Step 2

				$u_i \downarrow$
	(50) 1	(30) Δ	(220) −120 340	−170
	(90) 3	(45) 70 −25	(170) 80 250	−130
	(250) 220 30	(200) 2	(50) 2	0
$v_j \rightarrow$	200	200	50	

TABLE 3.39

Step 3

				$u_i \downarrow$
	(50) 1	(30) 5 25	(220) −145 365	−195
	(90) 3	(45) Δ	(170) −105 275	−155
$v_j \rightarrow$	(250) 245 5	(200) 2	(50) 2	0
	245	200	50	

EXAMPLE 2:

Solve the cost-minimizing transportation problem for the data given in Table 3.40.
(BT level 3)

Solution:
By using Vogel's approximation method, Table 3.41 provides an initial BFS problem.

TABLE 3.40

Data for Example 2

	D_1	D_2	D_3	D_4	Available
O_1	5	3	6	5	15
O_2	10	7	12	4	11
O_3	7	5	8	4	13
Demand	8	12	13	6	

TABLE 3.41

BFS Using Vogel's Method

	D_1	D_2	D_3	D_4	
O_1	(5) 8	(3) 7	6	5	15
O_2	(10)	(7) 5	(12)	(4) 6	11
O_3	(7)	(5)	(8) 13	4	13
	8	12	13	6	

The total transportation cost

$$= 8 \times 5 + 7 \times 3 + 5 \times 7 + 6 \times 4 + 13 \times 8 = \text{Rs.}224.$$

To test the solution for optimality. Here the total number of allocations is five, which is one less than the number $m + n - 1 = 6$ (Table 3.42).

And now the solution is degenerate.

To resolve the degeneration, we allocate a small amount in the cell (3, 4) with the minimum cost among unoccupied cells and the independent position.

To check the solution's optimality, we find the set of ui and vj so that for each occupied cell (r, s), $crs = ur +$ versus normally taking $u1 = 0$.

Now we enter $ui + vj$ and $dij = cij - (ui + vj)$ at the appropriate corners in each unoccupied cell and observe that $d31 < 0$, $d32 < 0$ (see Table 3.42).

So the solution is also not optimal.

Since minimum $dij = -2 < 0$ is in both cells, we take in one of these two cells, say cell (3, 1), and test the optimality again by finding the set of ui and vj for all occupied cells (r, s) so that $crs = ui + vj$, and entering $ui + vj$ and $dij = cij - (ui + vj)$ at the appropriate corner of each unoccupied cell, we get Table 3.43 in which we observe that all $dij \geq 0$.

TABLE 3.42

Step 1

	D_1	D_2	D_3	D_4	$u_i \downarrow$
	(5) 8	(3) 7	6 (4) (2)	(5) (0) (5)	0 (u_1)
	(10) (9) (1)	(7) 5	(12) (8) (4)	(4) 6 (0)	4 (u_2)
	(7) (9) (−2)	(5) (7) (−2)	(8) 13	4 Δ	4 (u_3)
$v_j \rightarrow$	5	3	4	0	
	V_1	V_2	V_3	V_4	

TABLE 3.43

Step 2

				$u_i \downarrow$
(5)	(3)	6	(5)	0 (u_1)
8	7	(6)	(0)	
		(0)	(5)	
(10)	(7)	(12)	(4)	4 (u_2)
(9)	5	(10)	6	
(1)		(2)		
(7)	(5)	(8)	(4)	2 (u_3)
Δ	(5)	13	(2)	
	(0)		(2)	
$v_j \rightarrow$ 5	3	6	0	

This solution is optimal and is given as follows:

$$x_{11} = 8, x_{12} = 7, x_{22} = 5, x_{24} = 6, x_{33} = 13,$$

and the minimum transportation cost = Rs. 224.

3.5 Unbalanced Transportation Problem

If in a transport problem the amount of all available quantities is not equal to the sum of all requirements, then this problem is called an unbalanced transportation problem.

Unbalanced transportation issues can arise in two different manners:

(i) *Shortage of the supply problem:* To change this form of unbalanced transportation issue to a balanced type, we add a dummy source row in the transportation table. The unit transport costs from this dummy source to any destination are set at zero. The availability of this dummy source is assumed equal to the gap.

(ii) *Excess of the supply problem:* To change this form of unbalanced transport issue to a balanced type, we insert a dummy destination column in the transportation table. The unit transport costs from any source to this dummy destination are set at zero. The necessity for this dummy destination is supposed to equal the gap.

Examples based on an unbalanced transportation problem

EXAMPLE 1:

Solve the following unbalanced transportation problem (Table 3.44; symbols have their usual meanings). *(BT level 3)*

TABLE 3.44

Data for Example 1

	D_1	D_2	D_3	$a_i \downarrow$
O_1	4	3	2	10
O_2	2	5	0	13
O_3	3	8	6	12
$b_j \rightarrow$	8	5	4	

Solution:

Here $\sum a_i = 35$, $\sum b_j = 17$. Since $\sum a_i$ is greater $\sum b_j$, the problem is of the unbalanced type. We convert this problem to a balanced one by introducing a fictitious destination D_4 with the requirement $35 - 17 = 18$, with all the transportation costs equal to zero. The balanced transportation table is given next (Table 3.45).

Applying the Vogel's method in the usual manner, the initial BFS is obtained as given next. This gives the transportation cost (Table 3.46): $= 3.5 + 0.5 + 2.8 + 0.4 + 0.1 + 0.12 =$ Rs. 31.

Since all $d_{ij} \geq 0$, the solution under assessment is optimal (Table 3.47).

Thus the solution of the given problem is $x_{12} = 5$, $x_{21} = 8$, $x_{23} = 4$ and min. cost = Rs. 31. (allocations in the dummy column are not considered).

TABLE 3.45

Balanced Transportation Problem

	D_1	D_2	D_3	D_4	$a_i \downarrow$
O_1	4	3	2	0	10
O_2	2	5	0	0	13
O_3	3	8	6	0	12
$b_j \rightarrow$	8	5	4	18	

TABLE 3.46

BFS Using Vogel's Method

	D_1	D_2	D_3	D_4	$a_i \downarrow$
O_1	(4)	(3) 5	(2)	(0) 5	10
O_2	(2) 8	(5)	(0) 4	(0) 1	13
O_3	(3)	(8)	(6)	(0) 12	12
$b_j \rightarrow$	8	5	4	18	

TABLE 3.47

BFS Using Vogel's Method

				vi ↓
(4)	(3)	(2)	(0)	0
2	5	0	5	
2		2		
(2)	(5)	(0)	(0)	0
8	3	4	1	
	2			
(3)	(8)	(6)	(0)	0
2	3	0	12	
1	5	6		
u_j	2	3	0	0

3.6 Introduction to an Assignment Problem

In this section, we deal with a special type of LPP generally called an "assignment problem." Although simplex method is powerful enough to solve all the LPPs, the previous types of problems may be solved by a very interesting method called the "assignment technique," which is described in this chapter. The classical

LO 6. To understand assignment problems and the use of assignment models in industry and business.

problems of assigning a number of sources (jobs) to the same number of destinations (persons) at a minimum cost (or maximum profit) are called "assignment problems." Here we assume that every person can do every job but with different levels of efficiency. These problems can include assigning people to offices, classes in rooms, drivers in cars, carriage routes, or research teams, etc.

An assignment problem involves what happens to the efficiency function when we connect each of a number of "origins" with each of the same "destination" numbers. To optimize the total efficiencies of resources, only one job must be linked to each resource or facility, and associations must be made so that they cannot be divisible among jobs and jobs cannot be divisible among resources.

3.7 The Nature of an Assignment Problem

Let there be n jobs to be performed and for doing these jobs, n persons are available. Assume that each person can do each job at a time, although with varying degrees of efficiency. Let C_{ij} be the cost (payment) of assigning the ith person to the jth job. Then the problem is to find an assignment (which job should be assigned to which person) so that the total cost for performing all jobs is minimum.

The previous assignment problem can be stated in the form of $n \times n$ matrix $[c_{ij}]$ of real numbers called the cost matrix or effectiveness matrix, as shown in Table 3.48.

TABLE 3.48

Cost Matrix

				Jobs				
		1	2	3	.....	j	....	n
	1	C_{11}	C_{12}	C_{13}	....	C_{1j}	...	C_{1n}
	2	C_{21}	C_{22}	C_{23}	....	C_{2j}	...	C_{2n}
	3	C_{31}	C_{32}	C_{33}	....	C_{3j}	....	C_{3n}
	⋮	⋮	⋮	⋮	⋮	⋮		⋮
Persons	i	C_{i1}	C_{i2}	C_{i3}	....	C_{ij}	....	C_{in}
	⋮	⋮	⋮	⋮	⋮	⋮		⋮
	⋮	⋮	⋮	⋮	⋮	⋮		⋮
	n	C_{n1}	C_{n2}	C_{n3}	....	C_{nj}	....	C_{nn}

3.8 Mathematical Formulation of an Assignment Problem

Mathematically, an assignment problem can be stated as follows:

LO 7. Formulate an assignment problem

Minimize the total cost

$$Z = \sum_{i=1}^{n} Cij\, Xij,$$

where $Xij = \{$ 1, if ith person is assigned to the jth job 0, if not subject to the conditions

(i) $\sum_{i=1}^{n} Cij = 1$, (only one job is done by the ith person, I = 1 , 2, ..., n),

(ii) $\sum_{i=1}^{n} Xij = 1$, (only one person should be assigned to the jth job, J = 1,2 ..., n.

3.8.1 Fundamental Theorems

Theorem 1 (reduction theorem): *In an assignment problem, if we add (or subtract) a constant to every element of a row (or column) of the cost matrix [c_{ij}], then an assignment which minimizes the total cost for one matrix also minimizes the total cost for the other matrix.*

Now we shall prove two important theorems on which the solution to an assignment problem is fundamentally based.

Mathematical statement of the reduction theorem. If $x_{ij} = x_{ij}$,

$$\text{Minimizes } Z = \sum_{i=1}^{n} \sum_{j=1}^{n} CijXij \text{ over all } x_{ij} = 0 \text{ or } 1 \text{ such that}$$

$$\sum_{i=1}^{n} Xij = 1, \sum_{j=1}^{n} Xij = 1, \text{then } x_{ij} = x_{ij} \text{ also minimizes}$$

$$Z = \sum_{i=1}^{n} \sum_{j=1}^{n} Cij'Xij, \text{ where } Cij \pm a_i \pm b_j; a_i, b_j$$

are some real numbers for I, $j = 1, 2, \ldots., n$.

Theorem 2: If all $C_{ij} \geq 0$, and there exists a solution $X_{ij} = x_{ij}$ which satisfies

$$\sum_{i=1}^{n} \sum_{j=1}^{n} CijXij = 0,$$

then this solution is an optimal solution for the problem (i.e. minimizes the objective function).

3.9 Assignment Algorithm (Hungarian Assignment Method)

From the two theorems, we get a powerful method known as the Hungarian assignment method for solving an assignment problem. The Hungarian method, suggested by Mr. Koning of Hungary, or the reduced matrix method or the flood's technique, is used for solving assignment problems

LO 8. Understand the use of the Hungarian method

since it is quite efficient over the other techniques. It involves a rapid reduction of the original matrix and finding a set of n independent zeros, one in each row and column, which results in an optimal solution. The various steps of the computational procedure for obtaining an optimal assignment are as follows:

Step 1. Subtract the minimum element of each row of the cost matrix and form all the elements of the respective rows. Further, modify the resulting matrix by subtracting the minimum element of each column from all the elements of the respective columns. These operations create zeros.

Step 2. Make assignments using only zeros. If a complete assignment is possible, then this is the required optimal assignment plan, and if not, then we shall modify the cost matrix to create some more zeros in it.

Thus at the end of step 1, the question arises to decide whether a complete assignment is possible or not. It can be done easily in the case of smaller cost matrices, but it is not so easy in the case of larger cost matrices. For this, we apply the following procedure:

Starting with row 1 of the matrix obtained in step 1, examine rows successively until a row with exactly one zero element is found.

Mark "☐" at this zero, as an assignment will be made there. Mark "X" at all other zeros if they lie in the column containing the assigned zero. This eliminates the possibility of marking further assignments in that column. Continue in this manner until all the rows have been examined.

When the set of rows has been completely examined, an identical procedure is applied successively to the columns. Starting with column 1, examine all the column assignments in that position (indicated by ▢) and mark "x" at all zeros in the row containing this marked zero. Proceed in this way until the last column is examined.

Continue the previous operations on rows and columns successively until reaching any of the two following situations.

    i. All the zeros have been marked "▢" or "X."
    ii. There are at least two of the remaining zeros in each row and column.

In situation (*i*), we have a maximal assignment (assignment as much as we can), and in situation (ii), we still have some zeros to be treated. To work with such situations of zeros, there is, again, an algorithm, which is complicated enough. But to avoid this highly complicated algorithm, we use the trial-and-error method to break up such ties of zeros.

Now there are two possibilities:

  (i) If there is an assignment in every row and every column (i.e., total number of marked "▢" zeros is exactly *n*), then we have obtained a complete optimal assignment plan.
  (ii) If every row and every column does not contain an assignment (i.e., total number of marked "▢" zeros is less than *n*), then we shall modify the cost matrix by creating some more zeros in it.

    Step 3. If in step 2 every row and every column of the matrix does not contain an assignment, then draw the minimum number of horizontal and vertical lines to cover all the zeros at least once in the resulting matrix.

Rule to draw the minimum number of lines.

  1. Mark ($\sqrt{\ }$) all rows that do not have any marked "▢" zeros.
  2. Mark ($\sqrt{\ }$) columns which have zeros in marked rows.
  3. Mark ($\sqrt{}$) rows (not already marked) which have assignments in marked columns.
  4. Repeat steps (ii) and (iii) until the chain of marking ends.
  5. Draw lines through unmarked rows and through marked columns. This will give us the minimal system of lines.

    Step 4. Select the smallest of the elements that do not have a line through them, subtract it from all the elements that do not have a line through them, add it to every element that lies at the intersection of two lines, and leave the remaining elements of the matrix unchanged. In the modified matrix, the number of zeros is increased (never decrease) as opposed to those in step 2.

Now apply step 2 to this new matrix. If a complete optimal assignment is still not possible in this matrix, then repeat steps 3 and 4 iteratively. Continue the process until the minimum number of lines is *n*. Thus exactly one marked "▢" zero in each row and each column of the matrix is obtained. The assignment corresponding to these marked "▢" zeros will give the optimal assignment.

Examples based on an assignment problem

EXAMPLE 1:

A computer centre has three expert programmers. The centre needs three application programmes to be developed. We estimate the computer time in minutes required by the experts to the application programmers as follows (Table 3.49).

Assign the programmers to the programmes in such a way that the total computer time is least. *(BT level 3)*

Solution:
We shall solve this problem stepwise so that the readers may understand the method described earlier.

Step 1. Subtracting the minimum element of each row from every element of the corresponding row, the matrix reduces to Table 3.50.

Again, subtract the minimum element of each column from every element of the corresponding column. The previous matrix reduces to Table 3.51.

TABLE 3.49

Data for Example 1

		Programme		
		A	B	C
Programmer	1	120	100	80
	2	70	90	110
	3	110	140	120

TABLE 3.50

Step 1

		Programme		
		A	B	C
Programmer	1	40	20	0
	2	0	20	40
	3	0	30	10

TABLE 3.51

Step 1

		Programme		
		A	B	C
Programmer	1	40	0	0
	2	0	0	40
	3	0	10	10

TABLE 3.52

Step 2

	A	B	C		A	B	C		A	B	C
1	40	0	0	1	40	⊗	[0]	1	40	⊗	[0]
2	⊗	0	40	2	⊗	0	40	2	⊗	[0]	40
3	[0]	10	10	3	[0]	10	10	3	[0]	10	10

Step 2. Now we examine rows successively until we get the row having one zero only and make a mark "▢" at this zero. Here such a row is the third row in which there is only one zero we mark "▢" and cross all zeros in the first column corresponding to this mark "▢" Again, we examine columns successively and get the third column with only one zero and mark "▢" at this zero and cancel all other zeros in the first row corresponding to this mark "▢." Now again, we start with rows if some zeros are left unmarked and continue the process from row to column, and then to rows and columns until all the zeros are marked either by "▢" or are crossed. Proceeding in this way, we get the following allocations.

Successive allocations (making of ▢) are shown in the following matrices in order (Table 3.52).

In the final table, all the zeros have been either assigned (marked ▢) or crossed out. In this table, we observe that every row and every column have one assignment, so we have reached the solution of the problem. And thus the final assignment which is optimal for the given problem is

$$1 \to C, 2 \to B, 3 \to A$$

(i.e., programmer 1 is allotted programme C, programmer 2 is allotted programme B, and programmer 3 is allotted programme A). In this case, the total computer time (from the given table)

= 80 + 90 + 110 = 280 minutes, which is the least amount of time.

EXAMPLE 2:

Solve the following minimal assignment problems (Table 3.53). *(BT level 3)*

Solution:

Step 1. Subtracting the minimum element in each row from every element of the row and then in the resulting matrix and subtracting the minimum element of each column from every element of the corresponding column, we get the following two matrices in order (Table 3.54).

Step 2. Now we examine rows in the second table successively until we reach the second row with only one zero and mark "[0]" at this zero and cancel all other zeros

TABLE 3.53

Data for Example 2

		Job				
		1	2	3	4	5
	A	8	4	2	6	1
	B	0	9	5	5	4
Persons	C	3	8	9	2	6
	D	4	3	1	0	3
	E	9	5	8	9	5

TABLE 3.54

Step 1

	1	2	3	4	5		1	2	3	4	5
A	7	3	1	5	0	A	7	3	0	5	0
B	0	9	5	5	4	B	0	9	4	5	4
C	1	6	7	0	4	C	1	6	6	0	4
D	4	3	1	0	3	D	4	3	0	0	3
E	4	0	3	4	0	E	4	0	2	4	0

in first column corresponding to this " $\boxed{0}$ ". Again, we proceed to the third row with only one zero and mark " $\boxed{}$ " at this zero and cancel all other zeros in the fourth column corresponding to this " $\boxed{}$ ".

Again, proceeding to the fourth row, we get only one unmarked zero in this row, so we mark " $\boxed{}$ " at this zero and cancel all other zeros in the third column corresponding to this " $\boxed{}$ " . Proceeding again to the fifth row, we get two zeros in this row so that we do not mark " $\boxed{}$ " at any zero and proceed to the columns. Proceeding successively in the columns, we get only one zero in the second column at which we mark " $\boxed{}$ " and cancel all other zeros in the fifth row corresponding to it. Proceeding similarly, we reach the fifth column, which has only one unmarked zero, and mark it " $\boxed{}$ " and cancel all other zeros in the first row corresponding to it. All the markings are shown in Table 3.55 successively.

In Table 3.55 (v), we observe that all zeros are either marked " $\bigotimes$ " or are cancelled. Also, there is an assignment " $\bigotimes$ " in every row and every column; hence, we have reached the solution of the problem. The optimal solution to the problem is

$$A \to 5, B \to 1, C \to 4, D \to 3, E \to 2.$$

TABLE 3.55

Step 2

	(i)						(ii)			
7	3	0	5	0		7	3	0	5	0
[0]	9	4	5	4		[0]	9	4	5	4
1	6	6	0	4		1	6	6	[0]	4
4	3	0	0	3		4	3	0	⊗	3
4	0	2	4	0		4	0	2	4	0

	(iii)						(iv)			
7	3	⊗	5	0		7	3	⊗	5	0
[0]	9	4	5	4		[0]	9	4	5	4
1	6	6	[0]	4		1	6	6	[0]	4
4	3	[0]	⊗	3		4	3	[0]	0	3
4	0	2	4	0		4	[0]	2	4	0

	(v)			
7	3	⊗	5	0
[0]	9	4	5	4
1	6	6	[0]	4
4	3	[0]	⊗	3
4	0	2	4	⊗

EXAMPLE 3:

Solve the following minimal assignment problem (Table 3.56). *(BT level 3)*

Solution:

Step 1. Subtracting the minimum element in each row from every element of that row we get the following matrix (Table 3.57).

Again, subtracting the minimum element in each column of Table 3.57, the resulting matrix from the corresponding column forms the matrix in Table 3.58.

Step 2. Examining the rows successively and giving assignment "☐" in the rows containing only one unmarked row and cancelling all zeros in the columns corresponding to the assignment "☐", we get Table 3.59.

TABLE 3.56

Data for Example 3

		Man				
		1	2	3	4	5
	I	12	8	7	15	4
	II	7	9	17	14	10
Job	III	9	6	12	6	7
	IV	7	6	14	6	10
	V	9	6	12	10	6

TABLE 3.57

Step 1

	1	2	3	4	5
I	8	4	3	11	0
II	0	2	10	7	3
III	3	0	6	0	1
IV	1	0	8	0	4
V	3	0	6	4	0

TABLE 3.58

Step 1

	1	2	3	4	5
I	8	4	0	11	0
II	0	2	7	7	3
III	3	0	3	0	1
IV	1	0	5	0	4
V	3	6	3	4	0

TABLE 3.59

Step 2

	1	2	3	4	5
I	8	4	0	11	0
II	$\boxed{0}$	2	7	7	3
III	3	0	3	0	1
IV	1	0	5	0	4
V	3	6	3	4	$\boxed{0}$

Again, examining the columns successively and giving assignment "$\boxed{0}$" in the columns containing only one unmarked zero and cancelling all zeros in the row corresponding to the assignment "$\boxed{0}$", we get the following matrix in Table 3.60.

Again, examining rows successively, we observe that all rows contain two unmarked zeros, and then examining columns successively, we observe that every column also contains two unmarked zeros. Thus we examine rows again, and upon reaching the third row, we mark "$\boxed{0}$" one of the zeros and cancel zeros in the corresponding columns. Selecting one of the two zeros in the third row and continuing the process from rows to columns and then from columns to rows in succession, we get the following two matrices, which provide the optimal solutions, as at this stage, all zeros have been either assigned or crossed, and every row and every column have one assignment (Table 3.61).

Hence the problem has two optimum solutions:

   (i) I→3, II→1, III →2, IV→4, V→5, and
  (ii) I→3, II→1, III →4, IV→2, V→5.

In both cases, the total minimum cost will be equal.

TABLE 3.60

Step 2

	1	2	3	4	5
I	8	4	$\boxed{0}$	11	0
II	$\boxed{0}$	2	7	7	3
III	3	0	3	0	1
IV	1	0	5	0	4
V	3	6	3	4	$\boxed{0}$

TABLE 3.61

Step 2

	1	2	3	4	5		1	2	3	4	5
I	8	4	⊗	11	⊗	I	8	4	$\boxed{0}$	11	⊗
II	$\boxed{0}$	2	7	7	3	II	$\boxed{0}$	2	7	7	3
III	3	$\boxed{0}$	3	⊗	1	III	3	$\boxed{0}$	3	⊗	1
IV	1	⊗	5	$\boxed{0}$	4	IV	1	⊗	5	$\boxed{0}$	4
V	3	6	3	4	$\boxed{0}$	V	3	6	3	4	$\boxed{0}$

EXAMPLE 4:

A department head has four subordinates, and four tasks have to be performed. Subordinates differ in efficiency, and tasks differ in their intrinsic difficulty. The time each subordinate would take to perform each task is given in the effectiveness matrix in Table 3.62. How should the tasks be allocated, one to each subordinate, to minimize the total work hours? *(BT level 3)*

Solution:
We shall solve this problem step by step to understand the method described earlier.

   Step 1. Subtracting the minimum element of each row from every element of the corresponding row, the matrix reduces to Table 3.63.

   Now subtracting the minimum element of each column from every element of the corresponding column, the matrix reduces to Table 3.64.

TABLE 3.62

Data Example 4

		Subordinates			
		I	II	III	IV
Tasks	A	8	26	17	11
	B	13	28	4	26
	C	38	19	18	15
	D	19	26	24	10

TABLE 3.63

Step 1

	I	II	III	IV
A	0	18	9	3
B	9	24	0	22
C	23	4	3	0
D	9	16	14	0

TABLE 3.64

Step 1

	I	II	III	IV
A	0	14	9	3
B	9	20	0	22
C	23	0	3	0
D	9	12	14	0

Step 2. Starting with row 1 of this reduced matrix, we examine the rows one by one until a row containing only one zero element is found. We mark "$\boxed{0}$" at this zero (i.e., make an assignment and mark a cross "X" over all zeros lying in the column containing the assigned zero). Continue in this manner until all the rows have been examined. Thus we get the following matrix (Table 3.65).

Now starting with column 1, examine all the columns until a column containing only one zero is found. We mark "$\otimes$" at this zero (i.e., make an assignment and cross out other zeros if lying in the row containing this marked zero). Continue in this manner until all the columns have been examined. Thus we get the following matrix (Table 3.66).

At this stage, we see that all zeros have been either assigned or crossed out and every row and every column have an assignment. Hence an optimal solution has been obtained. The optimal solution is

$$A \rightarrow I, B \rightarrow III, C \rightarrow II, D \rightarrow IV.$$

And the minimum total work hours (from the original matrix)

$$= 8 + 4 + 19 + 10 = 41.$$

TABLE 3.65

Step 2

	I	II	III	IV
A	$\boxed{0}$	14	9	3
B	9	20	$\boxed{0}$	22
C	23	0	3	$\otimes$
D	9	12	14	$\boxed{0}$

TABLE 3.66

Step 2

	I	II	III	IV
A	$\boxed{0}$	14	9	3
B	9	20	$\boxed{0}$	22
C	23	$\boxed{0}$	3	$\otimes$
D	9	12	14	$\boxed{0}$

EXAMPLE 5:

Five wagons are available at five stations 1, 2, 3, 4, and 5. They are required at five stations I, II, III, IV, and V. The mileages between various stations are given in Table 3.67.

How should the wagons be transported so as to minimize the total mileage covered? *(BT level 3)*

Solution:

Step 1. Subtracting the smallest element in each row from every element of the corresponding row and then subtracting the smallest element of each column from every element of the corresponding column, we get the matrix in Table 3.68.

Step 2. Giving zero assignment in the usual manner, we get the following table in which we observe that rows 4 and 5 in columns 4 and 5 have no assignments. So we proceed to the next step (Table 3.69).

TABLE 3.67

Data for Example 5

	I	II	III	IV	V
1	10	5	9	18	11
2	13	9	6	12	14
3	3	2	4	4	5
4	18	9	12	17	15
5	11	6	14	19	10

TABLE 3.68

Step 1

4	0	0	11	3
6	3	0	4	5
0	0	2	0	0
8	0	3	6	3
4	0	8	11	1

TABLE 3.69

Step 2

4	0	0	11	3
6	3	[0]	4	5
[0]	⊗	2	[0]	⊗
8	⊗	3	6	3
4	⊗	8	11	1

TABLE 3.70

Step 3

	L_1				
4	[0]	0	11	3	√(4)
L_2 6	3	[0]	4	5	
L_3 [0]	⊗	2	[0]	⊗	
8	⊗	3	6	3	√(1)
4	⊗	8	11	1	√(2)
	√(3)				

TABLE 3.71

Step 4

3	0	3	10	2
6	4	0	4	5
0	1	2	0	0
7	0	2	5	2
3	0	7	10	0

Step 3. Here we draw a minimum number of lines to cover all the zeros at least once. The number of such lines is three (Table 3.70).

Step 4. The smallest element among all the uncovered elements is 1. Subtracting this element 1 from all the uncovered elements, adding to every element that lies at the intersection of two lines, and leaving remaining elements unchanged in the table of step 3 reduces to Table 3.71.

Step 5. Giving zero assignments in the usual manner, we observe that row 4 and column 4 have no assignments. So we draw a minimum number of lines to cover all zeros at least once. The number of such lines is four (Table 3.72).

TABLE 3.72

Step 5

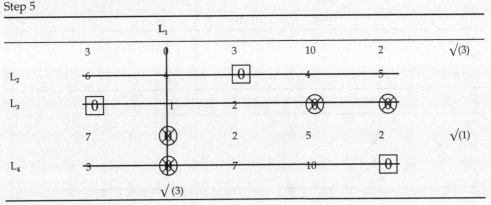

TABLE 3.73

Step 6

1	0	1	8	0
6	6	0	4	5
0	3	2	0	0
5	0	0	3	0
3	2	7	10	0

Step 6. The smallest element among all the uncovered elements is two. Subtracting element 2 from all the uncovered elements, adding it to every element that lies at the intersection of two lines, and leaving the remaining elements unchanged in the table of step 5 reduces the matrix to Table 3.73.

Step 7. Giving zero assignments in the usual manner, we observe that row 4 and column 4 have no assignments. So we draw a minimum number of lines to cover all zeros at least once. The number of such lines is four (Table 3.74).

Step 8. The smallest element among all the uncovered elements is one. Subtracting element 1 from all the uncovered elements, adding it to every element that lies at the intersection of two lines, and leaving the remaining elements unchanged in the table of step 7 reduces the matrix to Table 3.75.

TABLE 3.74

Step 7

TABLE 3.75

Step 8

0	0	1	7	0
5	6	0	3	5
0	4	3	0	1
4	0	0	2	0
2	2	7	9	0

TABLE 3.76

Step 9

	I	II	III	IV	V
1	$\boxed{0}$	⊗	1	7	⊗
2	5	6	$\boxed{0}$	3	5
3	⊗	4	3	$\boxed{0}$	1
4	4	$\boxed{0}$	⊗	2	⊗
5	2	2	7	9	$\boxed{0}$

Step 9. Giving zero assignments as usual, we observe that each row and each column has an assignment. Hence the optimal assignment is Table 3.76.

$$1 \rightarrow I, 2 \rightarrow III, 3 \rightarrow IV, 4 \rightarrow II, 5 \rightarrow V$$

The minimum total mileage from the first matrix = $10 + 6 + 4 + 9 + 10 = 39$ miles.

3.10 The Maximal Assignment Problem

We have discussed the problem of minimizing the total cost. Sometimes the assignment problem deals with the maximization of an objective function rather than minimizing it.

LO 9. Understand the solving of the maximization assignment problem

For example, the problem may be to assign persons to jobs in such a way that the expected profit is the maximum. This problem may be solved easily by first converting it to a minimization problem, and then we can use the usual procedure of the assignment algorithm. This conversion can be very easily done by modifying the given profit matrix to the cost matrix in either of the following two ways.

i. *Select the greatest element of the given profit matrix and then subtract each element of the matrix from the greatest element to get the modified matrix.*

For if $[c_{ij}]$ is the given profit matrix and c_{rk} is the greatest element of this matrix, then the modified matrix will be $[c_{ij}]$, where $c_{ij} = c_{rk} - c_{ij}$. It can be shown that

$$x_{ij} = x_{ij} \text{ maximizes } Z = \sum\sum C_{ij}x_{ij} = \sum\sum(C_{rk} - c_{ij})x_{ij} = \sum\sum c_{rk}x_{ij} = nc_{rk} - Z.$$

ii. *Place the minus sign before each element of the profit matrix to get the modified matrix.* In this case, if $[c_{ij}]$ is the given profit matrix, then the modified matrix will be $[c_{ij}]$, where $c_{ij} = -c_{ij}$. It can be shown that $x_{ij} = x_{ij}$ maximizes $Z = \sum\sum C_{ij}x_{ij}$ then minimizes $Z' = \sum\sum C_{ij}'x_{ij}$.

Examples based on maximization

EXAMPLE 1:

A company has five jobs that need to be done. The following matrix (Table 3.77) shows the return in rupees on assigning ith (i = 1, 2, 3, 4, 5) machine to the jth job (j = A, B, C, D, E). Assign the five jobs to the five machines so as to maximize the total expected profit. *(BT level 3)*

Solution:

First, we shall convert the problem from maximization to minimization. The greatest element of the given matrix is 14. By subtracting all the elements of the given matrix from 14, the modified matrix is obtained. Now we shall follow the usual procedure of solving an assignment problem (Table 3.78).

Step 1. Subtracting the minimum element of each row from every element of the corresponding row and then subtracting the minimum element of each column from every element of the corresponding column, the matrix reduces to Table 3.79.

TABLE 3.77

Data For Example 1

		Jobs				
		A	B	C	D	E
	1	5	11	10	12	4
	2	2	4	6	3	5
Machines	3	3	12	5	14	6
	4	6	14	4	11	7
	5	7	9	8	12	5

TABLE 3.78

Minimization Problem

9	3	4	2	10
12	10	8	11	9
11	2	9	0	8
8	0	10	3	7
7	5	6	2	2

TABLE 3.79

Step 1

3	1	2	0	7
0	2	0	3	0
7	2	9	0	7
4	0	10	3	6
1	3	4	0	6

Step 2. Giving zero assignments in the usual manner, we observe that rows 3 and 5 and columns 3 and 5 have no assignments. So we draw the minimum number of lines to cover all the zeros at least once. The number of such lines is three (Table 3.80).

Step 3. In Table 3.80, the smallest of the uncovered elements is one. Subtracting this element from all covered elements, adding it to each element (i.e., at the intersection of two lines), and leaving all remaining elements unchanged, the reduced matrix is in Table 3.81.

Step 4. Giving zero assignments in the usual manner, we observe that row 1 and column 5 have no zero assignments. So we again draw the minimum number of lines to cover all zeros at least once. The number of such lines is four (Table 3.82).

TABLE 3.80

Step 2

			L_1		
3	1	2	[0]	7	√(4)
L_2 [0]	2	(X)	5	(X)	
7	2	9	(X)	7	√(1)
L_3 4	[0]	10	3	6	
1	3	4	(X)	6	√(2)
			√(3)		

TABLE 3.81

Step 3

2	0	1	0	6
0	2	0	4	0
6	1	8	0	6
4	0	10	4	6
0	2	3	0	5

TABLE 3.82

Step 4

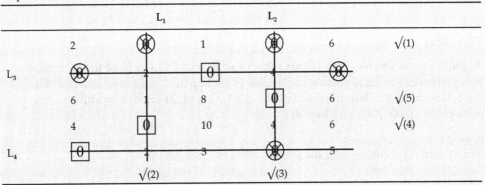

Step 5. In the last reduced table, the smallest uncovered element is one. Subtracting element 1 from all uncovered elements, adding it to each element that lies at the intersection of two lines, and leaving the remaining elements unchanged, the reduced matrix is in Table 3.83.

Step 6. Giving zero assignments in the usual manner, we observe that each row and each column has an assignment (Table 3.84).

Hence the optimal assignment of jobs to maximize the profit is

$$\text{Machine} \to \text{job}; 1 \to C, 2 \to E, 3 \to D, 4 \to B, 5 \to A.$$

From the given matrix, the maximum profit = 10 + 5 + 14 + 14 + 7 = Rs. 50.

TABLE 3.83

Step 5

1	0	0	0	5
0	3	0	5	0
5	1	7	0	5
3	0	9	4	5
0	3	3	1	5

TABLE 3.84

Step 6

	A	B	C	D	E
1	1	⊗	0	⊗	5
2	⊗	3	⊗	5	0
3	5	1	7	0	5
4	3	0	9	4	5
5	0	3	3	1	5

EXAMPLE 2:

Alpha Corporation has four plants which can manufacture any of the company's four products. Production costs differ from plant to plant, as do sales revenue. From the following data, determine which product each plant should produce to maximize profit (Table 3.85). *(BT level 3)*

Solution:

Since profit = revenue – cost, the profit matrix is Table 3.86.

TABLE 3.85

Data for Example 2

Sales revenue (Rs.' 1000) Product					Production cost (Rs.' 1000) Product				4
Plant ↓					Plant ↓				
	1	2	3	4		1	2	3	
A	50	68	49	6	A	49	60	45	61
B	60	70	51	74	B	55	63	45	69
C	55	67	53	70	C	52	62	49	68
D	58	65	54	69	D	55	64	48	66

TABLE 3.86

Profit Matrix

	1	2	3	4
A	1	8	4	1
B	5	7	6	5
C	3	5	4	2
D	3	1	6	3

This is the maximization problem. To convert this to a minimization problem, multiply each element of the given matrix by −1. Thus the resulting matrix is in Table 3.87.

Now we solve this assignment problem in the usual manner.

Step 1. Subtracting the smallest element of each row from every element of the corresponding row and then subtracting the smallest element of each column from every element of the corresponding column, we get Table 3.88.

Step 2. Giving zero assignments in the usual manner, we observe that there is zero assignment in each row and each column in Table 3.89.

Hence the optimal assignment for maximum profit is

$$\text{Plant} \rightarrow \text{Product}:$$

$$A \rightarrow 2, B \rightarrow 4, C \rightarrow 1, D \rightarrow 3.$$

TABLE 3.87

Minimization Problem

−1	−8	−4	−1
−5	−7	−6	−5
−3	−5	−4	−2
−3	−1	−6	−3

TABLE 3.88

Step 1

5	0	4	5
0	0	1	0
0	0	1	1
1	5	0	1

TABLE 3.89

Step 2

	1	2	3	4
A	5	☐0	4	5
B	⊗	⊗	1	☐0
C	☐0	⊗	1	1
D	1	5	☐0	1

3.11 Unbalanced Assignment Problem

In case the number of tasks (jobs) is not equal to the number of facilities (persons), the assignment problem is called an *unbalanced assignment problem*. Thus the cost matrix of an unbalanced assignment problem is not a square matrix.

LO 10. Understand and solve unbalanced assignment problems

For the solution of such a problem, we add the dummy (fictitious) rows or columns to the given matrix with zero costs to form it in a square matrix. Then the usual assignment algorithm can be applied to this resulting balanced assignment problem.

Examples based on an unbalanced assignment problem

EXAMPLE 1:

A department head has four tasks to be performed and three subordinates. The subordinates differ in efficiency. The estimates of the time each subordinate would take to perform a task are given in the matrix in Table 3.90. How should he allocate the tasks, one to each subordinate, to minimize the total work hours? *(BT level 3)*

Solution:
Since the matrix is not square, it is an unbalanced assignment problem. We introduce one fictitious subordinate (fourth column with zero costs) to get a square matrix. Thus the resulting matrix is shown in Table 3.91. Now the problem can be solved by the usual method.

TABLE 3.90

Data for Example 1

		Subordinates		
		1	2	3
Tasks	I	9	26	15
	II	13	27	6
	III	35	20	15
	IV	18	30	20

TABLE 3.91

Balanced Assignment Problem

	1	2	3	4
I	9	26	15	0
II	13	27	6	0
III	35	20	15	0
IV	18	30	20	0

Step 1. Subtracting the minimum element of each row from every element of the corresponding row and then subtracting the minimum element of each column from every element of the corresponding column, the matrix reduces to Table 3.92.

Step 2. Giving zero assignments in the usual manner, we observe that each row and each column have zero assignments (Table 3.93).

TABLE 3.92

Step 1

0	6	9	0
4	7	0	0
26	0	9	0
9	10	14	0

TABLE 3.93

Step 2

	1	2	3	4
I	[0]	6	9	⊗
II	4	7	[0]	⊗
III	26	[0]	9	⊗
IV	9	10	14	[0]

Hence the optimal assignment is as follows.

$$\text{Tasks} \rightarrow \text{subordinates}, I \rightarrow 1, II \rightarrow 3, III \rightarrow 2.$$

Task IV remains unassigned.
From the original matrix, the total time (work hours) = 9 + 6 + 20 = 35 hours.

EXAMPLE 2:

A company has four machines to do three jobs. Each job can be assigned to one and only one machine. The cost of each job on each machine is given in Table 3.94.
    What are the job assignments which will minimize the cost? *(BT level 3)*

Solution:
Since the matrix is not square, it is an unbalanced assignment problem. We introduce one fictitious job D (4th row with zero costs) to get a square matrix (Table 3.95).
    Now the problem can be solved by the usual method.
    Step 1. Subtracting the smallest element in each row from every element of the corresponding row and then subtracting the smallest element in each column from every element of the corresponding column, the matrix reduces to Table 3.96.
    Step 2. Giving zero assignments in the usual manner, we observe that rows 2 and 3 and columns 3 and 4 have no zero assignments. So we proceed to the next step (Table 3.97).

TABLE 3.94

Data for Example 2

		Machine			
		W	X	Y	Z
	A	18	24	28	32
Job	B	8	13	17	19
	C	10	15	19	22

TABLE 3.95

Balanced Assignment Problem

	W	X	Y	Z
A	18	24	28	32
B	8	13	17	19
C	10	15	19	22
D	0	0	0	0

TABLE 3.96

Step 1

0	6	10	14
0	5	9	11
0	5	9	12
0	0	0	0

TABLE 3.97

Step 2

Step 3. Here we draw the minimum number of lines to cover all the zeros at least once. The number of such lines is two.

Step 4. The smallest elements among all the uncovered elements is five. Subtracting element 5 from all uncovered elements, adding it to every element that lies at the intersection of two lines, and leaving the remaining elements unchanged, the table of step 3 reduces to the following form of Table 3.98.

Step 5. Giving zero assignments in the usual manner, we observe that row 3 and column 4 have no zero assignments (Table 3.99).

So, again, we draw the minimum number of lines to cover all the zeros at least once. The number of such lines is three.

Step 6. Now the smallest element among all uncovered elements is four. Subtracting element 4 from all uncovered elements, adding it to all the elements that lie at the intersection of two lines, and leaving the remaining elements unchanged, the table of step 5 reduces to the following form in Table 3.100.

TABLE 3.98

Step 3

0	1	5	9
0	0	4	6
0	0	4	7
5	0	0	0

TABLE 3.99

Step 4

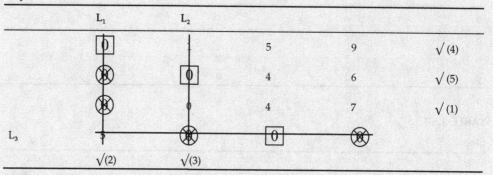

	L_1		L_2			
	$\boxed{0}$			5	9	$\sqrt{}$ (4)
	⊗		$\boxed{0}$	4	6	$\sqrt{}$ (5)
	⊗		0	4	7	$\sqrt{}$ (1)
L_3	5		⊗	$\boxed{0}$	⊗	
	$\sqrt{}$(2)		$\sqrt{}$(3)			

TABLE 3.100

Step 5

0	1	1	5
0	0	0	2
0	0	0	3
9	4	0	0

TABLE 3.101

Step 6

		W	X	Y	Z			W	X	Y	Z
	A	$\boxed{0}$	1	1	5		A	$\boxed{0}$	1	1	5
	B	⊗	$\boxed{0}$	⊗	2		B	⊗	⊗	$\boxed{0}$	2
Job	C	⊗	⊗	$\boxed{0}$	3	Job	C	⊗	$\boxed{0}$	⊗	3
	D	9	4	⊗	$\boxed{0}$		D	9	4	⊗	$\boxed{0}$

Step 7. Giving zero assignments in the usual manner, we get the following two assignments (Table 3.101).

Hence the optimal assignment of jobs to machines for minimum cost are as follows.

$$\text{Job } A \to W, B \to X, C \to Y, \text{ and Job } A \to W, B \to Y, C \to X.$$

In both cases, no job is assigned to machine Z.

EXAMPLE 3:

There are three persons P_1, P_2, and P_3 and five jobs J_1, J_2...., J_5. Each person can do only one job, and a job is to be done by one person only. Using the Hungarian method, find which two jobs should be left undone in the following cost-minimizing assignment problem (Table 3.102). *(BT level 3)*

Solution:
Introducing two fictitious persons P_4, P_5 (fourth and fifth rows with zero costs), the given unbalanced problem reduces to the following balanced assignment problem (Table 3.103).

Now the problem can be solved by the usual method.

Step 1. Subtracting the smallest element in each row from every element of the corresponding row and then subtracting the smallest element of each column from every element of the corresponding column, the matrix reduces to Table 3.104.

Step 2. Giving zero assignments in the usual manner, we observe that row 3 and column 5 have no zero assignments. So we draw the minimum number of lines to cover all zeros at least once. The number of such zeros is four (Table 3.105).

TABLE 3.102

Data for Example 3

	J_1	J_2	J_3	J_4	J_5
P_1	7	8	6	5	9
P_2	9	6	7	6	10
P_3	8	7	9	5	6

TABLE 3.103

Balanced Assignment Problem

	J_1	J_2	J_3	J_4	J_5
P_1	7	8	6	5	9
P_2	9	6	7	6	10
P_3	8	7	9	5	6
P_4	0	0	0	0	0
P_5	0	0	0	0	0

TABLE 3.104

Step 1

2	3	1	0	4
3	0	1	0	4
3	2	4	0	1
0	0	0	0	0
0	0	0	0	0

TABLE 3.105

Step 2

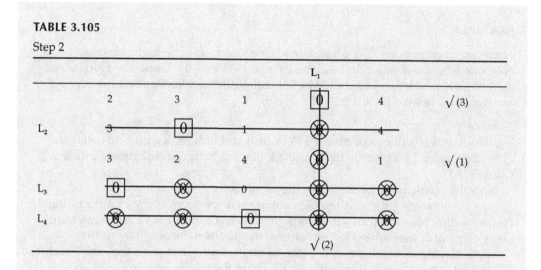

Step 3. The smallest of the uncovered elements is one. Subtracting element 1 from all uncovered elements, adding it to all the elements that lie at the intersection of two lines, and leaving the remaining elements unchanged, we get Table 3.106.

Step 4. Giving zero assignments in the table of step 5, we get the following two assignments (Table 3.107).

TABLE 3.106

Step 3

1	2	0	0	3
3	0	1	1	4
2	1	3	0	0
0	0	0	1	0
0	0	0	1	0

TABLE 3.107

Step 4

	J_1	J_2	J_3	J_4	J_5		J_1	J_2	J_3	J_4	J_5
P_1	1	2	⊡0	⊗	3	P_1	1	2	0	⊡0	3
P_2	3	⊡0	1	1	4	P_2	3	⊡0	⊗	1	4
P_3	2	1	3	⊡0	⊗	P_3	2	1	3	⊗	⊡0
P_4	⊡0	⊗	⊗	1	⊗	P_4	⊡0	⊗	⊗	1	⊗
P_5	⊗	⊗	⊗	1	⊡0	P_5	⊗	⊗	⊡0	1	⊗

Hence the two optimal assignment are

(i) $P_1 \rightarrow J_3, P_2 \rightarrow J_2, P_3 \rightarrow J_4$, ; Jobs J_1 and J_5 left undone, and

(ii) $P_1 \rightarrow J_4, P_2 \rightarrow J_2, P_3 \rightarrow J_5$, ; Jobs J_1 and J_3 left undone.

In both cases, the minimum cost is 17.

EXAMPLE 4:

Use the Hungarian method to find which of the two jobs should be left undone when each of the four persons will do only one job in the following cost-minimizing assignment problem (Table 3.108). *(BT level 3)*

Solution:

Introducing two fictitious persons P_5 and P_6 (fifth and sixth rows with zero costs), the given unbalanced problem reduces to the following balanced assignment problem (Table 3.109).

Now the problem can be solved by the usual method.

Step 1. Subtracting the smallest element in each row from every element of the corresponding row and then subtracting the smallest element in each column from every element of the corresponding column, we get Table 3.110.

Step 2. Giving zero assignments in the usual manner, we observe that rows 2, 3, and 4 and columns 3, 4, and 5 have no zero assignments. So we draw the minimum number of lines to cover all zeros at least once and get Table 3.111.

TABLE 3.108

Data for Example 4

		Job					
		J_1	J_2	J_3	J_4	J_5	J_6
	P_1	10	9	11	12	8	5
	P_2	12	10	9	11	9	4
Person	P_3	8	11	10	7	12	6
	P_4	10	7	8	10	10	5

TABLE 3.109

Balanced Assignment Problem

		J_1	J_2	J_3	J_4	J_5	J_6
	P_1	10	9	11	12	8	5
	P_2	12	10	9	11	9	4
	P_3	8	11	10	7	12	6
Person	P_4	10	7	8	10	10	5
	P_5	0	0	0	0	0	0
	P_6	0	0	0	0	0	0

TABLE 3.110

Step 1

5	4	6	7	3	0
8	6	5	7	5	0
2	5	4	1	6	0
5	2	3	5	5	0
0	0	0	0	0	0
0	0	0	0	0	0

Step 3. The smallest of the uncovered elements is one. Subtracting element 1 from all the uncovered elements, adding it to each element that lies at the intersection of two lines, and leaving the remaining elements unchanged, we get Table 3.112.

Step 4. Giving zero assignments in the usual manner, we observe that rows 2 and 4 and columns 3 and 5 have no zero assignments (Table 3.113).

So we draw the minimum number of lines to cover all zeros at least once and get Table 3.113.

TABLE 3.111

Step 2

TABLE 3.112

Step 3

4	3	5	6	2	0
7	5	4	6	4	0
1	4	3	0	5	0
4	1	2	4	4	0
0	0	0	0	0	1
0	0	0	0	0	1

TABLE 3.113

Step 4

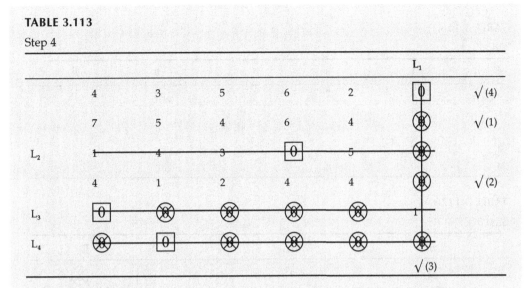

Step 5. The smallest of the uncovered elements is one. Subtracting element 1 from all the uncovered elements, adding it to each element that lies at the intersection of two lines, and leaving the remaining elements unchanged, we get Table 3.114.

Step 6. Giving zero assignments in the usual manner, we observe that row 2 and column 5 have no zero assignments (Table 3.115).

TABLE 3.114

Step 5

3	2	4	5	1	0
6	4	3	5	3	0
1	4	3	0	5	1
3	0	1	3	3	0
0	0	0	0	0	2
0	0	0	0	0	2

TABLE 3.115

Step 6

						L₁	
	3	2	4	5	1	0	√ (3)
	6	4	3	5	3	0	√ (1)
L₂	1	4	3	0	5	1	
L₃	3	0	1	3	3	0	
L₄	0	0	0	0	0	2	
L₅	0	0	0	0	0	2	
						√ (2)	

TABLE 3.116.

Step 7

2	1	3	4	0	0
5	3	2	4	2	0
1	4	3	0	5	2
3	0	1	3	3	1
0	0	0	0	0	3
0	0	0	0	0	3

TABLE 3.117.

Step 8

	J_1	J_2	J_3	J_4	J_5	J_6
P_1	2	1	3	4	$\boxed{0}$	⊗
P_2	5	3	2	4	2	$\boxed{0}$
P_3	1	4	3	$\boxed{0}$	5	2
P_4	3	$\boxed{0}$	1	3	3	1
P_5	$\boxed{0}$	⊗	⊗	⊗	⊗	3
P_6	⊗	⊗	$\boxed{0}$	⊗	⊗	3

So we again draw the minimum number of lines to cover all zeros at least once and get Table 3.115.

Step 7. The smallest of the uncovered elements is one. Subtracting element 1 from all the uncovered elements, adding it to each element that lies at the intersection of two lines, and leaving the remaining elements unchanged, we get Table 3.116.

Step 8. Giving zero assignments in the usual manner, we observe that each row and each column has an assignment (Table 3.117).

Hence the optimal assignment of persons is

$$P_1 \rightarrow J_5, P_2 \rightarrow J_6, P_3 \rightarrow J_4, P_4 \rightarrow J_2; \text{Jobs } J_1 \text{ and } J_3 \text{ left undone.}$$

CHAPTER SUMMARY

- The transportation problem deals with the distribution of goods from several supply sources to the number of destinations.
- The following are the three methods to develop an initial solution to the transportation problem:

North-west corner method

Low-cost method

Penalty cost method (Vogel's approximation method)

- The following are the two methods to obtain an optimum solution of the initial solution:

  Stepping-stone method

  U–V method

- Degeneracy and the unbalanced problem are the special cases of the transportation problem.

- The assignment problem can be considered a special case of the transportation problem.

- The Hungarian method is the most popular method for solving the assignment problem.

Questions

1. Explain the balanced and unbalanced transportation problem. *(BT level 2)*

2. Explain when and how we use dummy sources in solving a transportation problem. *(BT level 2)*

3. Explain degeneracy and how to resolve it in a transportation problem. *(BT level 2)*

4. Explain the BFS of a transportation problem. What are the methods to determine BFS? *(BT level 2)*

5. Name and compare methods to obtain an optimum solution for the transportation problem. *(BT level 1 and 5)*

6. Explain an assignment problem and give its application. *(BT level 2)*

7. Compare a transportation problem and an assignment problem. *(BT level 4)*

8. How could an assignment problem be solved using the transportation approach? *(BT level 2)*

4

Network Models

Learning Outcomes: After studying this chapter, the reader should be able to

LO 1. Understand a network, draw activity on an arc or arrow, and create activity on node networks. (BT level 1 and 2)

LO 2. Determine a critical path. (BT level 3)

LO 3. Compute forward and backward passes for a project. (BT level 3)

LO 4. Calculate the variance and standard deviation of activity times. (BT level 3)

LO 5. Conduct cost analysis of projects. (BT level 3 and 4)

LO 6. Describe a shortest path problem. (BT level 1 and 2)

LO 7. Understand and apply different algorithms for a shortest path problem. (BT level 2 and 3)

LO 8. Understand and apply an algorithm to solve the travelling salesman problem. (BT level 2 and 3)

LO 9. Understand and apply an algorithm to solve the Chinese postman problem (BT level 2 and 3)

LO 10. Become acquainted with the uses of network models. (BT level 3 and 4)

4.1 Introduction

In modern technological society, networks pervade everyday life. We do this over a transport network when we travel to work or to a place to shop. If we call or watch television, we receive electronic signals that are transmitted to us via a telecommunications network. Since the initial development of those disciplines, network models have played a key role in the development of operations research and management science. Networks occur in a variety of settings and guises. Networks of transportation, energy, and connectivity permeate our daily lives. Network models are also commonly used for topics in such diverse areas as manufacturing, distribution, project planning, location of facilities, resource management, and financial planning, to name only a few. In reality, a network representation provides such a powerful visual and conceptual help to depict the relationships between system components that it is used in virtually every area of science, as well as social and economic endeavours.

A network is a system of lines or channels which connect various points. In networks, we will be interested in sending some specified products from certain points of supply to some points of demand. For example, we may like to send water, oil, or gas from the supply station to demanding customers in a pipeline network system. In complex, interrelated business activities, the manager or administrator is constantly looking forward to those techniques or methods that will help them plan, schedule, and control such activities. We have been greatly aided by the idea of network planning. Much of the problem of network flow can be conceived as a problem of linear programming, and their solutions can be obtained using the simplex method. But a number of special network flow techniques have been developed that are generally more efficient than the simplex approach. This unit will be devoted to addressing some of the special network flow problems and their methods for solving them.

4.1.1 Network

LO 1. Understand a network, draw activity on an arc or arrow, and create activity on node networks

A network is a collection of points connected by arcs or branches called nodes. Each connection is linked to a pair of nodes and is usually drawn as a line that joins two points. If node 1 to node 2 is an arc, then node 1 is said to be a node 2 predecessor, and node 2 is a node 1 successor. The arranged pair also designates the arc (1, 2). The network in Figure 4.1 has two arcs and is labelled by naming the nodes at either end; for example, *1–2* is the arc between nodes *1* and *2* in Figure 4.1. A network's arcs may have some sort of flow through them. Flow can be considered as the total amount of an object originating at the source, making its way through the various arcs and passing through intermediate nodes, and finally reaching (or being consumed by) the destination node.

There are two methods to creating a network from a project: activity on node (AON) and activity on arc or arrow (AOA). Under the AON convention, activities are defined by nodes. Arrows under AOA reflect activities. Activities (except dummy activity) consume time and resources. The fundamental difference between AON and AOA is that the nodes in an AON diagram are activity representation. The nodes in an AOA network reflect an activity's start and end times and are also called an event. Therefore nodes in AOA consume no time or resources.

4.2 Dummy Activity

A dummy activity in the network is added only to determine one activity's dependence on the other but does not consume any time. It is necessary when (i) two or more project activities have the same head and tail events or (ii) two or more activities share some of

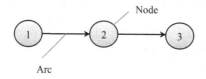

FIGURE 4.1
Network Components.

their immediate predecessors' activities. In the network diagram, dummy activities are usually represented through the dotted line. The use of dummy activity is illustrated in Figure 4.2.

While drawing the network, the following points have to be kept in mind:

- Arc represents activities in the network diagram, and the circle represents the events. The length of an arc is insignificant.
- Each activity is defined by one arc and only one. The tail of activity represents the beginning and the head represents the work to completion.
- In some situations where the activity is further subdivided into segments, each segment will be represented by a separate arc.
- No two or more activities may have the same events at the tail and head.
- Until all incoming activities have been completed, an event cannot occur.
- The arc should stay straight and should not be curved.
- The network should not have a loop.

After a logical network sequence is drawn, a number is assigned to each event. The sequence of numbers should reflect the flow of the network. The rule designed by D. R. Fulkerson (American mathematician) is used for numbering purposes and involves the following steps:

1. The initial event that has all arcs outgoing without an incoming arc is numbered '1.'
2. Delete all arcs from the '1' node. This will convert a few more nodes into initial events. Number these events '2,' '3,' …
3. To generate further initial events and allocate the next numbers to these events, delete all the arcs coming out of the numbered events obtained in step 2.
4. Continue until the final node is numbered with all the arcs coming in and no arcs going out.

To demonstrate the aforementioned numbering technique, steps let us consider the network shown in Figure 4.3.

Event A is the first event, numbered '1.' Delete arc 1 and arc 2. It produces two more B and C events. Number these events '2' and '3.' Now delete arc 3, arc 4, and arc 5. It turns D and F into events. Number these events as '4' and '5.' Delete arc 6, arc 7, and arc 8. It turns E and G into events. Number these events '6' and '7.' Event '8' is the last event.

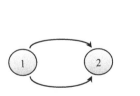

Incorrect Network

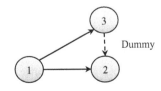

Correct Network with Dummy

FIGURE 4.2
Dummy Activity.

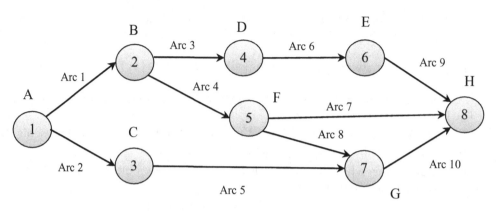

FIGURE 4.3
Network Numbering.

Examples based on drawing a network diagram

EXAMPLE 1:

A job required six operations A, B, C, D, E, and F with the following sequence: Operation A precedes B and C. Operation C and D precede E. Operation B precedes D. Operation E precedes F. Construct the network for a sequence of operations. *(BT level 6)*

Solution:
In this example, consider six operations as an activity A, B, C, D, E, and F. The network is shown in Figure 4.4.

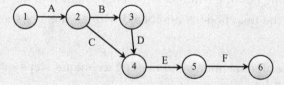

FIGURE 4.4
Network Diagram for Example 1.

EXAMPLE 2:

Construct the network for the activities shown in Table 4.1. *(BT level 6)*

TABLE 4.1

Activity and Immediate Predecessors for Example 2

Activity	Immediate Predecessors
A	–
B	A
C	B
D	C
E	C

Activity	Immediate Predecessors
F	E
G	D
H	G
I	C
J	F, I
K	H
L	J, K

Solution: The network is shown in Figure 4.5.

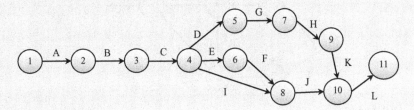

FIGURE 4.5
Network Diagram for Example 2.

EXAMPLE 3:

The XYZ company would like to have one more boiler installed in the factory. The XYZ company's plant manager has identified the eight tasks that need to be carried out to complete the project. When the project starts, two tasks can be started simultaneously: constructing the boiler's internal components (activity A) and making the necessary modifications for the floor and roof (activity B). Construction of the collection stack (activity C) will begin upon completion of the internal components. The frame installation (activity D) can be started as soon as the internal components have been completed and the roof and floor modified. Following the construction of the collection stack, two tasks may begin: building the burner (activity E) and installing the boiler (activity F). After the frame has been mounted, the chimney can be installed (activity G) and the burner built. Finally, the device may be checked and tested (activity H) after the boiler and chimney have been mounted.

Based on the previous information, create a table showing the relationships preceding the operation and construct the network. *(BT level 6)*

Solution:
Activities and precedence relationships is shown in Table 4.2. The network is shown in Figure 4.6.

TABLE 4.2

Activity and Immediate Predecessors for Example 3

Activity	Description	Immediate Predecessors
A	Building the internal components for the boiler	–
B	Modifications necessary for the floor and roof	–
C	Construction of the collection stack	A
D	Installation of the frame	A, B
E	Building the burner	C
F	Installing the boiler	C
G	Chimney installation	D, E
H	Inspection and testing	F, G

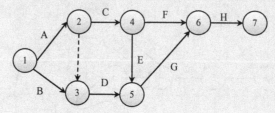

FIGURE 4.6
Network Diagram for Example 3.

EXAMPLE 4:

Construct the AON network for Example 3. *(BT level 6)*
Solution:
In the AON method, each activity is denoted by a node. The arcs reflect the relationship of the precedence between the activities. The network is shown in Figure 4.7.

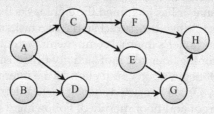

FIGURE 4.7
Network Diagram for Example 4.

4.3 Critical Path Method (CPM)/Program Evaluation and Review Technique (PERT)

PERT and CPM were each developed in the 1950s to help managers plan, manage, and control large and complex projects. CPM first appeared as a method designed to assist in the construction and maintenance of DuPont chemical plants. In 1958, PERT was developed independently for the US Marines. The best aspects of these two methods have appeared over the years to be incorporated into what is now commonly known as the PERT/CPM method. This project management network methodology continues to be widely used today.

CPM is based on deterministic task duration and is, therefore, ideal for production projects where previous experience with the subtasks helps management to make accurate estimates of the time. PERT, on the other hand, is based on probabilistic estimates of the duration of the projects and is, therefore, most useful in a research and development context where completion times cannot be determined beforehand. Common uses of these techniques include the following:

1. Researching and developing new products and processes
2. Plant development, building construction, and highways
3. Maintenance of large and complex equipment
4. New systems design and implementation

PERT and CPM can be of benefit to project managers to answer the following questions:

1. What is the total time needed to finish the project?
2. What are the scheduled dates for each specific activity to start and finish?
3. What are the critical tasks and needs to be done exactly as planned to keep the project on schedule?
4. How long can the "non-critical" tasks be postponed before the overall completion period of the project is increased?
5. What is the probability of the project finishing by a specific date?

PERT and CPM both have six basic steps to follow:

LO 2. Determine a critical path

1. Define the project and divide the work into activities.
2. Develop the relationships among the activities. Decide which activities are to precede and which others must follow.
3. Draw the connecting network of all the activities.
4. Assign estimates of the time and/or costs for each activity.
5. Calculate the network's longest time path. This is called the critical path.
6. Use the network to help plan, track, and manage the project.

4.3.1 CPM Computations

In CPM, the end result is a project time schedule. We conduct the critical path analysis for the network to find out just how long the project will take. In all network diagrams where there are some parallel relationships between the activities, there will be more than one sequence of activities that will lead from the start to the end of a project. Such activity sequences are called network paths. Each path has a total duration that is the aggregate of all of its activities. The critical path is the longest time path through the network (note that more than one critical path can be accessed). The critical path is named such because any delay in any of the activities along this path will disrupt the entire project. For each task, we measure two distinct start and end times to find the critical path. These are defined as follows:

Earliest start (ES) = earliest time at which an activity can start, assuming all predecessors have been completed.

Rule:

- If an activity has just one immediate predecessor, then its ES is equal to the predecessor's earliest finish (EF).

- If an activity has several immediate predecessors, its ES is the maximum of all EF values of its predecessors. That is,

$$ES = \max\{EF \text{ of all immediate predecessors}\}. \qquad (4.1)$$

Earliest finish (EF) = earliest time at which an activity can be finished.

Rule:

- The EF time of an activity is the sum of its ES time and its activity time. That is,

$$EF = ES + \text{Activity time}. \qquad (4.2)$$

Latest finish (LF) = latest time by which an activity has to finish so as to not delay the completion time of the entire project.

Rule:

- If an activity is an immediate predecessor of a single activity, its LF is equal to the latest start (LS) of the activity that follows it immediately.

- If an activity is an immediate predecessor of more than one activity, its LF is the minimum of all the LS values of all activities that follow it immediately. That is,

$$LF = \text{Min}\{LS \text{ of all immediately following activities}\}. \qquad (4.3)$$

Latest start (LS) = latest time at which an activity can start so as to not delay the completion time of the entire project.

Rule:

- The LS time of an activity is the difference between its LF time and its activity time. That is,

$$LS = LF - \text{Activity time}. \qquad (4.4)$$

These four times define the time slot that is available for the activity (illustrated in Figure 4.8).

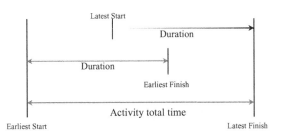

FIGURE 4.8
Activity Time.

4.3.2 Forward Pass and a Backward Pass

To determine time schedules for each operation, we use a two-pass process consisting of a forward pass and a backward pass. During the forward pass, the ES and EF times are determined. During the backward pass, the LS and LF times are determined. The forward pass starts with the project's first activity; the

LO 3. Compute forward and backward passes for a project

backward pass starts with the project's last activity. Although the forward pass helps us to determine the earliest completion time for the project, it does not identify the critical path. To identify this path to determine the LS and LF values for all activities, we must perform the backward pass. After we have calculated the earliest and most recent times for all activities calculating the amount of slack time of activity is a simple matter. *Slack* is the duration of an activity that can be postponed without delaying the entire project.

Mathematically,

$$\text{Slack} = \text{LS} - \text{ES or Slack} = \text{LF} - \text{EF}. \tag{4.5}$$

Zero slack activity is considered a critical activity and said to be on the critical path. The critical path, through the project network, is a continuous path that

- Begins at the first project activity,
- Finishes at the last project activity, and
- Only includes critical activities.

Often, the slack time of an activity is called its float. As listed next, there are four basic types of activity float:

Total Float: Total float is known as the amount of time that an activity can delay from its ES time without delaying the project's latest completion time. This refers to the amount of free time associated with an activity that can be used before, during, or after the activity is carried out. The total float of an activity is the difference between the latest completion time and the activity's earliest completion time or the difference between the LS time and the activity's ES time.

$$\text{Total Float} = \text{LF} - \text{EF} \tag{4.6}$$

Or

$$\text{Total Float} = \text{LS} - \text{ES} \tag{4.7}$$

Free Float: Free float is the amount of time that an activity can be postponed from its ES time without delaying any of its immediate successors' start times. It is that portion of the total float that can manipulate an activity without impacting the float of subsequent activities. Activity-free float is measured as the difference between the minimum ES time of the activity's successors and the EF time of the activity.

$$\text{Free Float} = \text{Min}\{\text{Succeeding ES}\} - \text{EF} \tag{4.8}$$

Interfering Float: Interfering float is the amount of time an activity interferes with (or obstructs) its successors while completely using its total float. The interference float is measured as the difference between the total float and the free float.

$$\text{Interfering Float} = \text{Total Float} - \text{Free Float} \tag{4.9}$$

Independent Float: Independent float is the amount of float an activity will always have irrespective of its predecessors' completion times or their successors' starting times. Independent floats are measured as

$$\text{Independent Float} = \text{Max}\{0, (\text{ES}j - \text{LF}i - t)\}, \tag{4.10}$$

where ESj is the earliest starting time of the preceding activity, LFi is the latest completion time of the succeeding activity, and t is the duration of the activity whose independent float is being calculated.

Examples based on CPM

EXAMPLE 1:

For the network shown in Figure 4.9, obtain the critical path and find the early and late start and completion times. The value in the bracket on the diagram shows the time (in days) of completion of each task. *(BT level 3)*

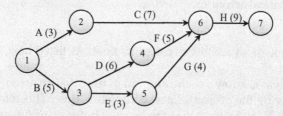

FIGURE 4.9
Network Diagram for Example 1.

Solution:
Paths on the network and their completion time are given in Table 4.3

TABLE 4.3

Completion Time and Path for Example 1

Path	Completion Time
A-C-H	$3 + 7 + 9 = 19$
B-D-F-H	$5 + 6 + 5 + 9 = 25$
B-E-G-H	$5 + 3 + 4 + 9 = 21$

From Table 4.3, we can conclude that the critical path is B-D-F-H with a completion time of 25 days.

Sample Calculation:

Consider the path A-C- H in Figure 4.9 and apply the forward pass.

Activity A has no predecessors. Using the ES time rule, the ES is equal to zero for activity A. Now using the EF time rule, the EF for A is 3 (= 0 + 3).

Since activity A precedes activity C, the ES of C equals the EF of A (= 3). The EF of C is therefore 10 (= 3 + 7).

We now come to activity H. Activities C, F, and G are immediate predecessors for H. C has an EF of 10, F has an EF of 16, and G has an EF of 12. Using the ES time rule, we compute the ES of activity H as follows:

$$ES \text{ of } H = Max\{EF \text{ of } C, EF \text{ of } F, EF \text{ of } G\} = \max(10,16,12) = 16$$
$$\text{The EF of H equals } 25 (= 16 + 9).$$

Now consider the path B-D-F-H in Fig 4.9 and apply for the backward pass

We begin by assigning an LF value of 25 days for activity H. That is, we specify that the LF time for the entire project is the same as its EF time. Using the LS time rule, the LS of activity H is equal to 16 (= 25 − 9).

Because activity H is the lone succeeding activity for activities C, F, and G, the LF for C, F, and G equals 16. This implies that the LS of F is 11 (= 16 − 5). The LF of D is 11, and its LS 5 (= 11 − 6).

Activity B is an immediate predecessor to two activities D and E. Using the LF time rule, we compute the LF of activity B as follows:

$$LF \text{ of } B = Min\{LS \text{ of } D, LS \text{ of } E\} = \min(5,9) = 5.$$

The LS of B is computed as 0 (= 5 − 5).

Proceeding in this manner, the ES and finish times, the LS and finish times for each activity are listed in Table 4.4

TABLE 4.4

ES, EF, LS, and LF for Example 1

Activity	Time	ES	EF	LS	LF
A	3	0	3	6	9
B	5	0	5	0	5
C	7	3	10	9	16
D	6	5	11	5	11
E	3	5	8	9	12
F	5	11	16	11	16
G	4	8	12	12	16
H	9	16	25	16	25

EXAMPLE 2:

For the data in Table 4.5, construct a network and indicate the critical path. Also, determine the ES and finish times, LS and finish times, and slack for each activity. (*BT level 6 and 3*).

TABLE 4.5

Activity and Immediate Predecessor for Example 2

Activity	Time (Days)	Immediate Predecessors
A	8	–
B	11	–
C	7	A
D	6	B
E	5	B
F	4	C, D
G	3	E, F

Solution:
The network is shown in Figure 4.10.

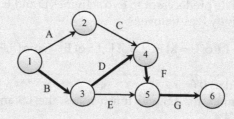

FIGURE 4.10
Network Diagram for Example 2.

TABLE 4.6

Completion Time and Path for Example 2

Path	Completion Time
A-C-F-G	8 + 7 + 4 + 3 = 22
B-D-F-G	11 + 6 + 4 + 3 = 24
B-E-G	11 + 5 + 4 = 20

TABLE 4.7

Slack for Example 2

Activity	Time	ES	EF	LS	LF	Slack
A	8	0	8	2	10	2
B	11	0	11	0	11	0
C	7	8	15	10	17	2
D	6	11	17	11	17	0
E	5	11	16	16	21	5
F	4	17	21	17	21	0
G	3	21	24	21	24	0

Paths on the network and their completion time are given in Table 4.6

From Table 4.6, we can conclude that the critical path is B-D-F-G with a completion time of 24 days.

As discussed in the sample calculation of Example 1, the ES and finish times, LS and finish times, and slack for each activity are given in Table 4.7

Table 4.7 summarizes the ES, EF, LS, LF, and slack time for all activities. Activity E, for example, has five days of slack time because its LS is 16 and its ES is 11 (alternatively, its LF is 21 and its EF is 16). This means that activity E can be delayed by up to five days, and the whole project can still be finished in 25 days.

On the other hand, activities B, D, F, G, and H have *no* slack time. This means that none of them can be delayed without delaying the entire project.

EXAMPLE 3:

For the network shown in Figure 4.11, determine the total, free, independent, and interfering floats for each activity. The value in the bracket on the diagram shows the time (in days) of completion of each activity. The critical path is shown by dark arcs. (BT level 3)

Solution:
As discussed in the sample calculation of Example 1, the ES and finish times and LS and finish times for each activity are given in Table 4.8

Table 4.8 also summarizes the total float, free float, interfering float, and independent float.

Sample Calculation:
For activity E of the network in Figure 4.11, the total available time (measured from the ES to the LF) is nine days while the time required to perform it is six days. Thus

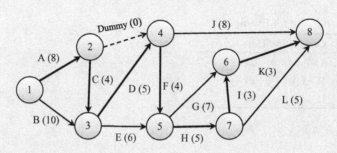

FIGURE 4.11
Network Diagram for Example 3.

the total float for this activity is LF − EF = 3 days (= 21 − 18). If an activity E is delayed by more than three days, it would delay completion of the project.

Free float of activity E is equal to three days because the ES of the succeeding activities G and H is 21, while the EF of this activity is 18.

$$\text{Interfering float of activity E} = \text{Total float} - \text{Free float} = 3 - 3 = 0$$

$$\text{Independent float of activity E} = \max\left\{0, \left(ESj - LFi - t\right)\right\}$$

$$= 21 - 12 - 6 = 3\,\text{days}$$

TABLE 4.8

Float for Example 3

Activity	Time	ES	EF	LS	LF	Total Float	Free Float	Interfering Float	Independent Float
A	8	0	8	0	8	0	0	0	0
B	10	0	10	2	12	2	2	0	2
C	4	8	12	8	12	0	0	0	0
Dummy		8	8	17	17	9	9	0	9
D	5	12	17	12	17	0	0	0	0
E	6	12	18	15	21	3	3	0	3
F	4	17	21	17	21	0	0	0	0
G	7	21	28	22	29	1	1	0	1
H	5	21	26	21	26	0	0	0	0
I	3	26	29	26	29	0	0	0	0
J	8	17	25	24	32	7	7	0	7
K	3	29	32	29	32	0	0	0	0
L	5	26	31	27	32	1	1	0	1

4.3.3 PERT Computations

In determining all the earliest and latest times so far, and the corresponding critical path(s), we followed the CPM method of assuming that all activity times were single values for all activity time estimates. By using only one estimation of the activity time, we are essentially assuming that the activity times are known with certainty (i.e., deterministic). Nonetheless, in practice, completion times of the activity are likely to vary depending on different factors. While deciding on a schedule for a project, we cannot ignore the impact of variation in activity times. PERT tackles this issue. Three time estimates for each activity are calculated in the PERT-type approach to estimating the activity times. For each activity, the three time estimates are the optimistic time (a), the most likely time (m), and the pessimistic time (b).

Optimistic time a = the shortest possible time to complete the activity if everything goes right.

Most probable time m = the most probable activity time under normal conditions. It is a subjective estimate of the activity time that would most frequently occur if the activity were repeated many times.

Pessimistic time b = the longest possible time to complete the activity assuming everything goes wrong.

These three time estimates are used to estimate the mean and variance of beta distribution as follows:

$$\text{Mean (expected time)} t = \frac{a + 4m + b}{6}. \tag{4.11}$$

With uncertain activity times, we can use the *variance* to describe the dispersion or variation in the activity time values. The variance of the activity time is given by the formula

$$\text{Variance } \sigma 2 = \left(\frac{b - a}{6}\right)^2. \tag{4.12}$$

For several factors, we assume that activity times can be represented by a beta distribution. With three time estimates, the beta distribution mean and variance can be approximated. The beta distribution is also continuous but it doesn't have a predetermined shape (such as the normal curve bell

> LO 4. Calculate the variance and standard deviation of activity times

shape). By the time estimates given, it will take on the specified shape – that is, be skewed. This is beneficial because we usually do not have prior knowledge of the shapes of the activity time distributions in a particular project network.

Figure 4.12 indicates two forms of the beta distribution. The left diagram represents a symmetrical distribution with a minimum of 0 and a maximum of 6. This distribution's most likely value is three minutes, at halfway between the extremes. The minimum and maximum durations in the asymmetric beta distribution in the right-side diagram are the same as in the left diagram, but the most likely value is rather lower.

4.3.4 Probability Computations

The probability of completing a given project in a defined time frame can be calculated using the following formula:

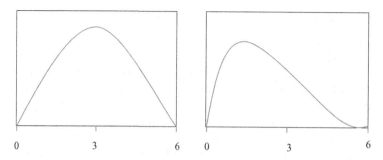

FIGURE 4.12
Beta Distribution Examples.

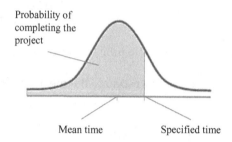

FIGURE 4.13
Probability Distribution of Project.

$$Z = \frac{Specified\ time - Project\ mean\ time}{standard\ deviation}.$$ (4.13)

The resulting value of z shows how many standard deviations of the project distribution the specified time is beyond the estimated length of the project. It is better if it is more of a positive value. A negative z value means the stated time is earlier than the estimated period. Once the z value is calculated, the probability that the project will be completed by the specified time from the area below the normal curve table may be used. Probability is proportional to that area below the normal curve to the left of z, as shown in Figure 4.13.

Examples based on PERT

EXAMPLE 1:

For the network shown in Figure 4.14, determine the expected time and variance for each activity and obtain the critical path. The value in the bracket on the diagram shows the three time estimates (in days) a, m, and b in the order a-m-b. *(BT level 3)*

Solution:
We put the activities in a tabular form and calculate the expected times and variance, as shown in Table 4.9.

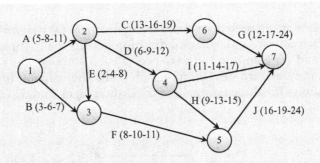

FIGURE 4.14
Network Diagram for Example 1.

TABLE 4.9

Expected Time and Variance for Example 1

Activity	Time			Expected Time $t = \dfrac{a + 4m + b}{6}$	Variance $= \left(\dfrac{b-a}{6}\right)^2$
	a	m	b		
A	5	8	11	8	1
B	3	6	7	5.67	0.44
C	13	16	19	16	1
D	6	9	12	9	1
E	2	4	8	4.33	1
F	12	17	24	17.33	4
G	9	13	15	12.67	1
H	11	14	17	14	1
I	8	10	11	9.83	0.25
J	16	19	24	19.33	1.78

Paths on the network and their completion time are given in Table 4.10.

From Table 4.10, we can conclude that the critical path is A-E-F-J with a completion time of 53.66 days.

TABLE 4.10

Paths and Completion Times for Example 1

Path	Completion Time
A-C-G	8 + 16 + 12.67 = 36.67
A-D-I	8 + 9 + 9.83 = 26.83
A-D-H-J	8 + 9 + 14 + 19.33 = 50.33
A-E-F-J	8 + 4.33 + 17.33 + 24 = 53.66
B-F-J	5.67 + 17.33 + 19.33 = 42.33

Example 2: For the network shown in Figure 4.15, (a) determine an expected time for each activity and expected duration for each path, (b) identify the critical path, and (c) compute the variance of each activity, the variance of each path, and the standard deviation of each path. The value in the bracket in the diagram shows the three time estimates (in days) a, m, and b in the order a-m-b. *(BT level 3)*

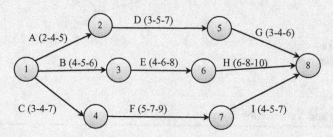

FIGURE 4.15
Network Diagram for Example 2.

Solution: (a) We put the activities in a tabular form and calculate the expected time for each activity and expected duration for each path, as shown in Table 4.11.

(b) From Table 4.11, we can conclude that the critical path is B-E-H with a completion time of 19 days.

(c) The variance of each activity and the variance and standard deviation of each path is given in Table 4.12.

TABLE 4.11

Expected Time and Path for Example 2

Path	Activity	Time			Expected Time $t = \dfrac{a + 4m + b}{6}$	Expected Path Duration
		a	m	b		
A-D-G	A	2	4	5	3.83	
	D	3	5	7	5	13
	G	3	4	6	4.17	
B-E-H	B	4	5	6	5	
	E	4	6	8	6	19
	H	6	8	10	8	
C-F-I	C	3	4	7	4.33	
	F	5	7	9	7	16.5
	I	4	5	7	5.17	

TABLE 4.12

Variance and Standard Deviation for Example 2

Path	Activity	Time a	Time m	Time b	Variance $=\left(\dfrac{b-a}{6}\right)^2$	Path Variance	Path Standard Deviation
A-D-G	A	2	4	5	0.25		
	D	3	5	7	0.44	0.94	0.97
	G	3	4	6	0.25		
B-E-H	B	4	5	6	0.11		
	E	4	6	8	0.44	1	1
	H	6	8	10	0.44		
C-F-I	C	3	4	7	0.44		
	F	5	7	9	0.44	1.13	1.06
	I	4	5	7	0.25		

EXAMPLE 3:

Using the information provided in Example 2 of PERT, calculate (a) the probability that the project may be completed within 20 days of its commencement, (b) the probability that the project will be completed within 18 days of its commencement, and (c) the probability that the project will not be completed within 18 days of its launch. *(BT level 3)*

Solution:
(a) Observe from the solution of Example 2 that paths A-D-G and C-F-I are well enough within the specified completion time of 20 days, so it is highly likely that both will be finished by day 20. Now find the probability of each path by computing the value of z using the following formula,

$$Z = \frac{Specified\ time - Project\ mean\ time}{standard\ deviation}.$$

Turning to Appendix A, Table B with the computed value of Z, find the area under the standardized normal curve. The computations are summarized in Table 4.13.

$$Probability\left(finish\ by\ day\ 20\right) = Probability\left(path\ A-D-G\right)$$
$$\times Probability\left(path\ B-E-H\right)$$
$$\times Probability\left(path\ A-D-G\right)$$

$$= 1.00 \times 0.8413 \times 1.00$$
$$= 0.8413$$

(b) For a specified time of 18 days, the z values are given in the Table 4.14.

$$\text{Probability}\left(\text{finish by day 18}\right) = \text{Probability}\left(\text{path A}-\text{D}-\text{G}\right)$$
$$\times \text{Probability}\left(\text{path B}-\text{E}-\text{H}\right)$$
$$\times \text{Probability}\left(\text{path A}-\text{D}-\text{G}\right)$$

$$= 1.00 \times 0.1587 \times 0.9192$$

$$= 0.1458$$

(c) The probability of not finishing before day 18 is the complement of the probability obtained in part (b).

$$= 1 - 0.1458$$

$$= 0.8542$$

TABLE 4.13

Probability of Paths for Example 3

Path	$Z = \dfrac{20 - \textit{Project mean time}}{\textit{standard deviation}}$	Probability of Completion in 20 Days
A-D-G	$\dfrac{20-13}{0.97} = +7.21$	1.00
B-E-H	$\dfrac{20-19}{1} = +1.00$	0.8413
C-F-I	$\dfrac{20-16.5}{1.06} = +3.30$	1.00

TABLE 4.14

Probability of Paths for Example 3

Path	$Z = \dfrac{18 - \textit{Project mean time}}{\textit{standard deviation}}$	Probability of Completion in 20 Days
A-D-G	$\dfrac{18-13}{0.97} = +5.15$	1.00
B-E-H	$\dfrac{18-19}{1} = -1.00$	0.1587
C-F-I	$\dfrac{18-16.5}{1.06} = +1.41$	0.9207

4.3.5 Project Crashing

To this extent, we have demonstrated the use of CPM and PERT network analysis to determine schedules for the project. This in itself is worthwhile for a project planning manager. The project manager is often confronted with the need to cut a

LO 5. Conduct cost analysis of projects

project's scheduled completion time to meet a deadline. In other words, the manager will complete the project earlier than the CPM/PERT network analysis implies. The duration of the project can often be shortened by allocating more labour to project activities in the form of overtime and allocating more resources (material, equipment, etc.). Nevertheless, extra labour and capital increase the cost of the project. Therefore, the decision to reduce the duration of the project must be based on an analysis of the trade-off between time and cost. The method by which we shorten a project's duration in the cheapest possible manner is called project crashing.

CPM is a method that has a normal or standard time for each and every activity that we use in our computations. The normal cost of the activity is associated with that normal time. Another time in project management, however, is the crash time, which is described as a reduction in the amount of time an activity takes. The crash cost of the activity is associated with that crash time. We may usually shorten an activity by adding additional resources (e.g. machinery, people) to it. Therefore, it is logical for an activity's crash cost to be higher than its normal cost.

The duration by which an activity (i.e., the difference between its normal time and crash time) can be shortened depends on the activity in question. There may be some activities that we cannot shorten at all. For example, adding more resources does not help shorten the time if a casting needs to be heat treated in the furnace for 48 hours. In contrast, we may be able to significantly shorten some activities (e.g., framing a house in three days instead of ten days using three times as many workers). The cost of crashing (or shortening) an activity likewise depends on the nature of the activity. Managers usually have an interest in speeding up a project at the least extra cost. Therefore, we need to ensure the following when choosing which activities to crash and by how much:

- The amount by which the activity is crashed is essentially permissible.
- The total cost of the crash is as low as possible.
- To make a rational decision about which activities to crash, if any, a manager needs certain information:
- Estimates of regular time and crash time for each activity
- Estimates of the regular costs and crash costs for each activity
- A list of activities that are on the critical path

For project crashing, the four-step procedure is as following:

Step 1. Find the critical path with the normal times and normal costs for the activities and identify the critical activities.

Step 2. Find out about the cost of the crash per unit time for each network activity. That is determined using the following formula:

$$\text{Crash cost / Time period} = \frac{Crash\ cost - Normal\ cost}{Normal\ time - Crash\ time}. \qquad (4.14)$$

Step 3. If there is only one critical path, then choose the activity on this critical path that may (a) still crash and (b) have the lowest crash cost per time. This activity crashes for one period. A critical activity is chosen for crashing based on the least additional cost. If there is a tie between two critical activities, the tie can be arbitrarily resolved.

Step 4. Find out which is the critical path with changed conditions after an activity crashes. Sometimes a reduction in the time of a critical path activity can cause a non-critical path to become critical. If the critical path with which we started is still the longest path, then go to step 3. Otherwise, define the new critical path, and then proceed to step 3.

Examples based on the crashing of the project

EXAMPLE 1:

A project has activities with the following normal and crash times and costs (Table 4.15). Determine a crash scheme for the previous project to reduce the total project period by two weeks. *(BT level 3)*

TABLE 4.15

Normal and Crash Times and Cost for Example 1

Activity	Immediate Predecessors	Normal Time (Week)	Normal Cost (Rs.)	Crash Time (Week)	Crash Cost (Rs.)
A	–	3	7,000	2	8,000
B	A	4	15,000	2	19,000
C	A	3	11,000	2	12,000
D	B	5	33,000	4	34,000
E	C	5	41,000	3	43,000
F	D	4	15,000	3	15,500
G	E	6	65,000	3	71,000
H	G	3	1,000	2	4,000

Solution:
The network diagram for the given project is shown in Figure 4.16.
    Paths on the network and their completion time are given in Table 4.16.

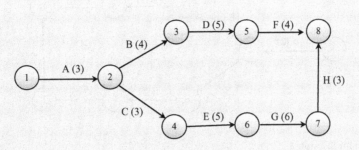

FIGURE 4.16
Network Diagram for Example 1.

TABLE 4.16

Paths and Completion Times for Example 1

Path	Completion Time
A-B-D-F	3 + 4 + 5 + 4 = 16
A-C-E-G-H	3 + 3 + 5 + 6 + 3 = 20

TABLE 4.17

Crash Cost per Week for Example 1

Activity	Crash Cost per Week (Rs)
A	8,000 – 7,000/3 – 2 = 1,000
B	19,000 – 15,000/4 –2 = 2,000
C	12,000 – 11,000/3 –2 = 1,000
D	34,000 – 33,000/5 –4 = 1,000
E	43,000 – 41,000/5 –3 = 1,000
F	15,500 – 15,000/4 – 3 = 500
G	71,000 – 65,000/6 – 3 = 2,000
H	4,000 – 1,000/3 – 2 = 3,000

From Table 4.16, we can conclude that the critical path is A-C-E-G-H with a completion time of 20 weeks.

The crash cost per week of the project is shown in Table 4.17. Note, for example, that activity B's normal time is four weeks, and its crash time is two weeks. This means that activity B can be shortened by up to two weeks if extra resources are provided. The cost of these additional resources is Rs. 4,000 (= difference between the crash cost of Rs 19,000 and the normal cost of Rs 15,000). If we assume that the crashing cost is linear over time (i.e., the cost is the same each week), activity B's crash cost per week is Rs. 2,000 (= Rs 4,000/2). Crash costs for all other activities can be computed in a similar fashion.

To begin crash analysis, the crash cost slope values for critical activities is Rs. 1,000 for A, Rs. 1,000 for C, Rs. 1,000 for E, Rs. 2,000 for G, and Rs. 3,000 for H.

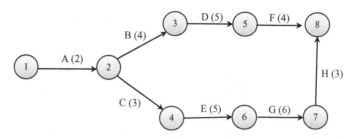

FIGURE 4.17
Network Diagram for Example 1.

TABLE 4.18

Paths and Completion Times

Path	Completion Time
A-B-D-F	2 + 4 + 5 + 4 = 15
A-C-E-G-H	2 + 3 + 5 + 6 + 3 = 19

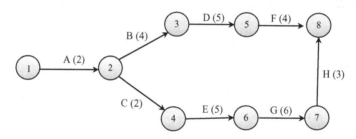

FIGURE 4.18
Network Diagram for Example 1.

The critical activity A with a cost slope of Rs 1,000 per week is the least expensive and can be crashed by one week. However, there is a tie between A, C, and E. The tie can be resolved arbitrarily. Let us select A for crashing. We reduce the time of A by one week by spending an extra amount of Rs. 1,000. After this step, we have the following network with the revised times for the activities (refer to Figure 4.17).

Revised paths on the network and their completion times are given in Table 4.18

From Table 4.18, we can conclude that the critical path is A-C-E-G-H with a completion time of 19 weeks.

The time for A cannot be reduced further. Therefore, we have to consider C, E, G, and H for crashing. Among them, C and E have the least crash cost per unit time (refer to Table 4.17). The tie between C and E can be resolved arbitrarily. Suppose we reduce the time of C by one week with an extra cost of Rs. 1,000. After this step, we have the following network with the revised times for the activities (refer to Figure 4.18).

Revised paths on the network and their completion times are given in Table 4.19

From Table 4.19, we can conclude that the critical path is A-C-E-G-H with a completion time of 18 weeks.

TABLE 4.19

Paths and Completion Times

Path	Completion Time
A-B-D-F	2 + 4 + 5 + 4 = 15
A-C-E-G-H	2 + 2 + 5 + 6 + 3 = 18

We have arrived at the following crashing scheme for the given project:

Reduce the time of A, and C by one week each.

Project time after crashing is 18 weeks.

Extra amount required = 1,000 + 1,000 = Rs. 2,000.

EXAMPLE 2:

A project has activities with the following normal and crash times and costs (Table 4.20). Construct the network and determine a crashing scheme for the project for minimum total time and the corresponding cost. Assume indirect cost per day as Rs. 10. *(BT level 6 and 3)*

TABLE 4.20

Normal and Crash Times and Costs for Example 2

Activity	Immediate Predecessors	Normal Time (Days)	Normal Cost (Rs.)	Crash Time (Days)	Crash Cost (Rs.)
A	–	6	60	4	100
B	–	4	60	2	200
C	A	5	50	3	150
D	A	3	45	1	65
E	B	6	90	4	200
F	C, E	8	80	4	300
G	D	4	40	2	100
H	F, G	3	45	2	80

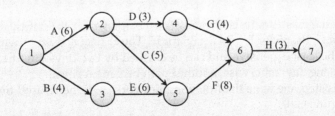

FIGURE 4.19
Network Diagram for Example 2.

Solution:
The network diagram for the given project is shown in Figure 4.19
  Paths on the network and their completion times are given in Table 4.21.

TABLE 4.21

Paths and Completion Times for Example 2

Path	Completion Time
A-D-G-H	$6 + 3 + 4 + 3 = 16$
A-C-F-H	$6 + 5 + 8 + 3 = 22$
B-E-F-H	$4 + 6 + 8 + 3 = 21$

TABLE 4.22

Crash Cost for Example 2

Activity	Crash Cost per Day (Rs)
A	$100 - 60/6 - 4 = 20$
B	$200 - 60/4 - 2 = 70$
C	$150 - 50/5 - 3 = 50$
D	$65 - 45/3 - 1 = 10$
E	$200 - 90/6 - 4 = 55$
F	$300 - 80/8 - 4 = 55$
G	$100 - 40/4 - 2 = 30$
H	$80 - 45/3 - 2 = 35$

From Table 4.21, we can conclude that the critical path is A- C-F-H with a completion time of 22 days.
  Crash cost per day of the project is shown in Table 4.22.

$$\text{Total cost} = \text{Direct normal cost} + \text{Indirect cost for } 22 \text{ days}$$

$$= 470 + 22 \times 10 = 690 / -$$

To begin the crash analysis, the crash cost slope values for critical activities is Rs. 20 for A, Rs. 50 for C, Rs. 55 for F, and Rs. 35 for H. The critical activity A with cost slope of Rs 20 per day is the least expensive and can be crashed by two days. But the time should only be reduced by one day; otherwise, another path becomes critical.
  After this step, we have the following network with the revised times for the activities (refer to Figure 4.20).
  Revised paths on the network and their completion times are given in Table 4.23
  From Table 4.23, we can conclude that the critical paths are A-C-F-H and B-E-F-H with a completion time of 21 days.

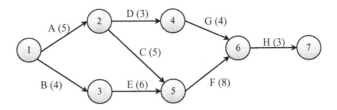

FIGURE 4.20
Network Diagram for Example 2.

TABLE 4.23

Paths and Completion Times for Example 2

Path	Completion Time
A-D-G-H	6 + 3 + 4 + 3 = 16
A-C-F-H	5 + 5 + 8 + 3 = 21
B-E-F-H	4 + 6 + 8 + 3 = 21

Total cost = Direct normal cost + Indirect cost for 22 days

$$= 470 + 21 \times 10 + 1 \times 20 = 700 / -$$

The total project cost for 21 weeks is more than the cost for 22 days. So further crashing is not desirable. Hence project optimum time is 22 days, and the cost is Rs 690.

4.4 Introduction to the Shortest Path Method

We will now consider a class of network problems in which we will attempt to find the shortest (or least expensive) path between two nodes. The path chosen does not necessarily need to move through all other nodes. A simple example of this type of

LO 6. Describe a shortest path problem

problem is illustrated by a vehicle travelling through the shortest route from a starting point to a final destination, passing through various points. Likewise is a distributed computer network that has to route data between specified pairs of processing nodes along the shortest path possible. Consider a salesman or a dairy seller or a postman who has to cover those areas already assigned to fulfil his daily routines. It is assumed that all places he wishes to visit are well connected to an appropriate mode of transport. He's managed to cover all those places. While doing so, if on the same day he visits the same place, again and again, there will be a loss of several resources, including time, money, etc. Therefore, he will place a constraint upon himself on the same day not to visit the same place again and again. He will be in a position to determine a route that would allow him to cover all the places, fulfilling the constraint. The shortest path method aims at discovering how a person can travel from one place to another, thereby reducing the total distance travelled. In other words, it seeks to define a set of destinations with the shortest route.

Although there are several other versions of the shortest path problem described in the next section, we will concentrate first on the following simple version. Consider a network that is undirected and connected with two special nodes called the origin and destination. There is a nonnegative distance associated with each of the links (undirected arcs). The goal is to find the shortest path from the origin to the destination (the path with the least total distance). To this problem, there is a relatively straightforward algorithm available. The essence of this procedure is that it fans from the origin, successively identifying the shortest path to each of the network nodes in the ascending order of their (shortest) distances from the origin, thus solving the problem when reaching the destination node.

The simple version operates by assigning a label to each node, which indicates the shortest distance to the destination from that node. A node is suitable for marking if it has labelled all its successors.

1. The destination node is initially given a zero label, which indicates that there is no cost or distance associated with going from that node to itself.
2. Choose any *eligible* node k, and assign it a label P_k as follows:
$P_k = \min \{d_{kj} + p_j\}$, the minimum taken over all successors j of node k.
3. Repeat step 2 until the source node is labelled. The label on the source *is* the shortest distance from the source to the destination. (For the illustration of this method, refer to Example 1.)

This reverse labelling procedure has an intuitive appeal when the problem is small enough to show the labels in a diagram. The procedure for larger problems consists of starting with a set containing a node and expanding the set by selecting a node in each subsequent step. This procedure works by switching to the destination node from the source node.

Step 1:

First, locate the origin. Then find the node nearest to the origin. Mark the distance between the origin and the nearest node in a box by the side of that node. In some cases, it may be necessary to check several paths to find the nearest node.

Step 2:

Repeat the previous process until the nodes in the entire network have been accounted for. The last distance placed in a box by the side of the ending node will be the distance of the shortest route. We note that the distances indicated in the boxes by each node constitute the shortest route to that node. These distances are used as intermediate results in determining the next nearest node. (For the illustration of this method, refer to Example 2.)

4.4.1 Dijkstra's Shortest Path Method

A more general algorithm that can be implemented to any network that has all nonnegative arc labels is known as the Dijkstra algorithm. Dijkstra's algorithm, published in 1959, was named after the Dutch computer scientist Edsger Dijkstra, its

LO 7. Understand and apply algorithms for a shortest path problem

author. This algorithm begins with the source node and calculates the shortest paths to every other node from the source node. During the operation of the Dijkstra algorithm, the nodes are divided into two sets: a set named S containing nodes known to be the shortest distance from the source and another set T containing nodes for which this shortest distance is not yet known. A label p_i is associated with every node i and specifies the length of the shortest path known *so far* from the source (node 1) to node i. Again, we let d_{ij} denote the direct distance from node i to node j.

1. Originally, only the source node is put in set S, and this node is labelled zero, indicating zero distance to itself from the source.

2. Initialize all other labels as follows:

$p_i = d_{1i}$ for $i \neq$ source node 1,

and $p_i = \infty$ if node i is not connected to the source.

3. Choose a node w, not in set S, whose label p_w is the minimum overall nodes not in S. Add node w to S and adjust the labels for all nodes v, not in set S, as follows:

$$p_v = \min\{p_v, p_w + d_{wv}\} \qquad (4.15)$$

Here we assume that p_v is the shortest distance from the source to node v directly through nodes in S. When we add node w to S, we check whether the new distance through w is shorter and update if necessary.

4. Repeat step 3 until all nodes belong to set S. (For the illustration of this method, refer to Example 3.)

4.4.2 Minimum Spanning Tree Algorithm

The problem of a minimal spanning tree bears some similarities to the main version of the shortest path problem presented in the previous section. In both cases, an undirected and linked network is considered, where the information given includes some measure of the positive length associated with each link (distance, cost, time, etc.). Both problems often include choosing a set of links that have the shortest total length between all sets of links that satisfy a particular property. For the shortest path problem, this property is that the links chosen must provide a path between the destination and the origin. The required property for the minimal spanning tree problem is that the links chosen must provide a path between each pair of nodes.

Consider a network connected with n nodes where all the arcs were removed. Then a tree can be "formed" by in some way adding one arc (or "branch") from the original network. The first arc can connect any pair of nodes anywhere. That new arc should then be between a node that is already connected to other nodes and a new node that was not previously connected to any other nodes. In this way, adding an arc prevents generating a loop and guarantees that the number of connected nodes is one bigger than the number of arcs. Each new arc generates a larger tree, which is a connected network that contains no undirected cycles (for some subset of the n nodes). Once the (n 1)st arc has been added, the process stops because of the resulting tree spans (connects) all n nodes. This tree is called a spanning tree (i.e., a connected network) that includes no undirected cycles for all n nodes.

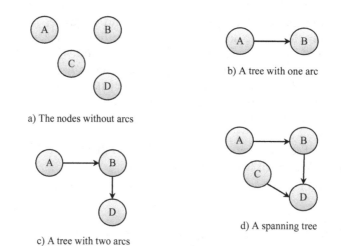

a) The nodes without arcs

b) A tree with one arc

c) A tree with two arcs

d) A spanning tree

FIGURE 4.21
Process of Growing a Tree.

Every spanning tree has exactly $n \_ 1$ arcs since this is the *minimum* number of arcs needed to have a connected network and the *maximum* number possible without having undirected cycles. Figure 4.21 uses the four nodes and some of the arcs to illustrate this process of growing a tree one arc (branch) at a time until a spanning tree has been obtained.

The minimum spanning tree problem can be solved with the help of the following steps.

Step 1: Select any node within the network first. This can be done arbitrarily. We'll begin with this node.

Step 2: Connect the selected node to the nearest node.

Step 3: Consider the nodes that are now connected. Consider the remaining nodes. If no node remains, then stop. On the other hand, if some nodes remain, figure out which one is closest to the already connected nodes among them. Choose this node, and go to step 2.

Thus the method involves the repeated application of steps 2 and 3. Since the number of nodes in the given network is finite, after a finite number of steps, the process must terminate. Stage 3 terminates the algorithm. A tie can be arbitrarily broken. We can end up with more than one optimal solution as a consequence of the tie. (For the illustration of this method, refer to Example 4.)

Examples based on the shortest path method

EXAMPLE 1:

Find the shortest path of the network shown in Figure 4.22. *(BT level 3).*

Solution:
In the illustration in Figure 4.22, initially label the destination node as $p_F = 0$. Next, node E is eligible and label as $p_E = 7 + 0 = 7$. The label for node D is computed as p_D

= min {6 + 0, 5 + 7} = 6. Node C is now eligible, and p_C = min {2 + 6, 3 + 7, 9 + 0} = 8. The label on node B is p_B = min {4 + 6, 4 + 8} = 10, and, finally, p_A = min {4 + 10, 5 + 8} = 13. Thus the length of the shortest path is 13, and the path itself is obtained by tracing back through the computations to find the path containing the arcs (A, C), (C, D), (D, F). The network with the label and shortest path is shown in Figure 4.23.

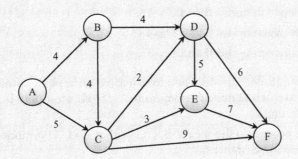

FIGURE 4.22
Network Diagram for Example 1.

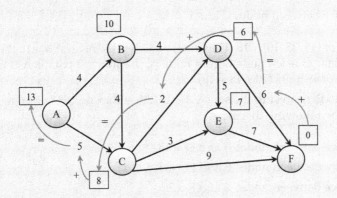

FIGURE 4.23
Shortest Path for example 1.

EXAMPLE 2:

Find the shortest path of the network shown in Figure 4.24. Number the of arcs represents the distance between nodes. *(BT level 3)*.

Solution:
Looking at Figure 4.24, we see that node A is the origin, and nodes B and C are neighbours to the origin. Node B is at 20 units of distance and node C is at 50 units of distance. The minimum of {20, 50} is 20. So select node B.

Now we search for the next node nearest to the set of nodes {A, B}. For this purpose, consider those nodes which are neighbours of either node A or node B.
The nodes C, D, and E fulfil this condition. We calculate the following
distances.

The distance between nodes A and C = 50,

The distance between nodes B and C = 18,

The distance between nodes B and D = 48,

The distance between nodes B and E = 33.

Minimum of {50, 18, 48, 33} = 18. Therefore, node C is the nearest one to the set {A,
B}. It means that to reach at node C from node A, the shortest path is A-B-C with a
total distance of 38 units.

The next node nearest to the set {A, B, C} is either node D or node E. We calculate the following distances.

The distance between nodes B and D = 48,

The distance between nodes B and E = 33,

The distance between nodes C and E = 10.

Minimum of {48, 33, 10} = 10. Therefore, node E is the nearest one to the set {A, B,
C}. It means that to reach node E from node A, the shortest path is A-B-C-E with a
total distance of 48 units (20 + 18 + 10).

The next node nearest to the set {A, B, C, E} is either node D or node F. We calculate the following distances.

The distance between nodes B and D = 48,

The distance between nodes E and D = 35,

The distance between nodes E and F = 20.

Minimum of {48, 35, 20} = 20. Therefore, node F is the nearest one to the set {A, B,
C, E}. It means that to reach node F from node A, the shortest path is A-B-C-E-F with
a total distance of 68 units (20 + 18 + 10 + 20). As node F is the destination node, we
have a path beginning from the source node A and terminating with the destination
node F. The network with the label and shortest path is shown in Figure 4.25.

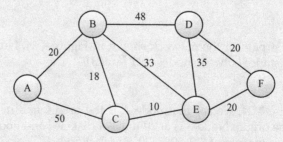

FIGURE 4.24
Network Diagram for Example 2.

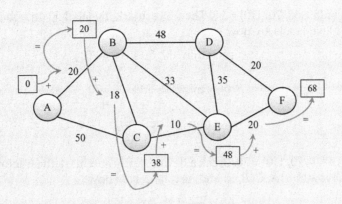

FIGURE 4.25
Shortest Path for Example 2.

EXAMPLE 3:

Find the shortest path of the network shown in Figure 4.26. *(BT level 3).*

Solution:

Initially, $S = \{A\}$, and $p_A = 0$, $p_B = 50$, $p_C = 30$, $p_D = 80$, $p_E = \infty$, and $p_F = \infty$. We then choose the minimum label 30 on node C, and $S = \{A, C\}$. Labels are now

$$p_B = \min\{50, 30 + \infty\} = 50,$$

$$p_D = \min\{80, 30 + \infty\} = 80,$$

$$p_E = \min\{\infty, 30 + 40\} = 70,$$

$$p_F = \min\{\infty, 30 + 80\} = 110.$$

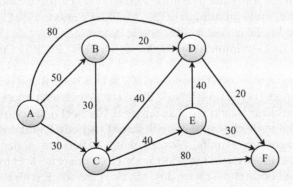

FIGURE 4.26
Network Diagram for Example 3.

Minimum of {50, 80, 70, 110} = 50. Therefore, mark the label 50 on node B so that S = {A, C, B}. Next, labels are now

$$p_D = \min\{80,\ 50+20\} = 70,$$

$$p_E = \min\{70,\ 50+\infty\} = 70,$$

$$p_F = \min\{110,\ 50+\infty\} = 110,$$

Minimum of {70, 70, 110} = 70. Break a tie arbitrarily and mark the minimum label 70 on node E. Now S = {A, C, B, E} and next labels are now,

$$p_D = \min\{70,\ 70+40\} = 70,$$

$$p_F = \min\{110,\ 70+30\} = 100,$$

Minimum of {70, 100} = 70. Therefore, mark the label 70 on node D so that S = {A, C, B, E, D}. Now the last label is

$$p_F = \min\{100,\ 70+30\} = 90.$$

Finally, node F is added to set S so that S = {A, C, B, E, D, F} with the shortest path of distance 310 (0 + 50 + 30 + 70 + 70 + 90).

EXAMPLE 4:

Determine the minimum spanning tree for the network, as shown in Figure 4.27. The thin lines represent potential links. *(BT level 3)*.

Solution:

Arbitrarily select node A to start. The unconnected node closest to node A is node C with a distance of 20 units (minimum of [30, 20, 40]). Connect node C to node A.

The unconnected node closest to either node A or node C is node D (closest to C) with a distance 25 units (minimum of [30, 30, 40, 50, 60, 30, 25, 40]). Connect node D to node C.

The unconnected node closest to nodes A, C, or D is node B and node E with the same unit of distances (i.e. 30 units). Now there is a tie between nodes B and E. The tie can be broken arbitrarily, and we select node B. Connect node B to node C.

The unconnected node closest to node B, C, or D is node E and node H with the same unit of distances (i.e. 30 units). Now there is a tie between nodes E and H. The tie can be broken arbitrarily, and we select node H. Connect node H to node B.

The unconnected node closest to node C, D, or H is node E with a distance of 30 units. Connect node E to node C.

The unconnected node closest to node C, E, or H is node F with a distance of 15 units. Connect node F to node E.

The only remaining unconnected node is node G, and it is closest to node F. Connect node G to node F.

All nodes are now connected, so this solution to the problem is the desired (optimal) one. The total distance of the links is 175 units. We obtain the minimum spanning tree for the given network as shown in Figure 4.28

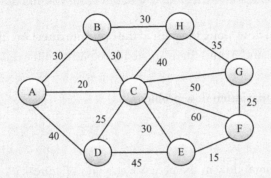

FIGURE 4.27
Network Diagram for Example 4.

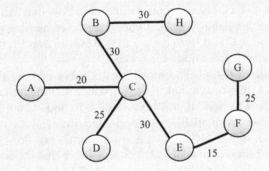

FIGURE 4.28
Spanning Tree for Example 4.

4.5 The Maximum Flow Problem

In the shortest route problem, we determined the shortest route from the origin to destinations. In the minimal spanning tree problem, we found the shortest connected network. In neither of these problems was the capacity of a branch limited to a specific number of items. However, there are network problems in which the branches of the network have limited flow capacities. The objective of these networks is to maximize the total amount of flow from an origin to a destination. These problems are referred to as maximal flow problems. Maximal flow problems can involve the flow of water, gas, or oil through a network of pipelines; the flow of forms through a paper processing system (such as a government agency); the flow of traffic through a road network; or the flow of products through a production line system. In each of these examples, the branches of the network have limited and often different flow capacities. Given these conditions, the decision maker wants to determine the maximum flow that can be obtained through the system.

The steps of the maximum flow solution method are as follows:

1. Arbitrarily select any path in the network from origin to destination.
2. Adjust the capacities at each node by subtracting the maximal flow for the path selected in step 1.
3. Add the maximal flow along the path in the opposite direction (if given) at each node.
4. Repeat steps 1, 2, and 3 until there are no more paths with available flow capacity.

Example based on the maximum flow problem

EXAMPLE 1:

Determine the maximal flow in the network shown in Figure 4.29. The number on the arcs represents the maximum flow capacity available between nodes. *(BT level 3).*

Solution:
First consider path A-C-F-G. The maximum number of flow that can be sent through this route is 6 (min of available 8, 6, 7). Notice that the remaining capacities of the arcs from node A to node C and from node F to node G are 2 and 1, respectively, and that no flows are available from node C to node F.

Now consider path A-B-E-G. The maximum number of flow that can be sent through this route is 4 (min of available 6, 4, 10). Notice that the remaining capacities of the arcs from node A to node B and from node E to node G are 2 and 6, respectively, and that no flows are available from node B to node E.

Now consider path A-B-C-E-G. The maximum number of flow that can be sent through this route is 2 (min of available 2, 2, 5, 6). Notice that the remaining capacities of the arcs from node C to node E and from node E to node G are 3 and 4, respectively, and that no flows are available from node A to node B and B to node C.

Now consider path A-C-E-G. The maximum number of flow that can be sent through this route is 2 (min of available 2, 3, 4). Notice that the remaining capacities

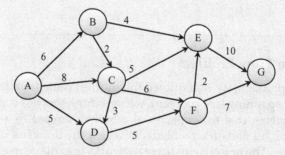

FIGURE 4.29
Network Diagram for Example 1.

of the arcs from node C to node E and from node E to node G are 1 and 2, respectively, and that no flows are available from node A to node C.

Now consider path A-D-F-E-G. The maximum number of flow that can be sent through this route is 2 (min of available 5, 5, 2, 2). Notice that the remaining capacities of the arcs from node A to node D and from node D to node F are 3 and 3, respectively, and that no flows are available from node F to node E and E to node G.

Next, consider path A-D-F-G. The maximum number of flow that can be sent through this route is 1 (min of available 3, 3, 1). Notice that the remaining capacities of the arcs from node A to node D and from node D to node F are 2 and 2, respectively, and that no flows are available from node F to node G.

Now there are no more paths with available flow capacity from source to destination so the current flow pattern is optimal with a maximum flow of 17. The previous steps are summarized in Table 4.24.

TABLE 4.24

Path and Flow Assign for Example 1

Path	Maximum Flow Assign from Available Capacities on the Path	Remaining Capacities of the Arcs	Max Flow Assigns to the Path
A-C-F-G	C-F = 6 (min of 8, 6, 7)	A-C = 2 C-F = 0 F-G = 1	6
A-B-E-G	B-E = 4 (min of 6, 4, 10)	A-B = 2 B-E = 0 E-G = 6	4
A-B-C-E-G	A-B = 2 (min of 2, 2, 5, 6)	A-B = 0 B-C = 0 C-E = 3 E-G = 4	2
A-C-E-G	A-C = 2 (min of 2, 3, 4)	A-C = 0 C-E = 1 E-G = 2	2
A-D-F-E-G	F-E = 2 (min of 5, 5, 2, 2)	A-D = 3 D-F = 3 F-E = 0 E-G = 0	2
A-D-F-G	F-G = 1 (min of 3, 3, 1)	A-D = 2 D-F = 2 F-G = 0	1

4.6 A Travelling Salesman Problem (TSP)

Perhaps the most famous classic case of computational optimization is called the travelling salesman problem. This picturesque name has been given because it can be described as a salesman (or saleswoman) who is expected to travel to several cities

LO 8. Understand and apply an algorithm to solve the travelling salesman problem

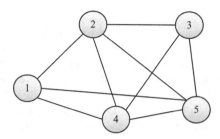

FIGURE 4.30
TSP Illustration.

during one tour. Beginning from his (or her) hometown, the salesman wishes to decide which route to follow to visit each city exactly once before returning to his hometown to reduce the total length of the tour. The basic idea is to try to find the shortest cycle in a network so that all nodes are visited and the minimum total distance is travelled.

Consider a network with N set of nodes, the set of arcs, and the matrix of costs C = [cij], where cij is the moving cost or the distance from node i to node j. The problem of the travelling salesman needs the Hamiltonian cycle in a network of minimal costs (a Hamiltonian cycle is a cycle which passes exactly once through each node i in N). Let us look at a situation in which five cities are represented as nodes, as shown in Figure 4.30. There is a person at node 1, and this person must reach only once per node and then return to the original (node 1) position. This process has to take place with minimal travelling distance or minimum cost or time. We are assuming the distance in either direction is the same. (This is referred to as a symmetric TSP.) The aim is to decide which route can minimize the total distance the salesman has to travel.

A variety of TSPs have been developed which have nothing to do with salesmen. For example, when a truck leaves a distribution centre to deliver goods to a number of locations, the problem of finding the shortest route to do this is a TSP. The manufacture of printed circuit boards for wiring chips and other components is another example. The problem of determining the most effective drilling sequence when many holes need to be drilled into a printed circuit board is a TSP.

4.6.1 TSP Solution by Branch and Bound Method

There are a large number of methods of solutions available for the TSP. The branch-and-bound process is discussed here.

Branch and bound is a general technique for enhancing the search process by enumerating all candidate solutions systematically. Branch and bound usually apply to those problems that have finite solutions in which the solutions can be represented as a sequence of options. In the first part of branch and bound, branching involves multiple choices to be made to branch the choices out into the space of the solution. The solution space is structured as a treelike structure in these approaches. Branching out to all possible choices means that no possible solutions are left uncovered. The solution space is often too vast to traverse. This problem is addressed by the branch-and-bound algorithm by bounding and pruning. Bounding refers to setting a bound on the quality of the solution (e.g., the length of the route for TSP), and pruning means cutting off branches in the solution tree whose quality is considered low. Bounding and pruning are the basic principles of the branch-and-bound method since they are used to minimize search space effectively.

The following are the steps for the branch-and-bound process to find an optimal TSP solution:

Step1: Compute a lower bound to the objective function and set the upper bound to infinity.

Step 2: Construct the branching tree and branch on possible values one of the variables can take.

Step 3: For each node, determine the lower bound. If the lower node bound is greater than the best upper bound, then the node must fathom.

Step 4: If the allocation results in a feasible solution at step 3, then evaluate the feasible solution. If the new solution is better, update the upper bound and fathom all nodes where the lower bound is greater than the present upper bound.

Step 5: When there is no open node, the process ends. The solution now is the optimal one.

For the illustration of this method, refer to Example 1.

4.7 Chinese Postman Problem (CPP)

In the CPP, a postman should visit a set of roads to deliver mail and return to the starting point. The roads are represented as network arcs. The postman will visit at least once each arc and return to the starting point travelling minimum total distance. A Chinese mathematician named Kuan Mei-Ko was interested in a postman delivering mail to a number of streets in 1962 so that the total distance the postman travelled was as short as possible.

If the given network has a Eulerian circuit, then it's optimal, and the postman will travel a minimal distance (equal to some of the distances of all the arcs), which is the Eulerian length. If the network given is not Eulerian, then some arcs must be travelled again, and this must be minimized. The problem then reduces the network to a minimum distance so that the resulting network is Eulerian. When the network has odd degree nodes, then it's not Eulerian.

For a network to be on Euler circuit or path, it must be traversable. Which means we can trace exactly once over all the arcs of a network without lifting our pencil. The degrees of a node are decided by counting the number of arcs that have that node as an end point. If it is even, the node is called even node. On the other hand, if a node's degree is odd, the node is considered an odd node.

The following are the steps for finding an optimal Chinese postman route:

Step 1: List all odd nodes.

Step 2: List any pairing of odd nodes possible.

Step 3: Find the arcs which connect the nodes to the minimum distance for each pairing.

Step 4: Find the pairings such that the sum of the distances is minimized.

Step 5: On the original network, add the arcs that have been found in step 4.

Step 6: The length of an optimal Chinese postman route is the sum of all the arcs added to the total found in step 4.

Step 7: A route corresponding to this minimum distance can then be easily found.

For the illustration of this method, refer to Example 2.

Examples based on TSP and CPP

EXAMPLE 1:

A salesman wants to visit five cities, starting with city 1 where he is stationed. The distances between various cities are given in Table 4.25. Determine a path through the four other cities and return to his home city in such a way that he has to travel the minimum distance. *(BT level 3)*

TABLE 4.25

TSP Data for Example 1

From City	To City				
	1	**2**	**3**	**4**	**5**
1	–	9	7	8	6
2	9	–	9	4	5
3	7	9	–	7	8
4	8	4	7	–	5
5	6	5	8	5	–

Solution:
Since we have to leave each city once, the minimum distance that will have to be travelled is the minimum in each row, and the sum of the five-row minimum values gives us a lower bound.

The lower bound of the total distance is LB = 6 + 4 + 7 + 4 + 5 = 26.

So the tour has to travel at least 26. The optimal solution will have to be more than or equal to the lower bound of 26.

Now create four more branches as $X_{1,2} = 1$, $X_{1,3} = 1$, $X_{1,4} = 1$, and $X_{1,5} = 1$.

For $X_{1,2} = 1$, the assignment gives a fixed value of nine. In addition, we need to have the minimum distance travelled from cities 2, 3, 4, and 5 without going to city 2 again. Leave out row 1 and column 2 from the original distance matrix, as shown in Table 4.26. We can't go 2→1, so set it to – (infinity).

The sum of the minimum in every row is 4 + 7 + 5 + 5 = 21. Add the distance of fixed allocation $D_{12} = 9$ and get a lower bound of 30 (21 + 9).

The lower bound corresponding to $X_{1,3} = 1$ is computed by creating a similar table (Table 4.27) with row 1 and column 3 left out and by fixing D_{31} (3→1) to infinity.

The sum of the minimum in every row is 4 + 7 + 4 + 5 = 20. Add the distance of fixed allocation $D_{13} = 7$ and get a lower bound of 27 (20 + 7).

TABLE 4.26

From City	To City			
	1	3	4	5
2	–	9	4	5
3	7	–	7	8
4	8	7	–	5
5	6	8	5	–

TABLE 4.27

From City	To City			
	1	2	4	5
2	9	–	4	5
3	–	9	7	8
4	8	4	–	5
5	6	5	5	–

TABLE 4.28

From City	To City			
	1	2	3	5
2	9	–	9	5
3	7	9	–	8
4	–	4	7	5
5	6	5	5	–

TABLE 4.29

From City	To City			
	1	2	3	4
2	9	–	9	4
3	7	9	–	7
4	8	4	7	–
5	–	5	8	5

The lower bound corresponding to $X_{1,4} = 1$ is computed by creating Table 4.28 with row 1 and column 4 left out and by fixing D_{41} (4→1) to infinity.

The sum of the minimum in every row is $5 + 7 + 4 + 5 = 21$. Add the distance of fixed allocation $D_{14} = 8$ and get a lower bound of 29 (21 + 8).

The lower bound corresponding to $X_{1,5} = 1$ is computed by creating Table 4.29 with row 1 and column 5 left out and by fixing D_{51} (5→1) to infinity.

The sum of the minimum in every row is $4 + 7 + 4 + 5 = 20$. Add the distance of fixed allocation $D_{15} = 6$ and get a lower bound of 26 (20 + 6).

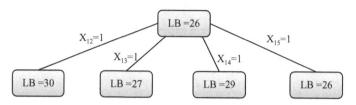

FIGURE 4.31
Step 1 Lower Bound for Example 1.

TABLE 4.30

From City	To City		
	2	3	4
3	9	–	7
4	4	7	–
5	–	8	5

TABLE 4.31

From City	To City		
	1	2	4
3	7	–	7
4	8	4	–
5	–	5	5

TABLE 4.32

From City	To City		
	1	2	3
3	7	9	–
4	8	–	7
5	–	5	8

All the previous computed lower bounds are shown in the branch-and-bound tree in Figure 4.31.

We want to minimize the total distance travelling, so evaluating up to this point 26 is minimum, so we branch further from this node and create three more branches as $X_{2,1} = 1$, $X_{2,3} = 1$, $X_{2,4} = 1$.

For $X_{2,1} = 1$, leave out rows 1 and 2 and columns 1 and 5 from the original distance matrix, as shown in Table 4.30. Until now, we traversed $2 \rightarrow 1 \rightarrow 5$, so we can't go $5 \rightarrow 2$, so set it to – (infinity).

The sum of the minimum in every row is $7 + 4 + 5 = 16$. Add the distance of fixed allocation $D_{21} = 9$ and $D_{15} = 6$ and get a lower bound of 31 (16 + 9 + 6).

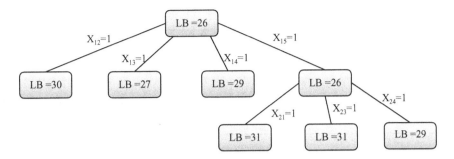

FIGURE 4.32
Step 2 Lower Bound for Example 1.

TABLE 4.33

From City	To City		
	2	4	5
3	–	7	8
4	4	–	5
5	5	5	–

For $X_{2,3} = 1$, leave out rows 1 and 2 and columns 3 and 5 from the original distance matrix, as shown in Table 4.31. Here we traversed 2→3→5, so we can't go 3→2, so set it to – (infinity).

The sum of the minimum in every row is $7 + 4 + 5 = 16$. Add the distance of fixed allocation $D_{23} = 9$ and $D_{15} = 6$ and get a lower bound of 31 (16 + 9 + 6).

For $X_{2,4}=1$, leave out rows 1 and 2 and columns 4 and 5 from the original distance matrix, as shown in Table 4.32. Here we traversed 2→4→5, so we can't go 4→2, so set it to – (infinity).

The sum of the minimum in every row is $7 + 7 + 5 = 19$. Add the distance of fixed allocation $D_{24} = 4$ and $D_{15} = 6$ and get a lower bound of 29 (19 + 4 + 6).

All the previous computed lower bounds are shown in the branch-and-bound tree in Figure 4.32.

We want to minimize the total distance travelling, so evaluating up to this point, 27 is minimum, so we branch further from this node and create three more branches as $X_{2,1} = 1$, $X_{2,4} = 1$, $X_{2,5} = 1$.

For $X_{2,1} = 1$, leave out rows 1 and 2 and columns 1 and 3 from the original distance matrix, as shown in Table 4.33. Here we traversed 2→1→3, so we can't go 3→2, so set it to – (infinity).

The sum of the minimum in every row is $7 + 4 + 5 = 16$. Add the distance of fixed allocation $D_{21} = 9$ and $D_{13} = 7$ and get a lower bound of 32 (16 + 9 + 7).

For $X_{2,4} = 1$, leave out rows 1 and 2 and columns 3 and 4 from the original distance matrix, as shown in Table 4.34. Here we traversed 2→4→3, so we can't go 4→2, so set it to – (infinity).

The sum of the minimum in every row is $8 + 5 + 5 = 18$. Add the distance of fixed allocation $D_{24} = 4$ and $D_{13} = 7$ and get a lower bound of 29 (18 + 4 + 7).

For $X_{2,5} = 1$, leave out rows 1 and 2 and columns 3 and 5 from the original distance matrix, as shown in Table 4.35. Here we traversed 2→5→3, so we can't go 5→2, so set it to – (infinity).

TABLE 4.34

From City	To City		
	1	2	5
3	–	9	8
4	8	–	5
5	6	5	–

TABLE 4.35

From City	To City		
	1	2	4
3	–	9	7
4	8	4	–
5	6	–	5

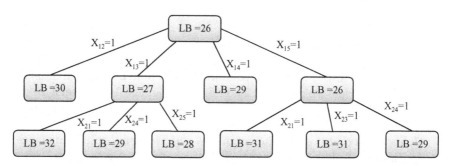

FIGURE 4.33
Step 3 Lower Bound for Example 1.

TABLE 4.36

From City	To City	
	1	4
4	8	–
5	–	5

The sum of the minimum in every row is $7 + 4 + 5 = 16$. Add the distance of fixed allocation $D_{25} = 5$ and $D_{13} = 7$ and get a lower bound of 28 ($16 + 5 + 7$).

All the previous computed lower bounds are shown in the branch-and-bound tree in Figure 4.33.

We want to minimize the total distance travelling, so evaluating up to this point, 28 is minimum, so we branch further from this node and create two more branches as $X_{3,2} = 1, X_{3,4} = 1$.

For $X_{3,2} = 1$, leave out rows 1, 2, and 3 and columns 2, 3, and 5 from the original distance matrix, as shown in Table 4.36. Here we traversed $1 \rightarrow 3 \rightarrow 2 \rightarrow 5$, so we can't go $5 \rightarrow 1$, so set it to – (infinity).

TABLE 4.37

From City	To City	
	1	2
4	8	4
5	6	–

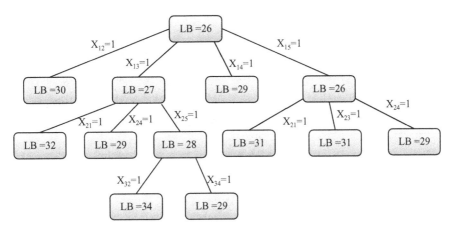

FIGURE 4.34
Step 4 Lower Bound for Example 1.

TABLE 4.38

From City	To City
	2
5	–

The sum of the minimum in every row is 8 + 5 = 13. Add the distance of fixed allocation $D_{13} = 7$, $D_{32} = 9$, $D_{25} = 5$ and get a lower bound of 34 (13 + 7 + 9+ 5).

For $X_{3,4} = 1$, leave out rows 1, 2, and 3 and columns 3, 4, and 5 from the original distance matrix, as shown in Table 4.37. We can't go 5→2, so set it to – (infinity).

The sum of the minimum in every row is 4+6=10. Add the distance of fixed allocation $D_{13} = 7$, $D_{34} = 7$, $D_{25} =5$ and get a lower bound of 29 (10+ 7+ 7+ 5).

All the above computed lower bounds are shown in the branch-and-bound tree in Figure 4.34

We want to minimize the total distance travelling, so evaluating up to this point, 29 is minimum, so we branch further from this node and create two more branches as $X_{4,1} = 1$, $X_{4,2} = 1$.

For $X_{4,1} = 1$, leave out rows 1, 2, 3, and 4 and columns 1, 3, 4, and 5 from the original distance matrix, as shown in Table 4.38. Here we traversed 1→3→5→4, so we can't go 5→2, so set it to – (infinity).

For $X_{4,2} = 1$, leave out rows 1, 2, 3, and 4 and columns 2, 3, 4, and 5 from the original distance matrix, as shown in Table 4.39.

Add the distance of fixed allocation $D_{13} = 7$, $D_{25} = 5$, $D_{34} = 7$, and $D_{42} = 4$ get a lower bound of 29 (6 + 7 + 5 + 7 + 4).

TABLE 4.39

From City	To City
	1
5	6

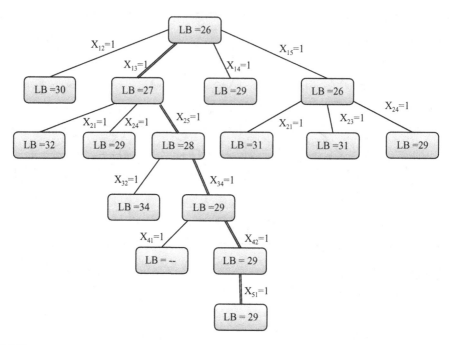

FIGURE 4.35
Branch-and-Bound Tree for Example 1.

We want to minimize the total distance travelling, so evaluating up to this point, 29 is minimum, so we branch further from this node and create one more branch as $X_{5,1} = 1$.

5 to 1 will be $7 + 5 + 7 + 4 + 6 = 29$.

The best optimal solution is $1 \to 3, 2 \to 5, 3 \to 4, 4 \to 2, 5 \to 1$

The city path is $1 \to 3 \to 4 \to 2 \to 5 \to 1$, which is 29.

The branch-and-bound tree for this optimum solution is shown in Figure 4.35.

EXAMPLE 2:

Find the length of an optimal Chinese postman route for the network shown in Figure 4.36. *(BT level 3)*

Solution:
The odd nodes in the network shown in Figure 4.36 are node A and node H. There is only one way of pairing these odd nodes – namely, AH. The following are the paths joining from node A to node H.

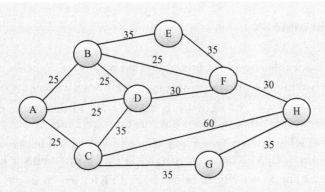

FIGURE 4.36
Network Diagram for Example 2.

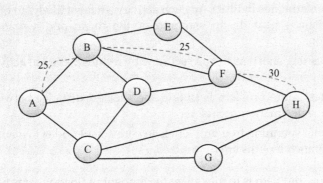

FIGURE 4.37
Path for Example 2.

A-B-E-F-H with a path length of 125.

A-B-F-H with a path length of 80.

A-B-D-F-H with a path length of 110.

A-D-F-H with a path length of 85.

A-C-D-F-H with a path length of 120.

A-C-H with a path length of 85.

A-C-G-H with a path length of 95.

The shortest way of joining A to H is using path A-B-F-H, a total length of 80.
Now draw the nodes on this path onto the original network, as shown in Figure 4.37.

The length of the optimal Chinese postman route is the sum of all the edges in the original network, which is 420 units, plus the answer found in step 4, which is 80 units. Hence the length of the optimal Chinese postman route is 500 units. One possible route corresponding to this length is ADCGHCABDFBEFHFBA, but many other possible routes of the same minimum length can be found.

CHAPTER SUMMARY

- A network is a collection of points connected by arcs or branches called nodes.
- There are two methods for creating a network from a project: AON and AOA.
- A dummy activity in the network is added only to determine one activity's dependence on the other but does not consume any time.
- CPM and PERT are two network analysis techniques used in planning and controlling projects. CPM is deterministic, and PERT is probabilistic in nature.
- CPM used a single activity time estimate. PERT used multiple activity time estimates.
- The critical path is the longest path through the network.
- Three time estimates in PERT for each activity – a most likely, an optimistic, and a pessimistic – provide an estimate of the mean and variance of a beta distribution.
- Project crashing shortens the project time by reducing critical activity times at a cost.
- The shortest route problem is to find the shortest distance between an origin and various destination points.
- The minimal spanning tree problem is to connect all nodes in a network so that the total branch lengths are minimized.

The maximal flow problem is to maximize the amount of flow of items from an origin to a destination.

Questions

1. What is a network, and why is it important is to study the network analysis? *(BT level 1 and 2)*
2. Compare an AOA network and an AON network. *(BT level 5)*
3. Explain why dummy activities are used in a CPM/PERT network? *(BT level 2)*
4. What is the critical path, and what is its importance in project planning? *(BT level 2)*
5. Explain how the *ES, EF, LF,* and *LS* times are computed. *(BT level 2)*
6. What are the three time estimates used with PERT? *(BT level 2)*
7. Which method for determining activity time estimates, deterministic, or probabilistic do you perceive to be preferable? Explain. *(BT level 5)*
8 Describe the meaning of slack and discuss how it can be determined. *(BT level 2)*
9. Would a project manager ever consider crashing a non-critical activity in a project network? Explain convincingly. *(BT level 4)*
10. What is a spanning tree? What is a minimum spanning tree? *(BT level 2)*

Practice Problem

1. The ABC Co. plans to erect a new hospital building. ABC has to set up the project activities; their immediate predecessor activities, along with their approximate duration of operation, are mentioned in Table 4.40. Construct AOA for construction activity. *(BT level 6).*

TABLE 4.40

Data for Practice Problem 1

Activity		Immediate Predecessor (S)	Duration (Days)
A	Purchase Land	–	50
B	Gain permission to build	A	40
C	Excavate	B	20
D	Building up	C	110
E	Do electrical work	D	20
F	Do plumbing work	D	30
G	Interior paint	D	20
H	Clean building	E, F, G	15
I	Shift in furniture	H	15

2. The project activities, their immediate predecessor activities, and their approximate duration of operation are mentioned in Table 4.41.
   a) Construct AOA network. *(BT level 6)*
   b) Determine the critical path *(BT level 3)*
   c) Can it delay activity D without delaying the whole project? If so how many weeks will it take? *(BT level 5)*

TABLE 4.41

Data for Practice Problem 2

Activity	Immediate Predecessor (S)	Duration (Week)
A	–	4
B	–	2
C	A	6
D	A	5
E	B	6
F	D, E	2
G	D, E	9
H	C, F	7

3. The activities and related information for a project are given in Table 4.42 (duration estimates measured in days).

   a) Draw up a network of programs. *(BT level 6)*
   b) Calculate the duration and variance expected for each activity. *(BT level 3)*
   c) Determine the critical path. *(BT level 3)*
   d) What is the expected duration and variance of the critical path? *(BT level 3)*
   e) What is the probability that the project will be completed within 31 days? *(BT level 3)*

TABLE 4.42

Data for Practice Problem 3

Activity	Immediate Predecessor (S)	Optimistic	Most Probable	Pessimistic
A	–	9	11	16
B	–	8	9	10
C	A	6	7	11
D	B	4	4	4
E	B, C	2	6	10
F	D	3	8	11
G	E	4	9	11

4. What is the minimum cost of crashing the following project (Table 4.43) by four days? *(BT level 3)*

TABLE 4.43

Data for Practice Problem 4

Activity	Predecessor (S)	Normal Time (Days)	Crash Time (Days)	Normal Cost	Crash Cost
A	–	5	4	Rs. 800	900
B	–	7	5	200	300
C	–	3	2	400	500
D	A	4	2	800	1,100
E	C	7	4	900	1,500

5. Find the shortest route in the network shown in Figure 4.38, from node 1 to node 6 if the objective is to minimize travel time (in a minute). *(BT level 3)*

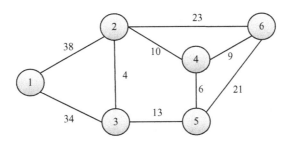

FIGURE 4.38
Network for Practice Problem 5.

6. Determine the maximum flow by vehicles per hour for the highway network system shown in Figure 4.39. Vehicle flow capacity is mentioned on the arc. *(BT level 3)*

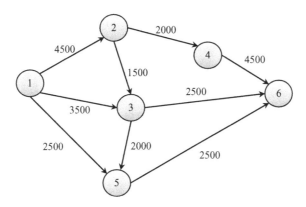

FIGURE 4.39
Network for Practice Problem 6.

7. Army battalion wants to set up a communications system that will connect its command to the eight camps. The following network (Figure 4.40) shows the distances (in meters) between the camps and the various paths over which a communication line can be built. Determine the minimum distance communication system which will link all eight camps using the minimal spanning tree approach. *(BT level 3)*

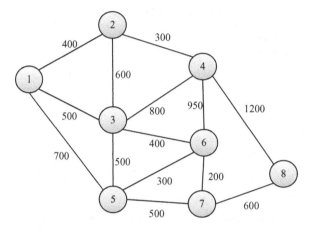

FIGURE 4.40
Network for Practice Problem 7.

5

Sequencing

Learning Outcomes: After studying this chapter, the reader should be able to

LO 1. *Understand the importance of sequencing. (BT level 2)*

LO 2. *Understand assumptions in the sequencing problem. (BT level 2)*

LO 3. *Understand various terminologies used in the sequencing problem. (BT level 2)*

LO 4. *Understand and construct a Gantt chart. (BT level 2 and 6)*

LO 5. *Name and describe different priority sequencing rules. (BT level 1 and 2)*

LO 6. *Use Johnson's rule. (BT level 3)*

LO 7. *Develop an understanding of sequencing on multiple machines/work centres. (BT level 2)*

5.1 Introduction

The sequencing problem arises when a series of tasks are to be performed in an appropriate order. In such cases, the pre-assigned order is known as sequencing. There can be many sequences that can be pursued to complete a task. The sequencing problem determines the optimal order in which tasks ought to be performed through a specific number of facilities to minimize total time and cost.

A shop production system is traditionally discovered in job-type machine shops, where a broad range of jobs are handled according to client requirements. The machine shops complex behaviour, consisting of several job centres, each with one or more of the same sort of machines. Usually, the job in the machine shop is started by obtaining customer orders. The order may include one or more jobs, each of which should be designed before releasing it to the shop. When the job is released to the shop, the related activities are routed through the work centres. Each of these activities occurs at only one work centre. It is natural that any work may include several units that can be determined or indicated in the customer order on an economical batch size basis.

When scheduling a set of jobs through the shop, the goal is to ensure that each job prescribed technological order is fulfilled and there are no conflicts between jobs while trying to fulfil job due dates or other requirements. In such a shop, the perfect scenario is that machines are continually busy; the workforce is just enough to manage the workload; materials and in-waiting inventories and end-product inventory, if any, are retained at reasonable rates; and work due dates are set to be met. However, in reality, the shop generally adapts to the present circumstances irrespective of scheduling processes or rules.

The sequencing issue is about determining the sequence in which a set of jobs on each of several machines is to be done. Of course, this issue is only part of the general problem of manufacturing control in a shop production system. The sequencing issue is preceded by the planning area that includes the functions as the acquisition of materials, technological procedures, production capability, routing specifications, time estimates for processing, and shop loading.

A sequencing problem is the assurance of "best" among all the sequences, where "best" is well defined with respect to performance measures for the concerned problem.

In terms of the following context, the classical shop scheduling method can be defined. A machine shop is fitted with different machine groups. Each group of machines consists of several identical machines in terms of their capacity to process the job anticipated in the shop. The amount of work constitutes several jobs to be performed on the machines. Each task may consist of a single unit or a batch of the same units, generally economically determined. Any job needs one or more operations, each on a single machine. The specification of the work and the machine concerned determines an operation ultimately. In a defined routing, the activities comprising each task are performed on the machines, hereafter referred to as a machine order. Machine orders are generally enforced by technological demands that always dictate the fundamental process of manufacturing and the kinds of machines, fixtures, and tools to be used.

Associated with each operation is a processing time that is required for the complete performance of this operation. The processing time of operation may consist of (1) the time needed to transport the work from one machine to another in accordance with the pre-specified ordering of the machine, (2) the time required to set up the machine to prepare it for that operation, (3) operation time—time needed to conduct the work on a specific machine, and (4) machine time—time required to reset the machine after completion of the procedure. Because of the limited capability of the available machines, the related activities must compete with each other for each of the machines as the jobs flow through the store. Upon completion of all activities involving a job, the job leaves the shop as a finished job.

The problem is to determine a sequence of jobs on machines such that it satisfies the constraints (if any) on the order in which jobs are to be performed on machines.

Although the issue of shop scheduling is illustrated in the context of manufacturing, it occurs quite naturally in almost all fields of activity, such as hospitals, colleges, banks, and government. By considering the following operating systems, the pervasive nature of shop scheduling issues can be illustrated:

1. A limited amount of mechanics (machines) in a car repair shop are scheduling vehicles (jobs) to be serviced.
2. The scheduling of courses (jobs) in an academic institution to classrooms.
3. Patient scheduling (jobs) in a hospital on a limited amount of test equipment (machines).
4. Scheduling of ships (jobs) in a harbour on a restricted amount of berths (machines).
5. Scheduling of shipments (jobs) in a loading dock on a limited amount of trucks (machines).

Many other operating systems that can be structured as a single-machine system include realistic situations. For example, the scheduling of programs (jobs) to be run on the computer (single machine) in a computer and data processing centre and the scheduling of cities (jobs) to be visited by a salesman (single machine), the well-known travelling salesman problem.

LO 1. Understand the importance of sequencing

The shop's performance efficiency can be evaluated in terms of one or more criteria, such as minimum price, minimum in-waiting inventory, the minimum completion time for all jobs, maximum profit, maximum machine and workforce usage, or the capacity to meet work due dates, whichever is most suitable.

The following is a general sequencing problem.

Let us consider that there are n jobs that need to be performed, each one in turn on every one of total m machines. The order of machines concerning which job ought to be performed is known. The expected completion time for each job on each machine is known. The objective is to determine the best sequence out of all possible sequences that minimize the total time elapsed between the start of the first job on the first machine and the completion of the last job on the previous machine.

5.1.1 Assumptions Made while Solving Sequencing Problems

LO 2. Understand assumptions in the sequencing problem

Assumptions Related to Jobs:

1. Each job is handled by a pre-set machine order, and no alternative ordering is allowed.
2. Each job, once it begins, must be completed (i.e., no cancellation of the job takes place).
3. Each job, once it begins on a machine, has to be completed before another work can begin on that machine (i.e., no preventive priorities).
4. Each job is an entity despite the fact that the job consists of individual units. This eliminates the division of jobs between two or more machines.
5. Not every job can be processed at a time by more than one machine.
6. Every work may have to wait between machines and is, therefore, allowed in-waiting inventory.
7. Each job may have a due date, which the customer and shop management generally determine. Such a date is fixed if any.
8. A job is processed on a machine as soon as possible.
9. The preceding operation must be completed before the succeeding one can proceed.

Assumptions Related to Machines:

1. No machine can execute more than one operation at a given point in time.
2. Processing time for each machine is known, and it does not depend on the order of processing various jobs.
3. The time that each job requires to move from one machine to another is negligible.
4. Each machine works separately in the shop, allowing each machine to operate at its own highest output speed.
5. During the planning period under the account, each machine is continually accessible for an assignment without disruption, such as machine breakdowns or maintenance.
6. The order of the jobs should be maintained over each machine.

Assumptions Related to Processing Time:

1. There are known and finite processing times.
2. Processing times may implicitly include transportation times between machines and, where appropriate, machine set-up and teardown times. Transport and exchange times can also be regarded as negligible.

5.1.2 Terms Used in Sequencing

Machines: It refers to the facilities on which a job is processed for completion.

Processing time: It indicated the time required to process the job on each machine. It includes both set-up time and actual processing time.

Idle time: It stands for the idle time of the machine (i.e., during this time, a machine is not assigned any job to process).

Total elapsed time: It is the total amount of time spent from processing of the first job on the first machine to the completion of the last job on the last machine. It includes both processing time and idle time.

Completion time: It is the time at which the job is completed. It is also known as flow time. The mean completion time, which is a common measure of system performance, is the arithmetic average of the completion times for all n jobs. The make-span is the time required to complete a group of jobs (all n jobs). Figure 5.1 shows the completion time and make-span for three jobs on two machines.

Due date: It is the time at which the processing of the job is to be completed.

Lateness: It is the difference between the actual completion time and due date of the job. It can be either positive or negative. As shown in Figure 5.2, if lateness is negative, it is earliness; when it is positive, it is tardiness.

Tardiness: It is the positive difference between the due date of the late job and its completion time.

Different books show the level of confusion arising from varying interpretations of the terms "sequencing" and "scheduling." We will, however, provide appropriate definitions

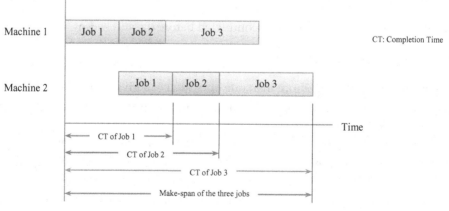

FIGURE 5.1
Completion Time and Make-Span.

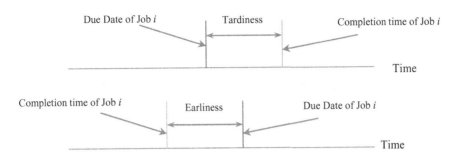

FIGURE 5.2
Tardiness and Earliness.

to distinguish between the two terms. Sequencing is about the arrangements and permutations in which all machines perform a set of jobs under consideration. Scheduling concerns the requirements on all machines for the start or completion times of specific jobs. Sequencing choices concentrate on events arrangement, whereas the time of occurrences is the focus of scheduling choices. Although distinguishable as the aforementioned, the terms sequencing and scheduling are often used as synonyms. The reason for the frequent use of both terms is that since it is always presumed that each task begins as soon as possible, a schedule is generated automatically when a specific sequence is searched.

The variety of circumstances in shop scheduling issues that can be experienced. Through sequencing and dispatching methods, shop scheduling issues are evaluated. A sequence of jobs for each machine is searched in advance in the sequencing strategy. Sequencing, therefore, relates to the method of specifying a set of jobs to be conducted on each of the concerned machines. On the other side, in the dispatching strategy, dispatching rules are sought, which can be applied to each machine to assign a job from a number of tasks waiting to be done. The choice of a dispatching rule must take into account its impact on the shop's activities. Examples of shipping rules are provided in this chapter's next section.

5.2 Gantt Chart

Gantt chart is a useful way to represent a shop scheduling issue of n jobs and m machines. A Gantt chart is a time-scale horizontal bar graph displaying the activities engaged in the operations. These operations' activities can be shown in a bar representing either a task or a machine, as shown in Figure 5.3.

LO 4. Understand and construct a Gantt chart

Notation:

Designate a job by an index n and a machine by an index m.

Processing of the jobs on any machine m in some sequence is given by

$$n_1, n_2, \ldots n_k \ldots n_J,$$

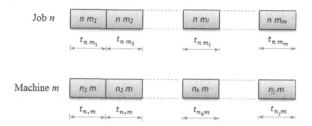

FIGURE 5.3
Representations of a Machine Ordering of Job *n* and a Job Sequence for Machine *m*.

where *J* is the total number of jobs.

For machines for any job *n* with respect to the preconceived order, the machines will be designated as

$$m_1, m_2, \ldots m_l \ldots m_M,$$

where *M* is the total number of machines.

The job and machine subscriptions are used, respectively, to indicate the place in a permutation sequence and permutation order.

Processing time $t_{n\,ml}$ is the time required to perform a job n on a particular machine m_l.

Gantt charts help to demonstrate the interrelations within the same job or machine and show the required coordination of the different operations. Gantt chart provides no guidelines for selecting but merely provides a graphical technique for displaying outcomes (schedule) and assessing outcomes (make-span, idle time, waiting time, machine usage, etc.). The first and likely most evident reason for using a Gantt chart is the capacity to impact the resource very efficiently. Gantt charts help to demonstrate the interrelations within the same job or machine and show the required coordination of the different operations. The Gantt chart provides no guidelines for selecting but merely offers a graphical technique for displaying outcomes (schedule) and assessing outcomes (make-span, idle time, waiting time, machine usage, etc.). The first and likely most evident reason for using a Gantt chart is the capacity to impact the resource very efficiently. A Gantt chart provides an apparent overview of all scheduled operations, how long each activity is scheduled to last, where operations overlap with other operations, and where each resource is put. With no more waste of resources, it is feasible to look ahead in the schedule and have all the required data to see when to allocate a particular resource. One of the biggest reasons to use a Gantt chart is that powerful and meaningful insights guide the scheduler. For example, the scheduler gets the total amount of time it takes for a project to be completed. And not only does it make it possible to visualize a series of scheduled occurrences but also a Gantt chart makes it possible to create a connection between operations, also known as "dependencies." A dependent task is linked to another task and starts after completion of the first task.

The above debate may be elucidated by an instance. Consider a sample problem on three machines having two jobs to be processed. The following (Table 5.1) are the orders in

TABLE 5.1

Sample Problem Data

Job	Machine Sequence (Processing Time)
1	$M_1(3) - M_3(5) - M_2(2)$
2	$M_3(4) - M_1(5) - M_2(6)$

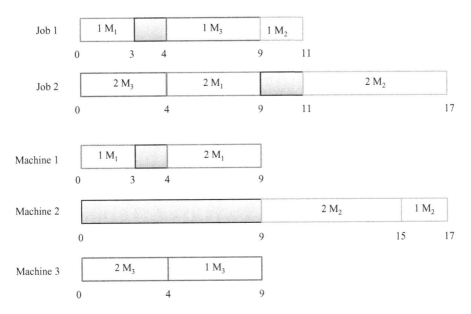

FIGURE 5.4
Gantt Chart for Sample Problem.

which jobs 1 and 2 are carried out on the three machines. It needs 3, 5, and 2 time units, respectively, to execute task 1 on machines 1, 3, and 2. Similarly, on machines 3, 1, and 2, task 2 needs 3, 4, and 5 units of time to be finished.

The combined machine ordering and processing time requirements of this problem result in the Gantt chart shown in Figure 5.4.

The first bar representing job 1 indicates that on machine 1 first, machine 3 seconds, and machine 2 last, this work has to be processed. Until the corresponding operation on machine 1 has been completed, machine 3 cannot be begun on job 1. Similarly, it is impossible to start machine 2 on the same job until the respective operating machine 3 is complete. Assuming the time scale begins at zero, job 1 completion times on machines 1, 3, and 2 are 3, 9, and 11, respectively. Job 2 is performed on machine 3 first, machine 1 second, and machine 2 last, represented by the second bar in Figure 5.4. Considering that the time scale also begins at zero, on all machines, the completion time of work 2 is 17. Machine 1 bar demonstrates that this machine must execute first job 1 and the second job 2. The unfilled portions in the bar indicate that machine 1 performs job 1 from zero to 3 and job 2 from 4 to 9, considering that the time scale starts at zero. The part filled from 3 to 4 reflects the idle time of the machine. From this Gantt chart, it should be observed that the scheduled time is 17 units of time as it represents the last operation's completion time.

5.3 Sequencing of *n* Jobs through One Machine

The sequencing of n jobs through one machine is the simplest sequencing problem. It consists of a queue of n jobs at a single facility or machine (see Figure 5.5). Some of the examples are n jobs to be processed on a lathe, n number of automobiles to be serviced at a service station, n number of patients waiting for a doctor in the clinic, and the like.

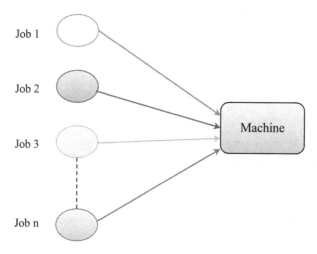

FIGURE 5.5
n Jobs through One Machine.

LO5. Name and describe different priority sequencing rules

In this sequencing problem, the completion time of each job will vary depending on its place in the sequence; however, the completion time for the group of jobs will not alter. Priority rules are used to select the sequence in which n jobs will be processed through one machine. Some common priority rules are listed next.

- ▶ FCFS (First Come, First Served): In this rule, jobs/tasks or entities are sequenced in the order in which they arrive at the machine or facility.
- ▶ SPT (Shortest Processing Time): In this rule, jobs/tasks or entities are sequenced according to the processing time required at a machine or facility, with the activity requiring the least processing time sequenced first.
- ▶ EDD (Earliest Due Date): In this rule, jobs/tasks or entities are sequenced according to the due date (i.e., in the order in which they are due to the customer for delivery).
- ▶ STR (Slack Time Remaining): In this rule, jobs/tasks or entities are sequenced according to least slack time first followed by the next smallest slack time, and so on. (Slack time is defined as the due date of the job minus its processing time.)

Examples based on sequencing of *n* jobs through one machine

EXAMPLE 1:

Six jobs are to be processed at a single work centre. Due dates and processing times for six jobs are given in Table 5.2 (assume the jobs arrived in the order given next).

TABLE 5.2

Data for Example 1

Job	Processing Time (Days)	Due Date (Days)
A	1	4
B	4	7
C	2	2
D	5	8
E	3	6
F	6	9

Determine the optimal sequence of jobs, average tardiness, average flow time, and average in-process inventory for each of the following sequencing rules. (*BT level 3*)

a. FCFS

b. SPT

c. EDD

Solution:

a. The optimal sequence of jobs using FCFS is simply A-B-C-D-E-F (Table 5.3).

TABLE 5.3

Processing Time

(1) Job Sequence	(2) Processing Time (days)	(3) Flow Time	(4) Due Date (Days)	(5) Days Tardy
A	1	1	4	0
B	4	5	7	0
C	2	7	2	5
D	5	12	8	4
E	3	15	6	9
F	6	21	9	12
	$\Sigma = 21$	$\Sigma = 61$		$\Sigma = 30$

In Table 5.3, column (3) represents cumulative processing time. The sum of all the processing times, when divided by the number of jobs, provides the average flow time (i.e. average time each job spends on a machine).

$$\therefore \text{Average flow time} = \frac{1+5+7+12+15+21}{6} = 10.17 \text{ days}$$

In Table 5.3, column (5) indicates job tardiness in days. It is calculated by subtracting column (4) from column (3) – i.e., (3) – (4). It is zero if the difference is negative. Adding all tardiness and then dividing it by the number of jobs provides the average tardiness.

$$\therefore \text{Average tardiness} = \frac{0+0+5+4+9+12}{6} = 5\,\text{days}$$

The average in-process inventory is calculated by dividing the sum of flow time by the total processing time.

$$\text{Average in} - \text{process inventory} = \frac{61}{21} = 2.91$$

b. The optimal sequence of jobs using SPT is A-C-E-B-D-F. Here the job with the least processing time is sequenced first followed by the one with the next smallest processing time, and so on (refer to Table 5.4).

In Table 5.4, column (3) represents cumulative processing time. The sum of all the processing times when divided by the number of jobs provides the average flow time (i.e., the average time each job spends on a machine).

$$\therefore \text{Average flow time} = \frac{1+3+6+10+15+21}{6} = 9.33\,\text{days}$$

In Table 5.4, column (5) indicates job tardiness in days. It is calculated by subtracting column (4) from column (3) – i.e., (3) – (4). It is zero if the difference is negative. Adding all tardiness and then dividing it by the number of jobs provides the average tardiness.

$$\therefore \text{Average tardiness} = \frac{0+1+0+3+7+12}{6} = 3.83\,\text{days}$$

The average in-process inventory is calculated by dividing the sum of flow time by the total processing time.

$$\text{Average in} - \text{process inventory} = \frac{56}{21} = 2.67$$

TABLE 5.4

Processing Time

(1) Job Sequence	(2) Processing Time (Days)	(3) Flow Time	(4) Due Date (Days)	(5) Days Tardy
A	1	1	4	0
C	2	3	2	1
E	3	6	6	0
B	4	10	7	3
D	5	15	8	7
F	6	21	9	12
	$\Sigma = 21$	$\Sigma = 56$		$\Sigma = 23$

TABLE 5.5

Processing Time

(1) Job Sequence	(2) Processing Time (Days)	(3) Flow Time	(4) Due Date (Days)	(5) Days Tardy
C	2	2	2	0
A	1	3	4	0
E	3	6	6	0
B	4	10	7	3
D	5	15	8	7
F	6	21	9	12
	$\Sigma = 21$	$\Sigma = 57$		$\Sigma = 22$

c. The optimal sequence of jobs using EDD is C-A-E-B-D-F. Here the job with the EDD is sequenced first followed by the jobs in the order of increasing due dates (refer to Table 5.5).

In Table 5.5, column (3) represents cumulative processing time. The sum of all the processing times when divided by the number of jobs provides the average flow time (i.e., the average time each job spends on a machine).

$$\therefore \text{Average flow time} = \frac{2+3+6+10+15+21}{6} = 9.5 \, \text{days}$$

In Table 5.5, column (5) indicates job tardiness in days. It is calculated by subtracting column (4) from column (3) – i.e., (3) – (4). It is zero if the difference is negative. Adding all tardiness and then dividing it by the number of jobs provides the average tardiness.

$$\therefore \text{Average tardiness} = \frac{0+0+0+3+7+12}{6} = 3.67 \, \text{days}$$

The average in-process inventory is calculated by dividing the sum of flow time by the total processing time.

$$\text{Average in-process inventory} = \frac{57}{21} = 2.71$$

EXAMPLE 2:

Seven jobs are to be processed in a single work centre. Due dates and processing times for jobs are given in Table 5.6 (assume jobs arrived in the order shown).

Determine the optimal sequence of jobs, average in-process inventory, average tardiness, and average flow time for STR sequencing rules. *(BT level 3)*

Solution:

First, calculate the slack time for seven jobs, as shown in Table 5.7.

Slack Time = Due Date – Processing Time

TABLE 5.6

Data for Example 2

Job	Processing Time (Days)	Due Date (Days)
A	3	19
B	11	28
C	1	14
D	10	15
E	9	17
F	2	4
G	5	8

TABLE 5.7

Slack Time

Job	Processing Time (Days)	Due Date (Days)	Slack Time (Days)
A	3	19	16
B	11	28	17
C	1	14	13
D	10	15	5
E	9	17	8
F	2	4	2
G	5	8	3

The optimal sequence of jobs using STR is F-G-D-E-C-A-B. Here the job with the least slack time is sequenced first followed by the jobs in the order of increasing slack times (refer to Table 5.8).

In Table 5.8, column (3) represents cumulative processing time. The sum of all the processing times, when divided by the number of jobs, provides the average flow time (i.e., average time each job spends on a machine).

$$\therefore \text{Average flow time} = \frac{2+7+17+26+27+30+41}{7} = 21.42 \text{ days}$$

In Table 5.8, column (5) indicates job tardiness in days. It is calculated by subtracting column (4) from column (3) – i.e. (3) – (4). It is zero if the difference is negative. Adding all tardiness and then dividing it by the number of jobs provides the average tardiness.

$$\therefore \text{Average tardiness} = \frac{0+0+2+9+13+11+13}{7} = 6.85 \text{ days}$$

TABLE 5.8

Processing Time

(1) Job Sequence	(2) Processing Time (Days)	(3) Flow Time	(4) Due Date (Days)	(5) Days Tardy
F	2	2	4	0
G	5	7	8	0
D	10	17	15	2
E	9	26	17	9
C	1	27	14	13
A	3	30	19	11
B	11	41	28	13
	$\Sigma = 41$	$\Sigma = 150$		$\Sigma = 48$

The average in-process inventory is calculated by dividing the sum of flow time by the total processing time.

$$\text{Average in} - \text{process inventory} = \frac{150}{41} = 3.65$$

EXAMPLE 3:

Seven jobs are to be processed at a single work centre. Due dates and processing times for jobs are given in Table 5.9 (assume the job arrived in the order shown).

If today is day 4 of the planning cycle, determine the optimal sequence of jobs, average flow time, and average tardiness for each of the following sequencing rules. Comment on your answer. *(BT level 3 and 5)*

a. FCFS

b. SPT

c. EDD

d. STR

TABLE 5.9

Data Example 3

Job	Processing Time (Days)	Due Date (Days)
A	2	9
B	9	11
C	1	24
D	3	7
E	4	14
F	7	17
G	6	19

Solution:

a. The optimal sequence of jobs using FCFS is simply A-B-C-D-E-F-G. As today is day 4, the start time of job A is 4 (refer to Table 5.10).

TABLE 5.10

Processing Time

(1) Job Sequence	(2) Processing Time (Days)	(3) Completion Time	(4) Due Date (Days)	(5) Days Tardy
A	2	4 + 2=6	9	0
B	9	15	11	4
C	1	16	24	0
D	3	19	7	12
E	4	23	14	9
F	7	30	17	13
G	6	36	19	17
	$\Sigma = 32$	$\Sigma = 145$		$\Sigma = 55$

In Table 5.10, column (3) represents cumulative processing time. The sum of all the processing times when divided by the number of jobs provides the average completion time.

$$\therefore \text{Average completion time} = \frac{6+15+16+19+23+30+36}{7} = 20.71 \text{days}$$

The average flow time is different from the average completion time for this problem since the start time was 4.

$$\therefore \text{Average flow time} = 20.71 - 4 = 16.71 \text{days}$$

In Table 5.10, column (5) indicates job tardiness in days. It is calculated by subtracting column (4) from column (3) – i.e. (3) – (4). It is zero if the difference is negative. Adding all tardiness and then dividing it by the number of jobs provides the average tardiness.

$$\therefore \text{Average tardiness} = \frac{0+4+0+12+9+13+17}{7} = 7.85 \text{days}$$

b. The optimal sequence of jobs using SPT is C-A-D-E-G-F-B. Here the job with the least processing time is sequenced first followed by the one with the next smallest processing time, and so on (refer to Table 5.11).

In Table 5.11, column (3) represents cumulative processing time. The sum of all the processing times when divided by the number of jobs provides the average completion time.

$$\therefore \text{Average completion time} = \frac{5+7+10+14+20+27+36}{7} = 17 \text{days}$$

The average flow time is different from the average completion time for this problem since the start time was 4.

$$\therefore \text{Average flow time} = 17 - 4 = 13 \text{days}$$

TABLE 5.11

Processing Time

(1) Job Sequence	(2) Processing Time (Days)	(3) Completion Time	(4) Due Date (Days)	(5) Days Tardy
C	1	4+1=5	24	0
A	2	7	9	0
D	3	10	7	3
E	4	14	14	0
G	6	20	19	1
F	7	27	17	10
B	9	36	11	25
	Σ =32	Σ =119		Σ = 39

In Table 5.11, column (5) indicates job tardiness in days. It is calculated by subtracting column (4) from column (3) – i.e., (3) – (4). It is zero if the difference is negative. Adding all tardiness and then dividing it by the number of jobs provides the average tardiness.

$$\therefore \text{Average tardiness} = \frac{0+0+3+0+1+10+25}{7} = 5.57 \text{ days}$$

c. The optimal sequence of jobs using EDD is D-A-B-E-F-G-C. Here the job with the EDD is sequenced first followed by the jobs in the order of increasing due dates (refer to Table 5.12).

In Table 5.12, column (3) represents cumulative processing time. The sum of all the processing times when divided by the number of jobs provides the average completion time.

$$\therefore \text{Average completion time} = \frac{7+9+18+22+29+35+36}{7} = 22.29 \text{ days}$$

TABLE 5.12

Processing Time

(1) Job Sequence	(2) Processing Time (Days)	(3) Completion Time	(4) Due Date (Days)	(5) Days Tardy
D	3	4+3=7	7	0
A	2	9	9	0
B	9	18	11	7
E	4	22	14	8
F	7	29	17	12
G	6	35	19	16
C	1	36	24	12
	Σ =32	Σ =156		Σ = 55

The average flow time is different from the average completion time for this problem since the start time was 4.

$$\therefore \text{Average flow time} = 22.29 - 4 = 18.29 \, \text{days}$$

In Table 5.12, column (5) indicates job tardiness in days. It is calculated by subtracting column (4) from column (3) – i.e., (3) – (4). It is zero if the difference is negative. Adding all tardiness and then dividing it by the number of jobs provides the average tardiness.

$$\therefore \text{Average tardiness} = \frac{0+0+7+8+12+16+12}{7} = 7.85 \, \text{days}$$

d. First, calculate the slack time for seven jobs, as shown in Table 5.13.

$$\text{Slack Time} = \text{Due Date} - \text{Processing Time}$$

The optimal sequence of jobs using STR is B-D-A-E-F-G-C. Here the job with the least slack time is sequenced first followed by the jobs in the order of increasing slack times (refer to Table 5.14).

TABLE 5.13

Processing Time

Job	Processing Time (Days)	Due Date (Days)	Slack Time (Days)
A	2	9	7
B	9	11	2
C	1	24	23
D	3	7	4
E	4	14	10
F	7	17	10
G	6	19	13

TABLE 5.14

Processing Time

(1) Job Sequence	(2) Processing Time (Days)	(3) Completion Time	(4) Due Date (Days)	(5) Days Tardy
B	9	4+9=13	11	2
D	3	16	7	9
A	2	18	9	9
E	4	22	14	8
F	7	29	17	12
G	6	35	19	16
C	1	36	24	12
	$\Sigma = 32$	$\Sigma = 169$		$\Sigma = 68$

In Table 5.14, column (3) represents cumulative processing time. The sum of all the processing times when divided by the number of jobs provides the average completion time.

$$\therefore \text{Average completion time} = \frac{13+16+18+22+29+35+36}{7} = 24.14\,\text{days}$$

The average flow time is different from the average completion time for this problem since the start time was 4.

$$\therefore \text{Average flow time} = 24.14 - 4 = 20.14\,\text{days}$$

In Table 5.14, column (5) indicates job tardiness in days. It is calculated by subtracting column (4) from column (3) – i.e., (3) – (4). It is zero if the difference is negative. Adding all tardiness and then dividing it by the number of jobs provides the average tardiness.

$$\therefore \text{Average tardiness} = \frac{2+9+9+8+12+16+12}{7} = 9.72\,\text{days}$$

From the summary in Table 5.15, SPT yields the lowermost average flow time, average tardiness, and a number of tardy jobs.

TABLE 5.15

Summary

Summary	FCFS	SPT	EDD	STR
Average Flow Time	16.71	13.00	18.29	20.14
Average Tardiness	7.85	5.57	7.85	9.72
No. of Tardy Jobs	5	4	5	7

EXAMPLE 4:

Mr. A tailor has eight jobs to be completed on one swing machine. Due dates and processing times for jobs are given in Table 5.16 (assume the jobs arrived in the order shown).

TABLE 5.16

Data Example 4

Job	Processing Time (Days)	Due Date (Days)
A	5	6
B	4	8
C	3	9
D	2	10
E	6	13
F	7	20
G	8	12
H	1	25

If today is day 2 of the planning cycle, determine the optimal sequence of jobs, average flow time, average tardiness, maximum tardiness, and a number of tardy jobs for each of the following sequencing rules. Comment on your answer. (*BT level 3 and 5*)

a. SPT

b. EDD

Solution:

a. The optimal sequence of jobs using SPT is H-D-C-B-A-E-F-G. Here the job with the least processing time is sequenced first followed by the one with the next smallest processing time, and so on (refer to Table 5.17).

In Table 5.17, column (3) represents cumulative processing time. The sum of all the processing times when divided by the number of jobs provides the average completion time.

$$\therefore \text{Average completion time} = \frac{3+5+8+12+17+23+30+38}{8} = 17 \text{ days}$$

The average flow time is different from the average completion time for this problem since the start time was 4.

$$\therefore \text{Average flow time} = 17 - 2 = 15 \text{ days}$$

In Table 5.17, column (5) indicates job tardiness in days. It is calculated by subtracting column (4) from column (3) – i.e., (3) – (4). It is zero if the difference is negative. Adding all tardiness and then dividing it by the number of jobs provides the average tardiness.

$$\therefore \text{Average tardiness} = \frac{0+0+0+4+11+10+10+26}{8} = 7.63 \text{ days}$$

TABLE 5.17

Processing Time

(1) Job Sequence	(2) Processing Time (Days)	(3) Completion Time	(4) Due Date (Days)	(5) Days Tardy
H	1	$2 + 1 = 3$	25	0
D	2	5	10	0
C	3	8	9	0
B	4	12	8	4
A	5	17	6	11
E	6	23	13	10
F	7	30	20	10
G	8	38	12	26
	$\Sigma = 36$	$\Sigma = 136$		$\Sigma = 61$

Maximum tardiness is 26 for job G.

The number of tardy jobs is five.

b. The optimal sequence of jobs using EDD is A-B-C-D-G-E-F-H. Here the job with the EDD is sequenced first followed by the jobs in the order of increasing due dates (refer to Table 5.18).

In Table 5.18, column (3) represents cumulative processing time. The sum of all the processing times when divided by the number of jobs provides the average completion time.

$$\therefore \text{Average completion time} = \frac{7+11+14+16+24+30+37+38}{8} = 22.13 \text{ days}$$

The average flow time is different from the average completion time for this problem since the start time was 4.

$\therefore$ Average flow time = 22.13 – 2 = 20.13 days

In Table 5.18, column (5) indicates job tardiness in days. It is calculated by subtracting column (4) from column (3) – i.e., (3) – (4). It is zero if the difference is negative. Adding all tardiness and then dividing it by the number of jobs provides the average tardiness.

$$\therefore \text{Average tardiness} = \frac{1+3+5+6+12+17+17+13}{8} = 9.25 \text{ days}$$

Maximum tardiness is 17 for jobs E & F.

The number of tardy jobs is eight.

TABLE 5.18

Processing Time

(1) Job Sequence	(2) Processing Time (Days)	(3) Completion Time	(4) Due Date (Days)	(5) Days Tardy
A	5	2+5=7	6	1
B	4	11	8	3
C	3	14	9	5
D	2	16	10	6
G	8	24	12	12
E	6	30	13	17
F	7	37	20	17
H	1	38	25	13
	$\Sigma = 36$	$\Sigma = 177$		$\Sigma = 74$

5.4 Sequencing of *n* Jobs through Two Machines

Let us consider *n* jobs, which will be processed on two work centres W_1 and W_2 in order W_1W_2. In other words, each job has to go through the same operation sequence (i.e., a job is first assigned to the W_1 and is assigned to the W_2 after completing processes at work centre W_1). If the W_2 work centre is not currently free to do the same job, the job must wait in a waiting line for the W_2 work centre for its turn (see Figure 5.6).

In this problem, the optimal sequence of jobs that minimizes T (the total elapsed time) from the start of the first job at the first work centre to the completion of the last job at the second work centre is to be determined. The total time elapsed includes the idle time, if any.

There are two advantages to determining a production sequence for a group of jobs to minimize the total elapsed time:

LO6. Use Johnson's rule

1. The job group is completed in the shortest possible time.

2. Maximizes the use of the work centres. By continuously using the first work centre until the last job is processed, the idle time on the second work centre is minimized.

Johnson's algorithm is useful in determining the sequence of n jobs being processed at two work centres. This algorithm minimizes idle time at the work centre and reduces the total time taken for completing all the jobs. All jobs are given equal priority, sequencing the jobs according to the time that is taken, which may minimize the idle time, which reduces total elapsed time.

Johnson's algorithm involves the following steps.

1. List the jobs in each work centre with their processing times.
2. Choose the job having the lowest time. If the smallest processing time is at the first work centre, sequence that job first; if the smallest processing time is at the second work centre, sequence that job last. Arbitrarily break ties.
3. Remove the job from further consideration.
4. Repeat steps 2 and 3 and place the remaining jobs next to first and next to the last until all the jobs are assigned.

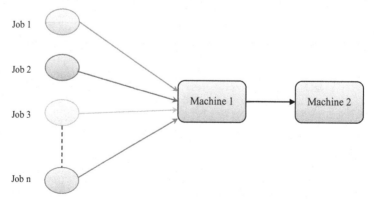

FIGURE 5.6
n Jobs through Two Machines.

Examples based on sequencing of *n* jobs through two machines

EXAMPLE 1:

Six jobs need to be processed on two work centres W_1 and W_2, in the order $W_1 W_2$. The processing times (in hours) for six jobs are given in Table 5.19.

Determine the optimal job sequence to minimize the total elapsed time for this job group. Also, determine the total elapsed time. *(BT level 3)*

Solution:

As per Johnson's algorithm, first, choose the job with the smallest processing time. It is job D at work centre W_1. So sequence job D at the beginning, as shown next.

D		

After eliminating job D and its time, the remaining set of jobs and processing times are shown in Table 5.20.

TABLE 5.19

Data Example 1

Job	W_1	W_2
A	7	7
B	6	5
C	10	11
D	4	9
E	8	10
F	14	17

TABLE 5.20

Processing Time

Job	W_1	W_2
A	7	7
B	6	5
C	10	11
E	8	10
F	14	17

The smallest processing time in the remaining set of jobs corresponds to Job B at work centre W_2. So sequence job B as last, as shown next.

D			B

TABLE 5.21

Processing Time

Job	W$_1$	W$_2$
A	7	7
C	10	11
E	8	10
F	14	17

After eliminating job D and its time, the remaining set of jobs and processing times are shown in Table 5.21.

The smallest processing time in the remaining set of jobs corresponds to Job A on both work centre W$_1$ and W$_2$. Here there is a tie. It doesn't matter if we put it at the start or end of the sequence. Assume it's placed at the end arbitrarily as shown next.

D		A	B

After eliminating job A and its time, the remaining set of jobs and processing times are shown in Table 5.22.

The smallest processing time in the remaining set of jobs corresponds to Job E at work centre W$_1$. So sequence job E at the beginning as shown next.

D	E		A	B

After eliminating job E and its time, the remaining set of jobs and processing times are shown in Table 5.23.

The smallest processing time in the remaining set of jobs corresponds to Job C at work centre W$_1$. So, sequence job C at the beginning and remaining Job F should be assigned next.

D	E	C	F	A	B

TABLE 5.22

Processing Time

Job	W$_1$	W$_2$
C	10	11
E	8	10
F	14	17

TABLE 5.23

Processing Time

Job	W$_1$	W$_2$
C	10	11
F	14	17

TABLE 5.24

Total Elapsed Time

Job	W_1		W_2	
	In	*Out*	*In*	*Out*
D	0	4	4	13
E	4	12	13	23
C	12	22	23	34
F	22	36	36	53
A	36	43	53	60
B	43	49	60	65

The previous is the optimum sequence of all six jobs. The total elapsed time is calculated as shown in Table 5.24.

In the table, job *out* time at work centre W_1 is determined as follows:

Job *out* time at work centre W_1 = Job *in* time at work centre W_1 + Processing time of job at work centre W_1.

In the previous table, job *in* time at work centre W_2 is determined as follows:

Job *in* time at work centre W_2 = max (Work centre W_1 time out of the current job, Work centre W_2 time out of the previous job)

In the previous table, job *out* time at work centre W_2 is determined as follows:

Job *out* time at work centre W_2 = Job *in* time at work centre W_2 + Processing time of job at work centre W_2.

From the previous table, the total elapsed time is 65 hours.

EXAMPLE 2:

Six jobs in a factory need to be processed on two machines M_1 and M_2, in the order M_1 M_2. The processing times (in hours) for six jobs are tabulated in Table 5.25.

Determine the optimal job sequence to minimize the total elapsed time for this job group. Also, determine the total elapsed time. *(BT level 3)*

TABLE 5.25

Data Example 2

Job	M_1	M_2
A	2	6
B	4	7
C	9	4
D	6	3
E	7	3
F	4	11

Solution:

As per Johnson's algorithm, first, choose the job with the smallest processing time. It is a Job A on M_1. So, sequence job A at the beginning as shown next.

A

After eliminating job A and its time, the remaining set of jobs and processing times are shown in Table 5.26.

TABLE 5.26

Processing Time

Job	M_1	M_2
B	4	7
C	9	4
D	6	3
E	7	3
F	4	11

The smallest processing time in the above remaining set of jobs corresponds to Job D and Job E on M_2. As the corresponding processing time for job E on M_1 is larger than that for Job D on M_1, Job E is sequenced last and Job D is sequenced second last as shown next.

A		D	E

After eliminating Job D and Job E and their time, the remaining set of jobs and processing times are shown in Table 5.27.

The smallest processing time in the above remaining set of jobs corresponds to Job B and Job F on M_1 and to Job C on M_2. Here there is a tie. As the corresponding processing time for Job F on M_2 is larger than that for Job B on M_2, Job F will be processed next to Job A and Job B should be sequenced next. Job C is sequenced on the remaining position as follows.

A	F	B	C	D	E

Above is the optimum sequence of all six jobs. The total elapsed time is calculated and shown in Table 5.28.

In the table, job *out* time on machine M_1 is determined as follows:

TABLE 5.27

Processing Time

Job	M_1	M_2
B	4	7
C	9	4
F	4	11

TABLE 5.28

Total Elapsed Time

Job	M_1		M_2	
	In	*Out*	*In*	*Out*
A	0	2	2	8
F	2	6	8	19
B	6	10	19	26
C	10	19	26	30
D	19	25	30	33
E	25	32	33	36

Job *out* time on machine M_1 = Job *in* time on machine M_1 + Processing time of job on machine M_1.

In the table, job *in* time on machine M_2 is determined as follows:

Job *in* time on machine M_2 = max (Machine M_1 time out of the current job, Machine M_2 time out of the previous job).

In the table, job *out* time on machine M_2 is determined as follows:

Job *out* time on machine M_2 = Job *in* time on machine M_2 + Processing time of job on machine M_2.

From the previous table, the total elapsed time is 36 hours.

EXAMPLE 3:

Nine jobs in a factory need to be processed on two machines M_1 and M_2, in the order M_1 M_2. The processing times (in hours) for nine jobs are tabulated in Table 5.29.

Determine the optimal job sequence to minimize the total elapsed time for this job group. Also, determine the total elapsed time. *(BT level 3)*

TABLE 5.29

Data Example 3

Job	M_1	M_2
A	2	6
B	5	8
C	4	7
D	9	4
E	6	3
F	8	9
G	7	3
H	5	8
I	4	11

Solution:

As per Johnson's algorithm, first, choose the job with the smallest processing time. It is Job A on M_1. So sequence Job A at the beginning, as shown next.

A

After eliminating job A and its time, the remaining set of jobs and processing times are shown in Table 5.30.

TABLE 5.30

Processing Time

Job	M_1	M_2
B	5	8
C	4	7
D	9	4
E	6	3
F	8	9
G	7	3
H	5	8
I	4	11

The smallest processing time in the previous remaining set of jobs corresponds to Job E and Job G on M_2. As the corresponding processing time for Job G on M_1 is larger than that for Job E on M_1, Job E is sequenced last, and Job G is sequenced second last as shown next.

A		G	E

After eliminating Job E and Job G and their time, the remaining set of jobs and processing times are shown in Table 5.31.

The smallest processing time in the previous remaining set of jobs corresponds to Job C and Job I on M_1 and to Job D on M_2. Here there is a tie. As the corresponding processing time for Job I on M_2 is larger than that for Job C on M_2, Job I will be processed next to Job

TABLE 5.31

Processing Time

Job	M_1	M_2
B	5	8
C	4	7
D	9	4
F	8	9
H	5	8
I	4	11

TABLE 5.32

Processing Time

Job	M$_1$	M$_2$
B	5	8
F	8	9
H	5	8

A, and Job C should be sequenced next. As the processing time of Job D is on M$_2$, sequence it before Job G, as shown next.

A	I	C				D	G	E

After eliminating Job C, Job D, and Job I and their time, the remaining set of jobs and processing times are shown in Table 5.32.

The smallest processing time in the previous remaining set of jobs corresponds to Job B and Job H on M$_1$. Also, their processing time on M$_2$ is the same. Here there are two sets of ties. Now choose arbitrarily Job B next to Job C and then job H or Job H next to Job C and then Job B as shown next.

A	I	C	B	H		D	G	E

OR

A	I	C	H	B		D	G	E

Job F is sequenced on the remaining position. Finally, there is two sequence order, as shown next.

A	I	C	B	H	F	D	G	E

OR

A	I	C	H	B	F	D	G	E

The total elapsed time for the first sequence is calculated and shown in Table 5.33.

The total elapsed time for the second sequence is calculated and shown in Table 5.34.

In the previous tables, job *out* time on machine M$_1$ is determined as follows:

Job *out* time on machine M$_1$ = Job *in* time on machine M$_1$ + Processing time of job on machine M$_1$.

In the table, job *in* time on machine M$_2$ is determined as follows:

Job *in* time on machine M$_2$ = max (Machine M$_1$ time out of the current job, Machine M$_2$ time out of the previous job).

In the table, job *out* time on machine M$_2$ is determined as follows:

Job *out* time on machine M$_2$ = Job *in* time on machine M$_2$ + Processing time of job on machine M$_2$.

From the previous tables, the total elapsed time is 61 hours. This elapsed time is the same for both sequencing order.

TABLE 5.33

Total Elapsed Time

Job	M_1		M_2	
	In	*Out*	*In*	*Out*
A	0	2	2	8
I	2	6	8	19
C	6	10	19	26
B	10	15	26	34
H	15	20	34	42
F	20	28	42	51
D	28	37	51	55
G	37	44	55	58
E	44	50	58	61

TABLE 5.34

Total Elapsed Time

Job	M_1		M_2	
	In	*Out*	*In*	*Out*
A	0	2	2	8
I	2	6	8	19
C	6	10	19	26
H	10	15	26	34
B	15	20	34	42
F	20	28	42	51
D	28	37	51	55
G	37	44	55	58
E	44	50	58	61

5.5 Sequencing of *n* Jobs through Three Machines

This is similar to the case before, but there are three machines instead of two (see Figure 5.7). The method developed by Johnson may solve problems falling within this category.

The two conditions of this approach are as follows:

1. The largest processing time for machine $M_2 \leq$ the smallest processing time for machine M_1

2. The largest processing time for machine $M_2 \leq$ the smallest processing time for machine M_3

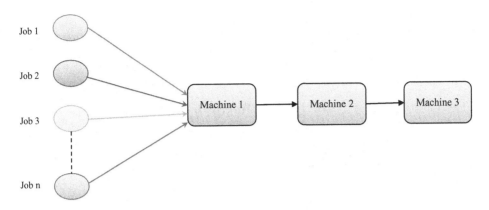

FIGURE 5.7
n Jobs through Three Machines.

At least one of the previous conditions must be satisfied. If so, replace the problem with n jobs and two machines (say, A and B) in an equivalent problem. Here two machines A and B are fictitious machines. The corresponding processing time of ith job on fictitious machines A and B are as follows:

$$A_i = (M_1)_i + (M_2)_i \, (i = 1, 2, \, ..., n),$$

$$B_i = (M_2)_i + (M_3)_i \, (i = 1, 2, \, ..., n).$$

Example based on sequencing of *n* jobs through three machines

EXAMPLE 1:

Five jobs are to be processed on three work centres W_1, W_2, W_3 and in the order W_1 W_2 W_3. The processing times in hours for five jobs are given in Table 5.35.

Determine the optimal job sequence to minimize the total elapsed time for this job group. Also, determine the total elapsed time. *(BT level 3)*

Solution:
Here minimum processing time at work centre W_1 is 2, maximum processing time at work centre W_2 is 5, and the minimum processing time at work centre W_3 is 5.

TABLE 5.35

Data Example 1

Job	W_1	W_2	W_3
A	3	3	5
B	8	4	8
C	7	2	10
D	5	1	7
E	2	5	6

Since the condition of the largest processing time for $W_2 \leq$, the smallest processing time for W_3 is satisfied, the problem can be solved by the procedure described in Section 5.4.

Now consider two fictitious work centre A and B such that

$$A_i = (W_1)_i + (W_2)_i \, (i = 1, 2, ..., 5),$$

$$B_i = (W_2)_i + (W_3)_i \, (i = 1, 2, ..., 5).$$

This calculation is shown in Table 5.36.

TABLE 5.36

Processing Time

Job	A	B
A	6	8
B	12	12
C	9	12
D	6	8
E	7	11

As per Johnson's algorithm, first, choose the job with the smallest processing time. It is Job A and Job D on machine A. So sequence Job A at the beginning, as shown next. Here there are two sets of ties. Now choose arbitrarily Job A first or Job D first, as shown next.

A	D

OR

D	A

After eliminating Job A and Job D and their time, the remaining set of jobs and processing times are shown in Table 5.37.

The smallest processing time in the remaining set of jobs corresponds to Job E on machine A. So sequence Job E as shown next

A	D	E

TABLE 5.37

Processing Time

Job	A	B
B	12	12
C	9	12
E	7	11

TABLE 5.38

Processing Time

Job	A	B
B	12	12
C	9	12

OR

D	A	E

After eliminating Job E and its time, the remaining set of jobs and processing times are shown in Table 5.38.

The smallest processing time in the remaining set of jobs corresponds to Job C on machine A. So sequence job C as shown next. Job b takes the last position.

A	D	E	C	B

OR

D	A	E	C	B

The total elapsed time for the first sequence is calculated and shown in Table 5.39.

In the above table, job *out* time at work centre W_1 is determined as follows:

Job *out* time at work centre W_1 = Job *in* time at work centre W_1 + Processing time of job at work centre W_1.

In the above table, job *in* time at work centre W_2 is determined as follows:

Job *in* time at work centre W_2 = max (Work centre W_1 time out of the current job, Work centre W_2 time out of the previous job).

In the table, job *out* time at work centre W_2 is determined as follows:

Job *out* time at work centre W_2 = Job *in* time at work centre W_2 + Processing time of job at work centre W_2.

In the table, job *in* time at work centre W_3 is determined as follows:

Job *in* time at work centre W_3 = max (Work centre W_2 time out of the current job, Work centre W_3 time out of the previous job)

In the table, job *out* time at work centre W_3 is determined as follows:

TABLE 5.39

Total Elapsed Time

Job	W_1		W_2		W_3	
	In	*Out*	*In*	*Out*	*In*	*Out*
A	0	3	3	6	6	11
D	3	8	8	9	11	18
E	8	10	10	15	18	24
C	10	17	17	19	24	34
B	17	25	25	29	34	42

Job *out* time at work centre W_3 = Job *in* time at work centre W_3 + Processing time of job at work centre W_3.

From the table, the total elapsed time is 42 hours. This elapsed time is the same for the second sequencing order.

5.6 Sequencing of *n* Jobs through *m* Machines

In problems involving n jobs and m machines, we are not able to reach optimal sequences using a general method. However, we have an applicable method, provided that no job transfers are permissible and either or both of the following conditions are met.

Let there are *n* jobs and six machines say $M_1, M_2, \ldots M_6$

1. The largest processing time for machines $(M_2, M_3, M_4, M_5) \leq$ the smallest processing time for machine M_1

2. The largest processing time for the machine $(M_2, M_3, M_4, M_5) \leq$ the smallest processing time for machine M_6

At least one of the aforementioned conditions must be satisfied. If so, replace the problem with n jobs and two machines (say, A and B) equivalent problem. Here two machines A and B are fictitious machines. The corresponding processing time of ith job on fictitious machines A and B are as follows:

$$A_i = \left(M_1\right)_i + \left(M_2\right)_i + \ldots + \left(M_{m-1}\right)_i,$$

$$B_i = \left(M_2\right)_i + \left(M_3\right)_i + \ldots + \left(M_m\right)_i.$$

Example based on the sequencing of *n* jobs through *m* machines

EXAMPLE 1:

Four jobs are to be processed on five work centres W_1, W_2, W_3, W_4, and W_5, and in the order $W_1 W_2 W_3, W_4, W_5$. The processing times in hours for four jobs are given in Table 5.40.

Determine the optimal job sequence to minimize the total elapsed time for this job group. Also, determine the total elapsed time. *(BT level 3)*

TABLE 5.40

Data Example 1

Job	W_1	W_2	W_3	W_4	W_5
A	6	4	1	2	8
B	5	5	3	4	9
C	4	3	4	5	7
D	7	2	2	1	5

Solution:

Here minimum processing time at work centre W_1 is 4 and minimum processing time at work centre W_5 is 5. The largest processing time for the work centre (W_2, W_3, W_4) is 5. Since the condition of the largest processing time for (W_2, W_3, W_4) ≤ the smallest processing time for W_5 is satisfied, the problem can be solved by the procedure described in Section 5.5.

Now consider two fictitious work centres A and B such that

$$A_i = \left(W_1\right)_i + \left(W_2\right)_i + \left(W_3\right)_i + \left(W_4\right)_i ,$$

$$B_i = \left(W_2\right)_i + \left(W_3\right)_i + \left(W_4\right)_i + \left(W_5\right)_i .$$

This calculation is shown in Table 5.41.

TABLE 5.41

Processing Time.

Job	A	B
A	13	15
B	17	21
C	16	19
D	12	10

As per Johnson's algorithm, first, choose the job with the smallest processing time. It is Job D on machine B. So sequence Job D as last, as shown next.

	D

After eliminating Job D and their time, the remaining set of jobs and processing times are shown in Table 5.42.

The smallest processing time in the remaining set of jobs corresponds to Job A on machine A. So sequence Job A at the beginning, as shown next.

A		D

TABLE 5.42

Processing Time

Job	A	B
A	13	15
B	17	21
C	16	19

TABLE 5.43

Processing Time

Job	A	B
B	17	21
C	16	19

After eliminating Job A and their time, the remaining set of jobs and processing times are shown in Table 5.43.

The smallest processing time in the above remaining set of jobs corresponds to Job C on machine A. So sequence Job C at the beginning and Job B will take a remaining position as shown next.

A	C	B	D

Above is the optimal sequence. The total elapsed time is calculated and shown in Table 5.44.

In the table, job *out* time at work centre W_1 is determined as follows:

Job *out* time at work centre W_1 = Job *in* time at work centre W_1 + Processing time of job at work centre W_1.

In the table, job *in* time at work centre W_2 is determined as follows:

Job *in* time at work centre W_2 = max (Work centre W_1 time out of the current job, Work centre W_2 time out of the previous job).

In the table, job *out* time at work centre W_2 is determined as follows:

Job *out* time at work centre W_2 = Job *in* time at work centre W_2 + Processing time of job at work centre W_2.

In the above table, job *in* time at work centre W_3 is determined as follows:

Job *in* time at work centre W_3 = max (Work centre W_2 time out of the current job, Work centre W_3 time out of the previous job)

In the above table, job *out* time at work centre W_3 is determined as follows:

Job *out* time at work centre W_3 = Job *in* time at work centre W_3 + Processing time of job at work centre W_3.

In the table, job *in* time at work centre W_4 is determined as follows:

Job *in* time at work centre W_4 = max (Work centre W_3 time out of the current job, Work centre W_4 time out of the previous job).

TABLE 5.44

Total Elapsed Time

Job	W_1		W_2		W_3		W_4		W_5	
	In	*Out*	*In*	*In*	*In*	*Out*	*In*	*Out*	*In*	*Out*
A	0	6	6	10	10	11	11	13	13	21
C	6	10	10	13	13	17	17	22	22	29
B	10	15	15	20	20	23	23	27	29	38
D	15	22	22	24	24	26	27	28	38	43

In the table, job *out* time at work centre W_4 is determined as follows:

Job *out* time at work centre W_4 = Job *in* time at work centre W_4 + Processing time of job at work centre W_4.

In the table, job *in* time at work centre W_5 is determined as follows:

Job *in* time at work centre W_5 = max (Work centre W_4 time out of the current job, Work centre W_5 time out of the previous job).

In the table, job *out* time at work centre W_5 is determined as follows:

Job *out* time at work centre W_5 = Job *in* time at work centre W_5 + Processing time of job at work centre W_5.

From the table, the total elapsed time is 43 hours.

Typical examples

EXAMPLE 1:

A, B, C, and D four people share a flat. Four newspapers are delivered to a flat every Saturday: the *Times of India* (ToI), the *Hindu* (TH), *Indian Express* (IE), and the *Economic Times* (ET). Each flat member insists on reading all the papers in their own specific order. A wants to start with ToI for 50 minutes. Then he takes 25 minutes to turn to TH, looks for 2 minutes at the IE, and ends with 5 minutes spent on the ET. B prefers to start with TH taking 1 hour 10 minutes, and then read the IE for 4 minutes, the ToI for 25 minutes, and the ET for 10 minutes. C starts reading the IE for 5 minutes, followed by TH for 20 minutes, the ToI for 10 minutes, and the ET for 30 minutes. Finally, D starts with the ET taking 1 hour and 20 minutes to spend 2 minutes each in that order on the ToI, TH, and the IE. Everyone is so insistent on their specific order of reading that they will wait for their next paper to be free instead of choosing another one. Also, no one is going to release a paper until it is completed. If A wakeup at 8:30 a.m., B and C wakeup at 8:45 a.m., and D at 9:30 a.m., and they can manage to wash, shower, shave, and eat breakfast while reading the newspaper, and since each of them insists on reading all the newspapers before leaving, what is the earliest time the four of them can leave together for the day. (*BT level 4*)

Solution:

First, write the data in a more compact form, as shown next in Table 5.45.

TABLE 5.45

Data Example 1

Reader	Morning Wake Up Time	Newspaper Reading Order (Times in Minutes)			
A	8:30 a.m.	ToI (50)	TH (25)	IE (2)	ET (5)
B	8:45 a.m.	TH (70)	IE (4)	ToI (25)	ET (10)
C	8:45 a.m.	IE (5)	TH (20)	ToI (10)	ET (30)
D	9:30 a.m.	ET (80)	ToI (2)	TH (2)	IE (2)

TABLE 5.46

Paper Reading Schedule

Newspaper	Read by			
	First	Second	Third	Fourth
ToI	A	D	C	B
TH	B	C	A	D
IE	C	B	A	D
ET	D	A	C	B

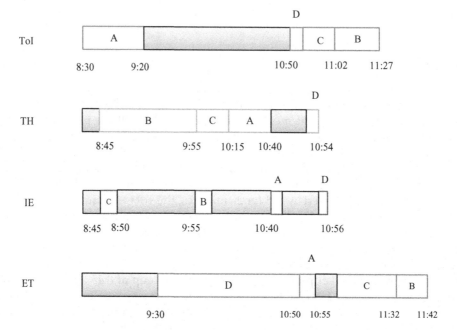

FIGURE 5.8
Gantt Chart.

Here the schedule for reading is a prescription of the order in which the papers rotate among readers. The following is one possible schedule (Table 5.46).

A has the ToI first, before it passes to D, then to C, and, finally, to B. Similarly TH passes between B, C, A, and D in that order, and so on.

It can be determined how long this reading schedule will take by plotting a Gantt chart (see Figure 5.8). To draw this chart, newspapers are rotated in the order given by Table 5.46 with the restriction that each reader follows his desire reading order.

Clearly, from the Gantt chart, the earliest the four of them can leave together for the day is 11:42 a.m.

EXAMPLE 2:

A, B, C, and D are products consisting of the order of a customer. Before the order can be delivered, they must all be finished. Figure 5.9 represents the activities to be

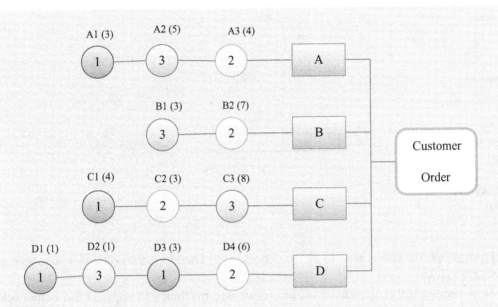

FIGURE 5.9
Activities for Example 2

conducted to produce each item. A1 is Product A's first procedure, A2 is the second, and so on. The numbers inside the circles are the machines used to complete each procedure, and the machining time is the numbers inside the brackets. Fifty units of each item were ordered by the customer. Use a 100-unit process batch and one transfer batch. Construct a Gantt chart for each machine to demonstrate how fast you can make the shipment. *(BT level 6)*

Solution:
First, find the bottleneck machine. It is determined by adding the machining time of all operations on the individual machine, as shown next (Table 5.47).

TABLE 5.47

Individual Machine Time

Machine 1:	A1	3	Machine 2:	A3	4	Machine 3:	A2	5
	C1	4		B2	7		B1	3
	D1	1		C2	3		C3	8
	D3	3		D4	6		D2	1
		11			20			17

From Table 5.47, machine 2 is the bottleneck.
Now find the fastest time to the bottleneck. It is determined by summing the machining time for all operations of individual products before machine 2.

A3	8	min
B2	3	min
C2	4	min
D4	5	min

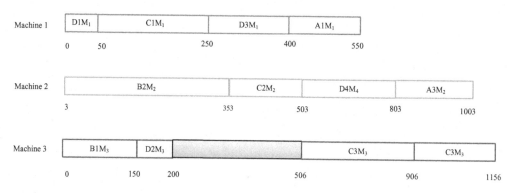

FIGURE 5.10
Gantt Chart

From above, the fastest time to the bottleneck is B2. Therefore, the bottleneck sequence is B2-C2-D4-A3.

Now sequence the operations on the remaining machines to support the bottleneck machine.

Sequence for Machine 1: D1-C1-D3-A1

Sequence for Machine 3: B1-D2-C3-A2

Considering the items are transferred one at a time, but processed in batches of 50, the Gantt chart for this problem is shown in Figure 5.10.

From the above Gantt chart, we can say that 50 units of the customer order can be assembled in 1,156 minutes.

CHAPTER SUMMARY

- Sequencing is the order of processing a set of tasks/jobs over available machines/work centres/facilities.
- Problem-solving approach:
  - ▶ Determine the priority rule to be used
  - ▶ List all jobs/tasks and their work centres
  - ▶ Using priority rule, determine which job should be worked on first and identify the sequence of all jobs/tasks
- Different priority rules:
  - ▶ FCFS
  - ▶ SPT
  - ▶ EDD
  - ▶ STR

- Different performance measures:

  Total elapsed time: It is the total time spent from processing of the first job on the first machine to the completion of the last job on the last machine. It includes both processing time and idle time.

  Lateness: It is the difference between actual the completion time and due date of the job.

  Completion time: It is the time at which the job is completed. It is also known as flow time.

  Tardiness: It is the difference between the due date of the late job and its completion time.

- Different sequencing problems:
  Sequencing of *n* Jobs through One Machine
  Sequencing of *n* Jobs through One Machine
  Sequencing of *n* Jobs through Three Machines
  Sequencing of *n* Jobs through *m* Machines

Questions

1 Explain the sequencing problem in detail. *(BT level 2)*

2 Explain the importance of the sequencing problem. *(BT level 2)*

3 What are the assumptions in the sequencing problem? *(BT level 2)*

4 Give different examples of sequence problems from your daily life. *(BT level 2)*

5 Explain the Gantt chart. *(BT level 2)*

6 Briefly explain the various method of solving sequencing problems. *(BT level 2)*

7 Explain the following priority rules: *(BT level 2)*

  ▶ FCFS

  ▶ SPT

  ▶ EDD

  ▶ STR

8 Explain different performance measures in sequence *(BT level 2)*

9 What are tardiness and lateness? *(BT level 2)*

10 Use the Johnson method to determine an optimal sequence on two machines for processing *n* jobs. *(BT level 3)*

11 How will you process *n* jobs through *m* machines? *(BT level 3)*

Practice Problem

1. The following jobs (Table 5.48) are waiting to be processed at the same machine centre. Determine the optimal sequence of jobs, average tardiness, average flow time, and average in-process inventory for each of the following sequencing rules. *(BT level 3)*

   a. FCFS

   b. SPT

   c. EDD

TABLE 5.48

Data Practice Problem 1

Job	Processing Time (Days)	Due Date (Days)
A	4	6
B	2	4
C	8	24
D	6	8
E	8	16
F	10	20

2. Six jobs in a factory need to be processed on two machines M_1 and M_2, in the order M_1 M_2. The processing times (in hours) for nine jobs are tabulated in Table 5.49.

TABLE 5.49

Data Practice Problem 2

Job	M_1	M_2
A	9	4
B	6	3
C	4	6
D	2	7
E	1	5
F	3	2

Determine the optimal job sequence to minimize the total elapsed time for this job group. Also, determine the total elapsed time and idle time for two machines M_1 and M_2. *(BT level 3)*

3. Five jobs are to be processed on three work centres M_1, M_2, M_3 and in the order M_1, M_2, M_3. The processing times in hours for five jobs are given in Table 5.50.

TABLE 5.50

Data Practice Problem 3

Job	M_1	M_2	M_3
A	7	4	3
B	9	5	8
C	5	1	7
D	6	2	5
E	10	3	4

Determine the optimal job sequence to minimize the total elapsed time for this job group. Also, determine the total elapsed time of all machines. *(BT level 3)*

6

Replacement Models

Learning Outcomes: After studying this chapter, the reader should be able to

LO 1. *Understand the replacement problem. (BT level 2)*

LO 2. *Understand and apply the replacement policy that deteriorates with time. (BT level 2 and 3)*

LO 3. *Understand and apply the group replacement policy. (BT level 2 and 3)*

6.1 Introduction

Organizations delivering goods and services use many resources for carrying out their activities, such as machinery and equipment. In addition to these services, there are a variety of other things that are important for facilitating the functioning of the organizations. All of those facilities should be monitored continuously for their efficient operation, as otherwise, the service quality of these facilities will be poor. Moreover, with the passage of time, the costs of their operation and maintenance would increase. It is, therefore, essential that the equipment is maintained in good working conditions at an economic cost. If a company wants to survive, it needs to decide whether to replace or retain the out-of-date equipment, taking account of the operating and maintenance costs.

Replacement problems are concerned with circumstances that occur when certain things, such as men, machines, and usable things, need to be replaced because of their reduced efficiency, deterioration, or breakdown. Such a reduced efficiency or complete breakdown may be either gradual or sudden. (In the event of a gradual failure, as the machine's life increases, its efficiency deteriorates and management can experience: (a) progressive increase in maintenance or operating costs, (b) decrease in equipment productivity, and (c) decrease in equipment value, i.e. decrease in equipment / facility resale value).

LO 1. Understand the replacement problem

In a sudden failure, after a certain time, the items fail. The equipment life is not predictable and is some kind of random variable. For any particular type of equipment, the period between installation and failure is not constant.

The need to replace the items is felt when

- Items that deteriorate with time (e.g., machine tools, vehicles, equipment buildings);
- Items become out of date because of new developments, such as ordinary weaving looms by automatic and manual tally accounting;

- Items that don't deteriorate but fail after certain usages, such as electronics components and streetlights; and
- An organization's current workforce is slowly reducing because of death, retirement, and retrenchment, and so on.

The solution to these problems of replacement is nothing but arriving at the best strategy that defines the period at which the replacement is most economical rather than continuing at an increased cost of maintenance.

Replacement problems can be broadly classified into the following two categories:

a) When the equipment/assets deteriorated with time and the value of money
   i. *Does not change with time*
   ii. *Changes with time*
b) When the equipment/assets fail completely all of a sudden.

6.2 Replacement of Items that Deteriorate with Time

LO 2. Understand and apply the replacement policy that deteriorates with time

Many items slowly deteriorate with usage and their efficiency declines over a time span. Generally, the maintenance costs of such items often escalate slowly over time until the stage comes where the maintenance costs are so large that it is easier and more economical to replace the item with a new one. There may be a variety of alternatives, and we may compare the different alternatives by considering the costs of waste, scrap, loss of production, damage to equipment, safety risks, etc.

These types of problems are solved by two methods. They are as follows:

(i) Replacement policy when the value of money does not change with time
(ii) Replacement policy when the value of money changes with time

6.2.1 Replacement Policy When the Value of Money Does Not Change with Time

The objective here is to decide the optimum replacement age of an equipment/item whose running/maintenance costs increase over time and the value of money remains static over that span.

Let,

C = Purchase cost or capital cost of the item,

S = Scrap value or resale value of the item; it is assumed that this cost will remain constant over time.

n = number of years that the equipment would be in use

CASE I:

Here we assume that the time 't' is a continuous variable. Let $R(t)$ be the maintenance or running cost at the time 't.'

The annual cost of the item at any time 't' $= R(t) + C - S$

Maintenance costs incurred during the 'n' years shall become $= \int_0^n R(t)\,dt$

Total costs incurred on the item $= \int_0^n R(t)\,dt + C - S$

The average annual total cost shall be determined by, $A(n) = \dfrac{Total\ Cost}{n}$

$$A(n) = \frac{1}{n}\int_0^n R(t)\,dt + \frac{C-S}{n} \tag{6.1}$$

Now, we find such time n for which $A(n)$ is minimum; the principle of minima will be employed.

Therefore, differentiating $A(n)$ with respect to 'n,' $\dfrac{dA(n)}{dn} = 0$ and $\dfrac{d^2A(n)}{dn^2} \geq 0$, we get

$$\frac{dA(n)}{dn} = \frac{1}{n}R(n) - \frac{1}{n^2}\int_0^n R(t)\,dt + \frac{C-S}{n^2} = 0,$$

which gives

$$R(n) = \frac{1}{n}\int_0^n R(t)\,dt + \frac{C-S}{n}, \tag{6.2}$$

$$\therefore R(n) = A(n).\,(\text{Comparing Eqs. 6.1 and 6.2})$$

Thus the maintenance cost at time 'n' = the average cost at time 'n.' Therefore, if time is continuously measured, the average annual cost of replacing the item is minimized if the average cost is equal to the current cost of maintenance.

CASE II:

Here time 't' is considered as a discrete variable. In this case, the time period is taken as one year, and 't' can take the values of 1, 2, 3 …etc.

Since the time is measured in discrete units, the cost Eq. (6.1) can be written as

$$A(n) = \sum_1^n \frac{R(t)}{n} + \frac{C-S}{n} \tag{6.3}$$

By using finite differences, $A(n)$ will be a minimum for that value of n for which

$$A(n+1) \geq A(n) \text{ and } A(n-1) \geq A(n), \text{ or }$$

$$A(n+1) - A(n) \geq 0 \text{ and } A(n) - A(n-1) \leq 0.$$

For this, we write

$$A(n+1) = \frac{C-S}{n+1} + \frac{1}{n+1}\sum_{1}^{n+1} R(t)$$

$$= \frac{1}{n+1}\left[C - S + \sum_{1}^{n+1} R(t)\right],$$

$$A(n+1) - A(n) = \frac{1}{n+1}\left[C - S + \sum_{1}^{n+1} R(t)\right] - \frac{1}{n}\left[C - S + \sum_{1}^{n} R(t)\right]$$

$$= \frac{1}{n+1} + \left[C - S + \sum_{1}^{n} R(t) + R(n+1)\right] - \frac{1}{n}\left[C - S + \sum_{1}^{n} R(t)\right]$$

$$= \frac{R(n+1)}{n+1} + \sum_{1}^{n} R(t)\left[\frac{1}{n+1} - \frac{1}{n}\right] + (C - S)\left[\frac{1}{n+1} - \frac{1}{n}\right]$$

$$= \frac{R(n+1)}{n+1} - \frac{1}{n(n+1)}\left[\sum_{1}^{n} R(t) + (C - S)\right]$$

$$A(n+1) - A(n) = \frac{1}{n+1}\left[R(n+1) - A(n)\right].$$

Thus A $(n + 1) -$ A $(n) \geq 0 \rightarrow$ R $(n + 1) \geq$ A(n).
Similarly, it can be shown as

$$A(n) - A(n-1) \leq 0 \rightarrow R(n) \leq A(n-1).$$

This suggests replacing the machine at the end of n years when maintenance costs in the $(n + 1)$ year are higher than the average total cost in the nth year and maintenance costs in the nth year are lower than the average total cost in the preceding year.

Examples based on the replacement policy when the value of money does not change with time

EXAMPLE 1:

The production manager of the manufacturing industry is considering replacing a small machine. A machine costs Rs.12,000, and its scrap value is only Rs.8,000. The maintenance costs are ascertained from experience as shown in Table 6.1.

At what year is the machine replaced? *(BT level 4)*

TABLE 6.1

Maintenance Cost for Example 1

Year	Maintenance Cost (Rs.)
1	2,000
2	5,000
3	8,000
4	12,000
5	18,000
6	25,000
7	32,000

Solution:

Here we are given the running cost, R(*t*), the scrap value S = Rs. 8,000, and the cost of machine, C = Rs. 1,22,000.

We first calculate an average cost per year during the machine's life to determine the optimal time *n* when the machine should be replaced.

The annual cost of the machine at any time '*t*' = R(*t*) + C − S

We calculate the average cost per year for the life of the machine, as shown in Table 6.2.

TABLE 6.2

Average Cost

Year (*n*) (1)	Running Cost R(*t*) (2)	Cumulative Running Cost (3)	Depreciation Cost (C-S) (4)	Total Cost (TC) (5) = (3)+(4)	Average Total Cost (6) = (5)/(1)
1	2,000	2,000	114,000	116,000	116,000
2	5,000	7,000	114,000	121,000	60,500
3	8,000	15,000	114,000	129,000	43,000
4	12,000	27,000	114,000	141,000	35,250
5	18,000	45,000	114,000	159,000	31,800
6	25,000	70,000	114,000	184,000	30,666.67
7	32,000	102,000	114,000	216,000	30,857.14

The calculations in Table 6.2 show that the average cost for the sixth year is the lowest (Rs. 30.666.67). Therefore, the machine should be replaced every six years; otherwise, the average cost per year of running the machine would start to rise.

EXAMPLE 2:

Machine M_1 costs Rs. 36,000. For the first year, the cost of annual operations is Rs. 800, and every year it increases by Rs. 8.000. Determine the best age at which the machine can be replaced. What will be the average annual cost of owning and operating the machine if the optimal replacement policy is followed?

Machine M_2 costs Rs. 40,000. For the first year, the annual operating costs are Rs. 1,600, and then each year they increase by Rs. 3,200. Now you've got a type M_1 machine that's one year old. (a) Will you replace it with M_2, or (b) if so, when? *(BT level 4)*

Solution:

Let the machine have no scrap (resale) value when replaced.

For M_1. Cost Price = 36,000, Operating Cost = 800, Increment By = 8,000
For M_2: Cost Price = 40,000, Operating Cost = 1,600, Increment By = 3,200

To determine the optimal time n when the machine M_1 should be replaced, we first calculate the average cost per year during the life of the machine M_1,
The annual cost of the machine at any time 't' = $R(t) + C - S$. Average cost is given in Table 6.3.

TABLE 6.3

Average Cost for Machine M1

Year (n) (1)	Running Cost R(t) (2)	Cumulative Running Cost (3)	Depreciation Cost (C-S) (4)	Total Cost (TC) (5) = (3)+(4)	Average Total Cost (6) = (5)/(1)
1	800	800	36,000	36,800	36,800
2	8,800	9,600	36,000	45,600	22,800
3	16,800	26,400	36,000	62,400	20,800
4	24,800	51,200	36,000	87,200	21,800
5	32,800	84,000	36,000	120,000	24,000
6	40,800	124,800	36,000	160,800	26,800

The table calculations show that the average cost of machine M_1 during the third year is the lowest (Rs. 20,800). The machine M_1 should, therefore, be replaced every third year.

To determine the optimal time n when the machine M_2 should be replaced, we first calculate the average cost per year during the life of the machine M_2 (Table 6.4).

TABLE 6.4

Average Cost for Machine M2

Year (n) (1)	Running Cost R(t) (2)	Cumulative Running Cost (3)	Depreciation Cost (C-S) (4)	Total Cost (TC) (5) = (3)+(4)	Average Total Cost (6) = (5)/(1)
1	1,600	1,600	40,000	41,600	41,600
2	4,800	6,400	40,000	46,400	23,200
3	8,000	14,400	40,000	54,400	18,133.33
4	11,200	25,600	40,000	65,600	16,400
5	14,400	40,000	40,000	80,000	16,000
6	17,600	57,600	40,000	97,600	16,266.67

Table calculations show that the average cost of machine M_2 during the fifth year is the lowest (Rs. 16,000). The machine M_2 should, therefore, be replaced every third year.

a) The lowest average running cost (Rs. 16,000) per year for machine M_2 is lower than the lowest average running cost (Rs. 20,800) per year for machine M_1. Machine M_1 should, therefore, be replaced by machine M_2.

b) Next, we compare the minimum average cost for M_2 (Rs. 16,000) with the annual cost of maintaining and using Machine M_1 to decide the time of replacement. Since there is no resale value, we only consider the cost of the maintenance. Machine M_1 is retained as long as the annual maintenance costs are lower than Rs. 16,000 and replaced if it exceeds Rs. 16,000. For maintenance in the next year, Rs. 8,800 would be required on a one-year-old M_1 machine and Rs. 16,800 would be needed for the following year. Since the maintenance cost of Machine M2 is more than the lowest average cost (Rs. 16,000) in the third year, after two years, Machine M1 should be replaced by Machine M2. Since Machine M1 is one-year-old, the replacement is expected to be one year from now.

EXAMPLE 3:

From his past records, the owner of an automobile finds that the cost per year of an automobile whose purchase price is Rs. 80,000 is, as shown in Table 6.5.
     Determine at what age it is due to be replaced. *(BT level 3)*

Solution:
     We first calculate an average cost per year during the automobile's life to determine the optimal time n when the automobile should be replaced.
     The annual cost of the machine at any time $'t' = R(t) + C - S$
     We calculate the average cost per year for the life of the machine, as shown in Table 6.6.

TABLE 6.5

Running and Resale Cost for Example 3

Year	Running Cost (Rs.)	Resale Cost (Rs.)
1	12,000	44,000
2	14,000	22,000
3	16,000	11,000
4	20,000	6,000
5	24,000	3,000
6	30,000	2,000
7	36,000	2,000

TABLE 6.6

Average Cost

Year (n) (1)	Running Cost R(t) (2)	Cumulative Running Cost (3)	Resale Value (S) (4)	Depreciation Cost (C-S) (5) = 80,000-(4)	Total Cost (TC) (6) = (3)+(5)	Average Total Cost (7) = (6)/(1)
1	12,000	12,000	44,000	36,000	48,000	48,000
2	14,000	26,000	22,000	58,000	84,000	42,000
3	16,000	42,000	11,000	69,000	111,000	37,000
4	20,000	62,000	6,000	74,000	136,000	34,000
5	24,000	86,000	3,000	77,000	163,000	32,600
6	30,000	116,000	2,000	78,000	194,000	32,333.33
7	36,000	152,000	2,000	78,000	230,000	32,857.14

The calculations in Table 6.6 show that the average cost for the sixth year is the lowest (Rs. 32,333.33). The automobile should, therefore, be replaced every six years; otherwise, the average annual cost of running the automobile would start to increase.

EXAMPLE 4:

Data are given in Table 6.7 in the running of a machine, which costs Rs. 1,60,000.
   Determine the optimum period for the replacement of the machine. *(BT level 3)*

Solution:
The running costs are determined jointly by the costs of spares and labour. In successive years, the running costs and resale price of the machine are given in Table 6.8.

TABLE 6.7

Data for Example 4

Year	Resale Cost (Rs.)	Cost of Spares (Rs.)	Labour Cost (Rs.)
1	100,000	8,000	20,000
2	80,000	10,000	25,000
3	65,000	12,000	30,000
4	50,000	18,000	35,000
5	35,000	24,000	40,000

TABLE 6.8

Resale and Running Cost

Year	Resale Cost (Rs.)	Running Cost (Rs.)
1	1,00,000	28,000
2	80,000	35,000
3	65,000	42,000
4	50,000	53,000
5	35,000	64,000

We first calculate an average cost per year during the machine's life to determine the optimal time n when the machine should be replaced.

The annual cost of the machine at any time 't' $= R(t) + C - S$.

We calculate the average cost per year for the life of the machine as shown in Table 6.9.

The calculations in Table 6.9 show that the average cost for the third year is the lowest (Rs. 66,666.67). The machine should, therefore, be replaced every third year; otherwise, the average cost per year of running the machine would start to increase.

TABLE 6.9

Average Cost

Year (n) (1)	Running Cost R(t) (2)	Cumulative Running Cost (3)	Resale Value (S) (4)	Depreciation Cost (C-S) (5) = 160,000-(4)	Total Cost (TC) (6) = (3)+(5)	Average Total Cost (7) = (6)/(1)
1	28,000	28,000	100,000	60,000	88,000	88,000
2	35,000	63,000	80,000	80,000	143,000	71,500
3	42,000	105,000	65,000	95,000	200,000	66,666.67
4	53,000	158,000	50,000	110,000	268,000	67,000
5	64,000	222,000	35,000	125,000	347,000	69,400

6.2.2 Replacement Policy When the Value of Money Changes with Time

We stated in the previous section that the value of the money does not change and stays constant, but it is well-known that as the equipment deteriorates and operating costs begin to increase, the money value begins to decrease over time. We, therefore, need to calculate the net present value of the money to be spent a few years from now. Otherwise, the resale value and the future operating costs will be unrealistic, and management will be unable to make optimal decisions.

The following assumptions are made when determining the replacement policy when the value of money is taken into account:

1. The equipment has no salvage value.
2. The maintenance costs are incurred at the beginning of time periods.

Let us understand the following terms first:

Money Value:

The time value of money is the value of money in a certain amount of interest earned over a given period of time. Since money has a value over time, the statement, "Money is worth 10% per annum" can, therefore, be explained in two ways:

a) Spending Rs.1,000 today would be, in one way, equivalent to spending Rs.1,100 in years. In other words, if we plan to spend Rs.1,100 a year from now, we could spend Rs.1,000 today and an investment worth Rs.1,100 the next year.

b) Alternatively, if we borrow Rs.1,000 at 10% interest per year and spend Rs.1,000 today, then after one year (next year), we have to pay Rs.1,100.

Therefore, we conclude that in a year, Rs.1,000 will be equal to Rs.1,100. Thus Rs. 1 is equivalent to $(1 + 0.1)^{-1}$ rupee now for a period of one year.

Present Worth Factor:

As we have seen, a rupee a year from now will be equivalent to $(1 + 0.1)^{-1}$ rupee today at the discount rate of 10% per year, one rupee spent two years from now is equivalent to $(1 + 0.1)^{-2}$ today. So in n years from now, one rupee will be equal to $(1 + 0.1)^{-n}$. Thus the amount $(1 + 0.1)^{-n}$ is called the present worth factor (PWF) or present value (PV) of one rupee that has been spent in n years from now. In general, if r is the interest rate, then $(1+r)^{-n}$ is called the PWF or the PV of one rupee spent in n years from now. The expression $(1+r)^{-n}$ is known as the compound quantity factor of one spent rupee over n years.

Discount Rate:

Let r be the interest rate. Therefore the PWF of the unit amount to be spent after one year is $v = 1/(1 + r)$. Then v is called the discount rate or the value of the depreciation.

The following method can determine the optimal replacement policy for items in which maintenance costs increase with time and money value changes at a constant rate:

Suppose the item (which may be a machine or equipment) is available for use at equal intervals over a series of time periods (say, in one year).

Let

C = Purchase price of the item to be replaced

$R(t)$ = Running (or maintenance) cost in the tth year

r = Rate of interest

$v = \dfrac{1}{1+r}$ is the present worth of a rupee to be spent in a year hence

Assuming that machines has no resale value at the time of replacement, the present worth of the machine in n years will be given by

$$P(n) = \left[(C+R_1) + R_2 v + R_3 v^2 + - - - - + R_n v^{n-1} \right] + \left[(C+R_1) v^n + R_2 v^{n+1} + R_3 v^{n+2} + - - - - - + \right.$$
$$R_n v^{2n-1} \left] + \left[(C+R_1) v^{2n} + R_2 v^{2n+1} + R_3 v^{2n+2} + - - - - - + R_n v^{3n-1} \right] + - - - \text{ and so on}$$

$$= (C+R_1)\left[1 + v^n + v^{2n} + - - \right] + R_2 v \left[1 + v^n + v^{2n} + - - \right] + - - - + R_n v^{n-1} \left[1 + v^n + v^{2n} + - - \right]$$

$$= (C+R_1 + R_2 v + - - + R_n v^{n-1}) \left[1 + v^n + v^{2n} + - - \right]$$

$$P(n) = (C + R1 + R2 v + - - - + Rn\, vn - 1)\left[\dfrac{1}{1-v^n} \right].$$

$$\text{Let } F(n) = (C + R_1 + R_2 v + - - - + R_n v^{n-1}),$$

$$P(n) = \left[\dfrac{F(n)}{1-v^n} \right] \text{ and } P(n+1) = \left[\dfrac{F(n+1)}{1-v^{n+1}} \right].$$

$F(n)$ and $F(n+1)$ given at $n = 0, 1, 2$, are called the weighted average cost of previous n years with weights $1, v, v^2, - - - - v^{n-1}$, respectively. $P(n)$ is the amount of money now required to pay all future costs of acquiring and operating the equipment once it is renewed every n years. However, if $P(n)$ is less than $P(n+1)$ it is preferable to replace the equipment every n year. Further, if the best policy is replacing every n years, then the two inequalities $P(n+1) - P(n) > 0$ and $P(n-1) - P(n) < 0$ must hold, but without giving the proof, we shall state the following two inequalities which hold good at n, the optimal replacement interval.

$$P(n) \le \dfrac{(C+R_1) + R_2 v + R_3 v^2 + - - - - + R_{n-1} v^{n-2}}{1 + v + v^2 + - - + v^{n-2}},$$

and

$$P(n+1) \ge \dfrac{(C+R_1) + R_2 v + - - - - + R_n v^{n-1}}{1 + v + v^2 + - - + v^{n-1}}.$$

As a result of these two inequalities, the rules for minimizing costs may be set out as follows:

1. Do not replace if the period's operating costs are below the weighted average of previous costs.

2. Replace if the operating costs for the next period exceed the weighted average of past costs.

Examples based on the replacement policy when value of money changes with time

EXAMPLE 1:

A manufacturing-industry manager is considering replacing a small machine. A machine's cost is Rs.45,000 and the running costs are given in Table 6.10 for the different years.

What is the replacement policy to be adopted if the capital is worth 10% and there is no salvage value? *(BT level 4)*

TABLE 6.10

Maintenance Cost for Example 1

Year	Maintenance Cost (Rs.)
1	4,000
2	5,000
3	8,000
4	11,000
5	14,000
6	17,000
7	21,500

Solution:

Since money is worth 10% per year, the discounted factor over a period of one year is given by

$$v = \frac{1}{1+0.1} = 0.9091.$$

It is also given that C = Rs. 45,000 The optimum replacement age must satisfy the condition Rn < W(n) < Rn+1.

The optimum replacement time is determined in Table 6.11.

TABLE 6.11

Weighted Average Cost

Year (1)	Running Cost (2)	Discounted Factor v^{n-1} (3) = 0.9091^{n-1}	Discounted Cost $Rn \times v^{n-1}$ (4) = (2)×(3)	$\Sigma R_i v^{i-1}$ (5) = Σ(4)	C + $\Sigma R_i v^{i-1}$ (6) = 45,000 + (5)	Sum of Discount Σv^{i-1} (7) = Σ(3)	Weighted Average Cost W(n) (8) = (6)/(7)
1	4,000	1	4,000	4,000	49,000	1	49,000
2	5,000	0.9091	4,545.45	8,545.45	53,545.45	1.9091	28,047.62
3	8,000	0.8264	6,611.57	15,157.02	60,157.02	2.7355	21,990.94
4	11,000	0.7513	8,264.46	23,421.49	68,421.49	3.4869	19,622.71
5	14,000	0.683	9,562.19	32,983.68	77,983.68	4.1699	18,701.72
6	17,000	0.6209	10,555.66	43,539.34	88,539.34	4.7908	18,481.17
7	21,500	0.5645	12,136.19	55,675.53	100,675.53	5.3553	18,799.37

The calculations in Table 6.11 show that the average cost is lowest in the sixth year (Rs. 18,481.17) and greater than the running cost of the current year (17,000 < 18,481.17), as well as lower than next year's running cost (18,481.17 < 21,500). Hence, the machine should be replaced after every sixth year; otherwise, the average cost per year of running the machine would begin to rise.

EXAMPLE 2:

The annual cost of the two machines M_1 and M_2 in rupees, where the value of money is neglected, is shown in Table 6.12. Find their cost pattern if the money value is 10% per year and find out which machine is more cost-effective. *(BT level 3)*

Solution:
Since money is worth 10% per year, the discounted factor over a period of one year is given by

$$v = \frac{1}{1+0.1} = 0.9091.$$

The discounted cost pattern of machines A and B are given in Table 6.13.

TABLE 6.12

Data for Example 2

Year	Machine M_1	Machine M_2
1	3,600	5,600
2	2,400	400
3	2,800	2,800

Table 6.13 shows that the total cost of the M1 machine is lower than that of the M2. Machine M1 is, therefore, more economical if the value of money changes.

TABLE 6.13

The Discounted Cost Pattern

Year	Machine M_1	Discounted Cost	Machine M_2	Discounted Cost
1	3,600	3,600	5,600	5,600
2	2,400	$2,400 \times 0.9091 = 2,181.84$	400	$400 \times 0.9091 = 363.64$
3	2,800	$2,800 \times (0.9091)^2 = 2,314.09$	2,800	$2,800 \times (0.9091)^2 = 2,314.09$
Total		8,095.93		8,277.73

EXAMPLE 3:

A machine costs Rs.50,000/-. Operation and maintenance costs are zero for the first year and increase by Rs. 10,000/- every year. If money is worth 5% every year, determine the best age at which the machine should be replaced. The machine's resale value is insignificantly small. What is the weighted average cost of owning the machine and operating it? *(BT level 3)*

Solution:
Since money is worth 5% per year, the discounted factor over a period of one year is given by

$$v = \frac{1}{1+0.05} = 0.9524.$$

The optimum replacement age must satisfy the condition $Rn < W(n) < Rn + 1$. The optimum replacement time is determined in Table 6.14.

TABLE 6.14

Weighted Average Cost

Year (1)	Running Cost (2)	Discounted Factor v^{n-1} (3) = 0.9524^{n-1}	Discounted Cost Rn x v^{n-1} (4) = (2)×(3)	$\Sigma R_i v^{i-1}$ (5) = Σ(4)	$C + \Sigma R_i v^{i-1}$ (6) = 50,000 + (5)	Sum of Discount Σv^{i-1} (7) = Σ(3)	Weighted Average Cost $W(n)$ (8) = (6)/(7)
1	0	1	0	0	50,000	1	50,000
2	10,000	0.9524	9,523.81	9,523.81	59,523.81	1.9524	30,487.8
3	20,000	0.907	18,140.59	27,664.4	77,664.4	2.8594	27,160.98
4	30,000	0.8638	25,915.13	53,579.53	103,579.53	3.7232	27,819.67
5	40,000	0.8227	32,908.1	86,487.63	136,487.63	4.546	30,024

The calculations in Table 6.14 show that the average cost is lowest during the third year (Rs. 27,160.98), and it is greater than the running cost of the current year (20,000 < 27,160.98) as well as less than the running cost of the next year (27,160.98 < 30,000). Hence, after every third year, the machine should be replaced; otherwise, the average cost of running the machine would start to increase.

EXAMPLE 4:

The equipment M is offered at a price of Rs.90,000 and the costs of operation and maintenance are estimated to be Rs.15,000 for each of the first five years, increasing by Rs. 5,000 annually in the sixth and subsequent years. What would the optimal replacement period be if money carries an interest rate of 10% annually? *(BT level 3)*

Solution:

Since money is worth 5% per year, the discounted factor over a period of one year is given by

$$v = \frac{1}{1+0.1} = 0.9091.$$

It is also given that C = Rs. 90,000. The optimum replacement age must satisfy the condition Rn < W(n) < Rn+1.

The optimum replacement time is determined in Table 6.15.

TABLE 6.15

Weighted Average Cost

Year (1)	Running Cost (2)	Discounted Factor v^{n-1} (3) = 0.9524^{n-1}	Discounted Cost Rn × v^{n-1} (4) = (2)×(3)	$\Sigma R_i v^{i-1}$ (5) = Σ(4)	C + $\Sigma R_i v^{i-1}$ (6) = 50,000 + (5)	Sum of Discount Σv^{i-1} (7) = Σ(3)	Weighted Average Cost W(n) (8) = (6)/(7)
1	15,000	1	15,000	15,000	105,000	1	105,000
2	15,000	0.9091	13,636.36	28,636.36	118,636.36	1.9091	62,142.86
3	15,000	0.8264	12,396.69	41,033.06	131,033.06	2.7355	47,900.3
4	15,000	0.7513	11,269.72	52,302.78	142,302.78	3.4869	40,811.25
5	15,000	0.683	10,245.2	62,547.98	152,547.98	4.1699	36,583.43
6	20,000	0.6209	12,418.43	74,966.41	164,966.41	4.7908	34,434.1
7	25,000	0.5645	14,111.85	89,078.26	179,078.26	5.3553	33,439.69
8	30,000	0.5132	15,394.74	104,473	194,473	5.8684	33,138.91
9	35,000	0.4665	16,327.76	120,800.76	210,800.76	6.3349	33,275.96
10	40,000	0.4241	16,963.9	137,764.66	227,764.66	6.759	33,697.86

The calculations in Table 6.15 show that the average cost is lowest during the eighth year (Rs. 33,138.91), and it is greater than the running cost of the current year (30,000 < 33,138.91) as well as less than the running cost of the next year (33,138.91 < 35,000). Hence, after every third year, the machine should be replaced; otherwise, the average cost of running the machine would start to increase.

6.3 Replacement When the Equipment/Assets Fail Completely All of a Sudden

There are certain items that do not deteriorate but fail completely after a certain amount of use. A large number of items fail due to their average life expectancy. This type of item may not have maintenance costs as such, but without any prior warning, it will suddenly fail. In addition, an immediate replacement cannot be possible in the event of sudden breakdowns. Fluorescent tubes, light bulbs, electronic chips, fuse, etc., are just a few examples.

Group replacement policy covers items that either perform perfectly or function partially or inefficiently or fail totally. This occurs generally when there are several identical low-cost items in the system which is becoming more susceptible to failure with age. In such cases, a set of costs would be incurred in replacing individual items, which is independent of the replaced number. It may, however, be advantageous to replace all items at a fixed interval at one time. This policy is known as a group replacement policy and is very attractive, particularly when the following occurs:

a) The value of every single item is so small.
b) The cost of keeping individual ages records is high and unjustified.
c) The purchase in the bulk of such identical items can be made at a discounted rate.
d) The average single replacement would be more expensive than the average group replacement.
e) If enough standby machines are available.
f) New machinery designs increase the output rate considerably.

In all of the above cases, the two types of replacement policies considered are as follows:

Individual Replacement
According to this policy, an item is replaced right after its failure.

LO 3. Understand and apply the group replacement policy

Group Replacement
Under this policy, a decision will be taken to replace all items, irrespective of the fact that the items have failed or not failed, provided that, if any item fails, it may be replaced individually before the optimum time.

We, therefore, have the following optimal policy:

Let all items in the system to be replaced after a 't' time interval so that an individual replacement can be made if and when any item fails during that time span.

Group replacement shall be made at the end of the tth period if the cost of individual replacement for the tth period exceeds the average cost per period by the end of the t period.

Group replacement at the end of the period t is not possible if, at the end of the period $t-1$, the cost of individual replacement is lower than the average cost per period at the end of the period t.

Examples based on the replacement policy when the equipment fails completely all of a sudden

EXAMPLE 1:

There are 500 bulbs in use, and it costs Rs. 50 to replace a single bulb that has burnt out. If all the bulbs were replaced at the same time, it would cost Rs. 20 per bulb. It is proposed that all bulbs be replaced at fixed intervals of time, whether they have burnt out or not, and that they continue to replace burnt-out bulbs as and when they fail. The failures rates for certain types of light bulbs are as follows (Table 6.16):

TABLE 6.16

Data for Example 1

Week	Percentage Fail at the End of the Week
1	10
2	20
3	50
4	80
5	100

At what intervals will all bulbs be replaced? At what group replacement price per bulb will the policy of purely individual replacement be preferable to the policy adopted? *(BT level 4)*

Solution:

Let p_i be the probability that a new bulb, when placed in a position to be used, will fail in the tth week of its life. Given 500 bulbs are in use, then we have

$$p1 = \text{the prob.of failure in 1st week} = \frac{10}{100} = 0.10,$$

$$p2 = \text{the prob.of failure in 2nd week} = \frac{20-10}{100} = 0.10,$$

$$p3 = \text{the prob.of failure in third week} = \frac{50-20}{100} = 0.30,$$

$$p4 = \text{the prob.of failure in 4th week} = \frac{80-50}{100} = 0.30, \text{and}$$

$$p5 = \text{the prob.of failure in 5th week} = \frac{100-80}{100} = 0.20.$$

The sum of all probabilities is 1, therefore, all further probabilities p_6, p_7, p_8 and so on will be zero.

Let N_i be the number of replacements made at the end of ith week, if all 500 bulbs are new initially.

$$N_0 = \text{number of items in the beginning} = 500.$$

$$N_1 = N_0 p_1 = 500 \times 0.10 = 50$$

$$N_2 = N_0 p_2 + N_1 p_1 = 500 \times 0.10 + 50 \times 0.10 = 55$$

$$N_3 = N_0 p_3 + N_1 p_2 + N_2 p_1$$

$$= 500 \times 0.30 + 50 \times 0.10 + 55 \times 0.10 = 161$$

$$N_4 = N_0 p_4 + N_1 p_3 + N_2 p_2 + N_3 p_1$$

$$= 500 \times 0.30 + 50 \times 0.30 + 55 \times 0.10 + 161 \times 0.10 = 187$$

$$N_5 = N_0 p_5 + N_1 p_4 + N_2 p_3 + N_3 p_2 + N_4 p_1$$

$$= 500 \times 0.20 + 50 \times 0.30 + 55 \times 0.30 + 161 \times 0.10 + 187 \times 0.10 = 167$$

$$N_6 = 0 + N_1 p_5 + N_2 p_4 + N_3 p_3 + N_4 p_2 + N_5 p_1$$

$$= 0 + 50 \times 0.20 + 55 \times 0.30 + 161 \times 0.30 + 187 \times 0.10 + 167 \times 0.10 = 111$$

$$N_7 = 0 + 0 + N_2 p_5 + N_3 p_4 + N_4 p_3 + N_5 p_2 + N_6 p_1$$

$$= 0 + 0 + 55 \times 0.20 + 161 \times 0.30 + 187 \times 0.30 + 167 \times 0.10 + 111 \times 0.10 = 144$$

And so on.

From the previous results, it is clear that the number of burnt-out bulbs increases to the fourth week and decreases to the sixth week and again begins to increase. The entire system arrives at a steady state where the proportion of failed bulbs in each week is the reciprocal of their average life.

As the mean age of bulbs

$$= 1 \times p_1 + 2 \times p_2 + 3 \times p_3 + 4 \times p_4 + 5 \times p_5,$$

$$= 1 \times 0.10 + 2 \times 0.10 + 3 \times 0.30 + 4 \times 0.30 + 5 \times 0.20,$$

$$= 3.4 \, \text{week}.$$

Therefore, numbers of failures in each week in the steady state become

$$= \frac{500}{3.4} = 147.$$

So the cost of replacing bulbs individually only upon failure

$$= 147 \times 50 = \text{Rs.7,350}.$$

Since replacing all 500 bulbs simultaneously costs Rs. 20 per bulb and replacing an individual bulb with failure costs Rs., 50; the average cost for various group replacement policies is shown in Table 6.17.

TABLE 6.17

Average Cost

End of Week (1)	Total Cost of Group Replacement in Rs. (2)	Average Cost per Week in Rs. (3) = (2) / (1)
1	50 × 50 + 500 × 20 = 12,500	12,500
2	55 × 50 + 12,500 = 15,250	7,625
3	161 × 50 + 15,250 = 23,300	7,767

In the third week, the cost of individual replacements exceeds the average cost of two weeks. It is, therefore, advisable to replace all the bulbs every two weeks; otherwise, the average cost will begin to rise.

EXAMPLE 2:

Find the cost per period of individual replacement policy of an installation of 600 compact fluorescent CFL bulbs. The cost of replacing an individual CFL is Rs. 20. The conditional probability of failure is given in Table 6.18. *(BT level 3)*

TABLE 6.18

Data for Example 2

Week	Percentage Fail at the End of the Week
0	0
1	0.20
2	0.40
3	0.80
4	1

Solution:

Let p_i be the probability that a new bulb, when placed in a position to be used, will fail in the tth week of its life. Given 500 bulbs are in use. Then, we have

$$p_0 = 0,$$

$$p_1 = \text{the prob.of failure in the first week} = 0.20,$$

p_2 = the prob.of failure in the second week = $0.4 - 0.2 = 0.20$,

p_3 = the prob.of failure in the third week = $0.8 - 0.4 = 0.40$, and

p_4 = the prob.of failure in the fourth week = $1 - 0.8 = 0.20$.

The sum of all probabilities is 1; therefore, all further probabilities p_5, p_6, p_7, and so on will be zero.

Let N_i be the number of replacements made at the end of ith week, if all 600 bulbs are new initially.

$$N_0 = \text{number of items in the beginning} = 600$$

$$N_1 = N_0 p_1 = 600 \times 0.20 = 120$$

$$N_2 = N_0 p_2 + N_1 p_1 = 600 \times 0.20 + 120 \times 0.20 = 144$$

$$N_3 = N_0 p_3 + N_1 p_2 + N_2 p_1$$

$$= 600 \times 0.40 + 120 \times 0.20 + 144 \times 0.20 = 293$$

$$N_4 = N_0 p_4 + N_1 p_3 + N_2 p_2 + N_3 p_1$$

$$= 600 \times 0.20 + 120 \times 0.40 + 144 \times 0.20 + 293 \times 0.20 = 256$$

It is clear from the above results that the number of CFL bulbs failing each week increases until the third week and then decreases during the fourth week.

As the mean age of bulbs

$$= 1 \times p_1 + 2 \times p_2 + 3 \times p_3 + 4 \times p_4,$$

$$= 1 \times 0.20 + 2 \times 0.20 + 3 \times 0.40 + 4 \times 0.20,$$

$$= 2.6 \, \text{week}.$$

Therefore, numbers of failures in each week in steady state become

$$= \frac{600}{2.6} = 231.$$

So the cost of replacing CFL bulbs individually only on failure

$$= 231 \times 20 = \text{Rs.4,620.}$$

EXAMPLE 3:

A unit of electrical equipment is subjected to failure. The probability distribution of its age at failure is given in Table 6.19.

When a unit fails in service, it costs Rs. 3 to replace it, but if all units in the same operation are replaced, it costs just Rs. 1 per unit. Assuming that the number of units is 10,000, decide on the best replacement policy and give reasons for this. *(BT level 4)*

Solution:

Let N_i be the number of replacements made at the end of ith week if all 10,000 electrical equipment units are new initially.

$$N_0 = \text{number of items in the beginning} = 10,000$$

$$N_1 = N_0 p_1 = 10,000 \times 0.10 = 1,000$$

$$N_2 = N_0 p_2 + N_1 p_1 = 10,000 \times 0.15 + 1,000 \times 0.10 = 1,600$$

$$N_3 = N_0 p_3 + N_1 p_2 + N_2 p_1$$

$$= 10,000 \times 0.25 + 1,000 \times 0.15 + 1,600 \times 0.10 = 2,810$$

$$N_4 = N_0 p_4 + N_1 p_3 + N_2 p_2 + N_3 p_1$$

$$= 10,000 \times 0.33 + 1,000 \times 0.25 + 1,600 \times 0.15 + 2,810 \times 0.10 = 4,071$$

$$N_5 = N_0 p_5 + N_1 p_4 + N_2 p_3 + N_3 p_2 + N_4 p_1$$

$$= 10,000 \times 0.13 + 1,000 \times 0.33 + 1,600 \times 0.25 + 2,810 \times 0.15 + 4,071 \times 0.10 = 2,859$$

$$N_6 = N_0 p_6 + N_1 p_5 + N_2 p_4 + N_3 p_3 + N_4 p_2 + N_5 p_1$$

TABLE 6.19

Data for Example 3

Week	Fail at the End of the Week
1	0.10
2	0.15
3	0.25
4	0.33
5	0.13
6	0.04

$$= 10,000 \times 0.04 + 1,000 \times 0.13 + 1,600 \times 0.33 + 2,810 \times 0.25 + 4,071 \times 0.15 + 2,859 \times 0.10$$

$$= 2,657$$

As the mean age of electrical equipment units

$$= 1 \times p_1 + 2 \times p_2 + 3 \times p_3 + 4 \times p_4 + 5 \times p_5 + 6 \times p_6,$$

$$= 1 \times 0.10 + 2 \times 0.15 + 3 \times 0.25 + 4 \times 0.33 + 5 \times 0.13 + 6 \times 0.04,$$

$$= 3.36 \, \text{week}.$$

Therefore, numbers of failures in each week in steady state become

$$= \frac{10000}{3.36} = 2,977$$

So the cost of replacing CFL bulbs individually only on failure

$$= 2,977 \times 3 = \text{Rs.8,931}.$$

In addition, since group replacement costs Rs. 12,500 after one week and individual replacement costs Rs. 7,350 after one week, then individual replacement is preferable.

TABLE 6.20

Average Cost

End of Week (1)	Total Cost of Group Replacement in Rs. (2)	Average Cost per Week in Rs. (3) = (2) / (1)
1	1,000 × 3 + 10,000 × 1 = 13,000	13,000
2	1,600 × 3 + 13,000 = 17,800	8,900
3	2,810 × 3 + 17,800 = 26,230	8,743.33
4	4,071 × 3 + 26,230 = 38,443	9,610.75

From Table 6.20, it follows that the optimal policy is to have group replacement on every third week. The average cost is Rs. 8.743.33, which is less than Rs. 8.931 for individual replacements. So group the replacement policy is preferred.

CHAPTER SUMMARY

- Maintenance and operating costs increase as the equipment ages while the salvage value decreases.

- Replacement models help in finding equipment's economic life and help decision makers make decisions about new acquisitions.
- Generally, there are two types of replacement problems: (1) items that deteriorate gradually and (2) items that fail suddenly.
- The economic life of items that gradually deteriorate over time is based on their average operating and maintenance costs and their resale value. The item is replaced when the total cost can be minimized.

In the case of equipment that suddenly fails, the decision maker must decide whether to replace components as they fail or at some point perform a group replacement.

Questions

1. What's the replacement problem? Describe some of the important replacement situations. *(BT level 1 and 2)*
2. What are the situations that make the replacement of items necessary? *(BT level 2)*
3. What are the categories to which replacements for items are classified? *(BT level 2)*
4. Briefly describe different replacement policies. *(BT level 2)*
5. Differentiate between individual and group replacement. *(BT level 4)*
6. Explain the discount factor. *(BT level 2)*
7. State the conditions under which group replacement is superior to individual replacement. *(BT level 1)*

Practice Problem

1. A company is considering replacing a machine that has a cost price of Rs. 25,000 and scrap value of Rs. 500. From experience, the maintenance costs are found to be as follows (Table 6.21):

TABLE 6.21

Data for Practice Problem 1

Year	Maintenance Cost (Rs.)
1	400
2	1,000
3	1,600
4	2,400
5	3,600
6	5,000
7	6,400
8	8,000

When should the machine be replaced? *(BT level 3)*

2. The price of a machine is Rs. 100,000. Operating costs for the first five years are Rs. 500 per annum. Operating costs increase by Rs. 1,000 per annum in the sixth and subsequent years. Find the optimum time needed to hold the machine before it is replaced. *(BT level 3)*

3. Machine M_1 costs Rs. 100,000/-. Annual operating costs are Rs. 4,000/- for the first year, and they increase by Rs. 8,000/- each year. Machine, M_2 which is one year old, costs Rs. 90,000/- and the annual operating costs are Rs. 2,000/- for the first year, and they increase by Rs. 20,000/- every year. Determine at what time it is profitable to replace machine M_1 with machine M_2. Assume that machines have no resale value and the future costs are not discounted. *(BT level 3)*

4. A vehicle with the first cost of Rs. 90,000 has a depreciation and service pattern as shown in Table 6.22:

TABLE 6.22

Data for Practice Problem 3

Year	Depreciation in Year	Service Cost
1	38,000	20,000
2	30,000	23,000
3	24,000	27,000
4	15,000	31,000
5	14,000	36,000
6	14,000	42,000

How many years should the vehicle be kept in service before replacement? *(BT level 3)*

5. The yearly cost of two machines M_1 and M_2, when money value is neglected is shown in Table 6.23. Find their cost patterns if the money is worth 10% a year and find out which machine is more economical. *(BT level 3)*

TABLE 6.23

Data for Practice Problem 5

Year	M_1 (Rs.)	M_2 (Rs.)
1	18,000	28,000
2	12,000	12,000
3	14,000	14,000

6. A piece of equipment cost Rs. 5,000. Operation and maintenance costs are zero for the first year and increase by Rs. 1,000 every year. If money is worth 10% every year, determine the best age at which the equipment should be replaced. The equipment resale value is negligibly small. What is the weighted average cost to own and operate the equipment? *(BT level 3)*

7. Find the cost per period of individual replacement policy of an installation of 600 electric bulbs. The cost of replacing an individual bulb is Rs. 15. The conditional probability of failure is given in Table 6.24. *(BT level 3)*

TABLE 6.24

Data for Practice Problem 7

Week	Prob. of Failure
0	0
1	0.1
2	0.33
3	0.67
4	1

8. A data processing firm has suggested that a company adopt the policy of periodically replacing all the tubes in a particular piece of equipment. It is known that a given type of tube has a distribution of mortality, as shown in Table 6.25.

TABLE 6.25

Data for Practice Problem 8

Week	Prob. of Failure
1	0.25
2	0.15
3	0.15
4	0.2
5	0.25

The cost of replacing the tubes on an individual basis is estimated to be Rs. 10 per tube and the cost of a group replacement policy average Rs. 3 per tube. Compare the cost of preventive replacement with the cost of remedial replacement. *(BT level 3)*

7

Game Theory

Learning outcomes: *After studying this chapter, you should be able to*

LO 1. *Understand the concept of a game. (BT level 2)*

LO 2. *Grasp the terms in the theory of games. (BT level 2)*

LO 3. *Understand the concept of a two-person, zero-sum game. (BT level 2)*

LO 4. *Have an awareness of the assumptions of a two-person, zero-sum game. (BT level 2)*

LO 5. *Understand the rules of minimax and maximin. (BT level 2)*

LO 6. *Understand the concept of a two-person, zero-sum game without saddle point. (BT level 2)*

LO 7. *Understand the principle of dominance. (BT level 2)*

LO 8. *Understand the graphical method of solving 2 × n and m × 2 games. (BT level 2)*

LO 9. *Interpret the results obtained from solving games. (BT level 3 and 4)*

7.1 Introduction

Life is filled with conflict and competitiveness. Parlour games, military wars, election campaigns, advertisement, and marketing strategies by competing business companies and so on are various examples that include adversaries in conflict. A basic characteristic in all of these cases is that the final outcome depends primarily on the combination of strategies chosen by the opponents. Game theory is a mathematical theory that deals formally and abstractly with the general features of competitive situations such as these. It puts a special focus on adversary decision-making processes.

Game theory is a technique for analysing conflict situations to determine the course of action to be taken and to take correct decisions in the long term. This theory, therefore, takes on value from the viewpoint of managers. Von Neumann and Morgenstern carried out the ground-breaking research on the theory of games through their publication entitled *The Theory of Games and Economic Behavior*, and, subsequently, other experts explored the subject matter. This principle will provide a manager with useful guidance in "strategic management," which can be used for acquisition, takeover, joint venture, etc., in the decision-making process. The results obtained by applying this theory will serve as an early warning to top management in meeting the challenges rose by competing business organizations and turning internal deficiencies and external threats into opportunities and strengths, thereby achieving the goal of benefit maximization. Although this theory does not explain any method for playing a game, it will encourage a player to achieve their goals by choosing the correct strategies to be pursued.

Game theory is, therefore, a body of expertise dealing with decision making when two or more intelligent and logical competitors are engaged in conflict or competition conditions.

LO 1. Understand the concept of a game

Through game theory, two or more decision makers, called players, are playing against each other. Every player chooses one of several strategies without understanding the strategy the other player or players have selected in advance. Combining the competing strategies gives the players the value of the game. Requirements for game theory have been developed for scenarios where the competing players are teams, businesses, political leaders, and contract bidders.

The emergence of games theory dates back to the 20th century. But John Von Neumann and Morgenstern dealt with the theory mathematically, and in 1944, they published a well-known paper called "Theory of Games and Economic Behaviour." Von Neumann's mathematical approach follow the minimax principle, which incorporates the basic concept of overall loss minimization. Game theory can address many competitive problems, but not all the competitive concerns can be addressed with game theory.

7.1.1 Terminology in Game Theory

Players: In a game, the competitors or decision makers are called the game players. The players can be any two firms, two countries, etc.

LO 2. Grasp the terms in the theory of games

Strategies: The alternative courses of action that a player can take are called his / her strategies. It's believed that competitors or circumstances (chance) cannot disrupt a strategy. Every player can have a set of strategies. There's no need to have the same amount of strategies for both players. Examples of strategies are when selling computer equipment, giving computer furniture away for free, giving 25% more equipment, giving special discounts, etc.

Strategies can be further grouped into pure and mixed strategies. If a player chooses a specific strategy with a probability of 1, then that strategy is referred to as pure strategy. This means the player only chooses a particular strategy and ignores the remaining strategies. When a player practices more than one strategy, then a mixed strategy implies the player's practices. Here the probability of the individual strategies being chosen will be less than one, and their sum will be equal to one.

Payoff: The outcome of playing a game is called the payoff for the player concerned. A player's payoff is the amount that the player receives for the outcome of the game (e.g. points, money, or anything that the player values). We say the goal of the player is to maximize his or her payoff. We should remember that the total payoff of the player can also be negative, in which case the player needs the least negative payoff (or nearest to 0). The tabular view of payoffs to players is called the game's payoff matrix under different alternatives. Any combination of Players A's and B's alternatives is associated with an a_{ij} outcome. If a_{ij} is negative it reflects Player A loss and Player B gain. When it's positive it's a win for A, and a loss for B. That means we can portray one player for the game in terms of payoff. Designating the two players as A and B with m and n strategies, respectively, the game is normally introduced to Player A, as shown in Table 7.1 in terms of the payoff matrix.

Optimal Strategy:

The optimum strategy for that player is called a strategy from which a player can obtain the best payoff.

TABLE 7.1

Payoff Matrix

	B_1	B_2	...	B_n
A_1	a_{11}	a_{12}	...	a_{1m}
A_2	a_{21}	a_{22}	...	a_{2m}
.	.	.	.	.
A_m	a_{m1}	a_{m2}	...	a_{mn}

Zero-Sum Game:
A game in which the net payoff at the end of the game to all players is null is referred to as a zero-sum game.

Non-Zero-Sum Game:
Non-zero-sum games are called games of "less than complete conflict of interest." It is under this umbrella that a significant number of business organizations face challenges. In these games, one player's benefit in terms of success doesn't need to be entirely at the cost of the other player.
N-person Game:
A game in which N-players take part is called an N-person game.

7.2 Two-Person, Zero-Sum Games (with Saddle Point)

A game with only two players, say, Player A and Player B, is called a two-person, zero-sum game if Player A's gain is equal to Player B's loss so that the total sum is null.

LO 3. Understand the concept of a two-person, zero-sum game

Assumptions of the Two-Person, Zero-Sum Game:
Some rational assumptions are very important for building any model. Some assumptions for constructing a two-person, zero-sum game model are described next.

a) Every player has available to him/her a finite number of possible action courses. Often, each player can have the same set of courses of action. Alternatively, certain courses of action may be available to both players, although each player may have other different courses of action that the other player cannot access.

LO 4. Have an awareness of the assumptions of a two-person, zero-sum game

b) Player A tries to maximize gains. Player B is attempting to reduce the losses.

c) Both players' decisions are taken independently before the play, without any communication between them.

d) The decisions are taken and announced simultaneously so that no player has a benefit resulting from the direct knowledge of the decision of the other player.

e) Both players know the possible payoffs for themselves and their opponents.

TABLE 7.2

Payoff Table

		Player B	
	Strategy	H	T
Player A	H	1	−1
	T	−1	1

To illustrate the basic characteristics of two-person, zero-sum games, consider the game called matching *head (H) and tail (T)* of a coin. This game consists simply of each Player (A and B) simultaneously showing either the head or tail of a coin. When the head or tail for each player matches (i.e., Player A shows H and Player B also shows H or Player A shows T and Player B also shows T), then the Player A wins the bet (say, Rs.1) from the Player B. If the H or T does not match (i.e., Player A shows H and Player B shows T or Player A shows T and Player B shows H), Player A loses the bet and pays Rs. 1 to Player B. Thus each player has two *strategies:* to show either head (H) or tail (T). The resulting payoff to Player A in Rs. is shown in the *payoff table* given in Table 7.2.

In general, a two-person game is characterized by the following:

1. A two-person game includes each player's strategies and the payoff matrix.

2. The payoff matrix displays the benefit for Player 1 (positive or negative) that would result from each combination of the two players' strategies. Note that the matrix for Player 2 is the negative of the matrix for Player 1 in a zero-sum game.

3. The payoff matrix entries may be in any unit, provided they represent the player's utility (or value).

7.3 The Maximin-Minimax Principle

The goal of game theory is to learn how these players have to pick their strategies to optimize their payoffs. Such a decision-making criterion is referred to as the principle of maximin-minimax. In pure strategy games, this principle leads to the best possible selection of a strategy for both players.

To illustrate this principle, let's consider a two-person, zero-sum game. The game's payoff matrix for Player A is set out in Table 7.3.

LO 5. Understand the rules of minimax and maximin

Let's evaluate this game first from Player A's point of view: Here payoff values in the payoff matrix are Player A's gains. So when Player A chooses a specific strategy, Player B will step in such a way that the payoff to Player A is minimal for that particular strategy since Player A's and Player B's interests conflict. Therefore, if Player A uses strategy I, he/she will receive Rs. 2 or Rs. 5 depending on the strategy Player B has adopted. Now, whatever strategy Player B

TABLE 7.3

Payoff Matrix

		Player B	
	Strategy	I	II
Player A	I	2	5
	II	3	4
	III	−2	6

adopts, Player A will at least earn Rs. 2 (min 2, 5). Likewise, if Player A uses strategy II, he/she will win at least Rs. 3 (min 3, 4) and if Player A uses strategy III, he/she will either lose Rs. 2 or gain Rs. 6. Player A clearly needs to go for the strategy, which maximizes his/her minimal gains. For a maximum of three {2, 3, −2} = 3 strategies, Player A will follow strategy II.

If Player A adopts strategy III, and Player B adopts strategy II, Player A's gain will be Rs. 6, greater than Rs. 3. But there's no guarantee B will adopt strategy II. He/she can employ strategy I, leading to a loss of Rs. 2 to Player A. So Player A will go for strategy II since we believe that both players are equally intelligent in game theory.

Now we're evaluating this game from Player B's point of view: If Player B uses strategy I, he/she can face a loss of Rs. 2, a loss of Rs. 3, or a gain of Rs. 2 depending on Player A's strategy. Currently, whatever strategy Player A adopts, Player B cannot incur more than a maximum loss of (2, 3, −2) = 3. Similarly, if Player B uses strategy II, he/she cannot incur a loss in excess of (5, 4, 6) = 6. Player B would obviously like to opt for the strategy which minimizes its maximum losses. Player B should adopt strategy I since min of (3, 6) = 3.

From the discussion above, we note that Player A will opt for the strategy that corresponds to the maximum value of payoff among the minimum row values (i.e., maximum among the minima). This is, therefore, known as the **maximin value**. Player B will opt for a strategy corresponding to the minimum payoff value between the column maximum. If maximin value = minimax value = v(say) v is known as the game value and Player A's and Player B's corresponding strategies are known as their optimal strategies. Also, the position of the element that corresponds to the two players' optimal strategies is known as the **saddle point**.

If the value of maximin ≠ minimax value, we say the game has no saddle point. So in terms of pure strategies, we can't get the game solution by applying the maximin-minimax principle. Games without a saddle point as their solutions have mixed strategies which are discussed in Section 7.4.

Examples based on the two-person, zero-sum games with saddle point

EXAMPLE 1:

For the game of payoff matrix for Player A given in Table 7.4, obtain the (i) optimal strategy for Player A, (ii) optimal strategy for Player B, and (iii) value of the game. *(BT level 3)*

TABLE 7.4

Payoff Matrix for Player A

	Strategy	Player B		
		I	II	III
Player A	I	6	8	5
	II	5	3	0
	III	7	1	4

TABLE 7.5

Maximin and Minimax

	Strategy	Player B			Row minimum	Maximin
		I	II	III		
Player A	I	6	8	5	5	
	II	5	3	0	0	5
	III	7	1	4	1	
Column maximum		7	8	5		
Minimax			5			

Solution:
Identify the minimum element of each row first, and then select the maximum among these minimum elements. This is called the value maximin, as shown in Table 7.5. Then identify each column's maximum element and choose the minimum among these maximum elements. This is called the value minimax, as shown in Table 7.5.

Now verify whether the values for the maximin and minimax are equal. From Table 7.5, maximin value = minimax value = 5. So the game is solved with the principle of maximin-minimax, and we get the following results:

i) The optimal strategy for Player A is I, as it corresponds to the maximin value.

ii) Optimal strategy for Player B is III, as it corresponds to the minimax value.

iii) The value of the game is 5 [maximin value = minimax value = 5].

EXAMPLE 2:

For the game of payoff matrix for Player A is given in Table 7.6, obtain the (i) optimal strategy for Player A, (ii) optimal strategy for Player B, and (iii) value of the game. (*BT level 3*)

Solution:
Identify the minimum element of each row first, and then select the maximum among these minimum elements. This is called the value maximin, as shown in

TABLE 7.6

Payoff Matrix for Player A

		Player B				
	Strategy	I	II	III	IV	V
	I	−3	6	−4	7	8
Player A	II	5	7	9	−2	7
	III	9	3	4	6	5
	IV	16	15	19	13	21

TABLE 7.7

Maximin and Minimax Value

		Player B					Row minimum	Maximin
	Strategy	I	II	III	IV	V		
Player A	I	−3	6	−4	7	8	−4	
	II	5	7	9	−2	7	−2	13
	III	9	3	4	6	5	3	
	IV	16	15	19	13	21	13	
Column maximum		16	15	19	13	21		
Minimax			13					

Table 7.7. Then identify each column's maximum element and choose the minimum among these maximum elements. This is called the value minimax, as shown in Table 7.7.

Now verify whether or not the values for the maximin and minimax are equal. From Table 7.7, maximin value = minimax value = 13. So the game is solved using the maximin-minimax principle, and we obtain the following results:

i) Optimal strategy for Player A is IV, as it corresponds to the maximin value.
ii) Optimal strategy for Player B is IV, as it corresponds to the minimax value.
iii) The value of the game is 13 [maximin value = minimax value = 13].

EXAMPLE 3:

Determine the range of values of M and N that will make the position (II, II) a saddle point for the game that has the payoff matrix given in Table 7.8. *(BT level 3)*
Solution:
Since it is given that the position (II, II) is the saddle point, the maximin value = minimax value = 5.

TABLE 7.8

Payoff Matrix

	Strategy	Player B		
		I	II	III
Player A	I	2	4	6
	II	9	5	M
	III	3	N	10

Now, maximin value = 5, so it must be the minimum element of the second row.

$$\therefore M \geq 5$$

Also, minimax value = 5 so it must be the maximum element of the second column.

$$\therefore N \leq 5$$

Hence, the range of M is $M \geq 5$ and the range of N is $N \leq 5$.

7.4 Two-Person, Zero-Sum Games (without Saddle Point)

LO 6. Understand the concept of a two-person, zero-sum game without saddle point

In Section 7.2, we learned how to use the maximin-minimax principle to solve two-person, zero-sum games with a saddle point. We have also understood that the presence of a saddle point depends on the equality of the value maximin and the value minimax. If they are equal, then the game has a saddle point and can be solved using the principle of maximin-minimax, which offers a pure strategy as a solution. But if they are not equal, there's no saddle point in the game. Then we need methods to solve the games without a saddle point. This section addresses one such approach for solving 2 × 2 two-person, zero-sum games without a saddle point. When there's no saddle point in the game, players respond to mixed strategies. No strategy for Player A can be considered his/her best strategy. Therefore, A will use all of its strategies. Similarly, no strategy for Player B can be considered the right strategy for him/her and he/she must use all strategies.

For any given payoff matrix without saddle point, the optimum mixed strategies are shown in Table 7.9.

Let p_1 and p_2 be the probability for Player A of playing strategy I and II, respectively.

Let q_1 and q_2 be the probability for Player B of playing strategy I and II, respectively.

Let v be the value of the game.

TABLE 7.9

Payoff Matrix

		Player B	
	Strategy	I	II
Player A	I	a_{11}	a_{12}
	II	a_{21}	a_{22}

The expected gain of Player A is

$$a_{11}p_1 + a_{21}p_2 \text{ (when Player B employs strategy I)},$$

$$a_{12}p_1 + a_{22}p_2 \text{ (when Player B employs strategy II)}.$$

Similarly, the expected loss of Player B is

$$a_{11}q_1 + a_{12}q_2 \text{ (when Player A employs strategy I)},$$

$$a_{21}q_1 + a_{22}q_2 \text{ (when Player A employs strategy II)}.$$

Since v is the value of the game and (p_1, p_2) a mixed optimal strategy for Player A, we have

$$a_{11}p_1 + a_{21}p_2 = v = a_{12}p_1 + a_{22}p_2,$$

$$\therefore a_{11}p_1 + a_{21}p_2 = a_{12}p_1 + a_{22}p_2.$$

Now as $p_1 + p_2 = 1, p_2 = 1 - p_1$,

$$a_{11}p_1 + a_{21}(1 - p_1) = a_{12}p_1 + a_{22}(1 - p_1),$$

$$\left(a_{11} - a_{21} - a_{12} + a_{22}\right)p_1 = a_{22} - a_{21},$$

$$\therefore p1 = \frac{a_{22} - a_{21}}{\left(a_{11} - a_{12}\right) + \left(a_{22} - a_{21}\right)}. \tag{7.1}$$

Now as $p_2 = 1 - p_1$,

$$\therefore p2 = 1 - \frac{a_{22} - a_{21}}{\left(a_{11} - a_{12}\right) + \left(a_{22} - a_{21}\right)},$$

$$= \frac{\left(a_{11} - a_{12}\right) + \left(a_{22} - a_{21}\right) - a_{22} + a_{21}}{\left(a_{11} - a_{12}\right) + \left(a_{22} - a_{21}\right)},$$

$$p2 = \frac{\left(a_{11} - a_{12}\right)}{\left(a_{11} - a_{12}\right) + \left(a_{22} - a_{21}\right)}. \tag{7.2}$$

The expected gain of Player A $= a_{11}\, p_1 + a_{21}\, p_2$.

$$= \frac{a_{11}\left(a_{22} - a_{21}\right)}{\left(a_{11} - a_{12}\right) + \left(a_{22} - a_{21}\right)} + \frac{a_{21}\left(a_{11} - a_{12}\right)}{\left(a_{11} - a_{12}\right) + \left(a_{22} - a_{21}\right)}$$

$$= \frac{a_{11}\, a_{22} - a_{11}\, a_{21} + a_{21}\, a_{11} - a_{12}a_{21}}{\left(a_{11} - a_{12}\right) + \left(a_{22} - a_{21}\right)}$$

$$= \frac{a_{11}\, a_{22} - a_{12}a_{21}}{\left(a_{11} - a_{12}\right) + \left(a_{22} - a_{21}\right)}$$

Processing similarly, since v is the value of the game and (q_1, q_2) is a mixed optimal strategy for Player B, we have

$$q1 = \frac{a_{22} - a_{12}}{\left(a_{11} - a_{12}\right) + \left(a_{22} - a_{21}\right)} \tag{7.3}$$

and

$$q2 = \frac{\left(a_{11} - a_{21}\right)}{\left(a_{11} - a_{12}\right) + \left(a_{22} - a_{21}\right)} \tag{7.4}$$

The expected gain of Player B $= a_{11}\, q_1 + a_{12}\, q_2$

$$= \frac{a_{11}\, a_{22} - a_{12}a_{21}}{\left(a_{11} - a_{12}\right) + \left(a_{22} - a_{21}\right)} \tag{7.5}$$

Examples based on the two-person, zero-sum game without saddle point

EXAMPLE 1:

For the game of payoff matrix for Player A is given in Table 7.10, obtain the (i) optimal strategy for Player A, (ii) optimal strategy for Player B, and (iii) value of the game. *(BT level 3)*

Solution:

We first check whether a saddle point exists or not. Identify the minimum element of each row first, and then select the maximum among these minimum elements. This is called the value maximin, as shown in Table 7.11. Then identify each column's maximum element and choose the minimum among these maximum elements. This is called the value minimax, as shown in Table 7.11.

TABLE 7.10

Payoff Matrix for Player A

		Player B	
	Strategy	I	II
Player A	I	4	3
	II	2	6

TABLE 7.11

Maximin and Minimax Value

		Player B		Row minimum	Maximin
	Strategy	I	II		
Player A	I	4	3	3	3
	II	2	6	2	
Column maximum		4	6		
Minimax		4			

Since the maximin value (=3) ≠ minimax value (=4), there is no saddle point. Hence, we have to solve the game in terms of mixed strategies.

Let,

p_1 and p_2 be the probability for Player A of playing strategy I and II, respectively,

q_1 and q_2 be the probability for Player B of playing strategy I and II, respectively, and

v be the value of the game.

From equations (7.1 to 7.5), we have

$$p1 = \frac{6-2}{(4-3)+(6-2)} = \frac{4}{5},$$

$$p2 = \frac{4-3}{(4-3)+(6-2)} = \frac{1}{5},$$

$$q1 = \frac{6-3}{(4-3)+(6-2)} = \frac{3}{5},$$

$$q2 = \frac{4-2}{(4-3)+(6-2)} = \frac{2}{5},$$

$$v = \frac{4 \times 6 - 3 \times 2}{(4-3)+(6-2)} = \frac{18}{5}.$$

Hence, the solution of the game is as follows:

The optimal strategy for Player A is ($\frac{4}{5}$, $\frac{1}{5}$), the optimal strategy for Player B is ($\frac{3}{5}$, $\frac{2}{5}$), and the value of the game is $\frac{18}{5}$.

EXAMPLE 2:

Player A and Player B take out one or two matches and guess how many matches the opponent has taken. When one of the players correctly guesses then the loser will pay twice as many rupees as the sum of the number owned by both players; otherwise, the payoff is zero. Make the payoff matrix and obtain the (i) optimal strategy for Player A, (ii) optimal strategy for Player B, and (iii) value of the game. *(BT level 6 and 3)*.

Solution:

Here strategy I for Player A is to guess one match and strategy II for Player A is to guess two matches. The same is for Player B. The payoff matrix is given in Table 7.12.

We first check whether a saddle point exists or not. Identify the minimum element of each row first, and then select the maximum among these minimum elements. This is called the value maximin, as shown in Table 7.13. Then identify each column's maximum element and choose the minimum among these maximum elements. This is called the value minimax, as shown in Table 7.13.

TABLE 7.12

Payoff Matrix

		Player B	
	Strategy	I	II
Player A	I	4	0
	II	0	8

TABLE 7.13

Minimax and Maximin Value

		Player B		Row minimum	Maximin
	Strategy	I	II		
Player A	I	4	0	0	0
	II	0	8	0	
Column maximum		4	8		
Minimax		4			

Since the maximin value (= 0) ≠ minimax value (= 4), there is no saddle point. Hence, we have to solve the game in terms of mixed strategies.

Let,

p_1 and p_2 be the probability for Player A of playing strategy I and II respectively,

q_1 and q_2 be the probability for Player B of playing strategy I and II respectively, and

v be the value of the game.

From equations (7.1–7.5), we have

$$p1 = \frac{8-0}{(4-0)+(8-0)} = \frac{2}{3}$$

$$p2 = \frac{4-0}{(4-0)+(8-0)} = \frac{1}{3}$$

$$q1 = \frac{8-0}{(4-0)+(8-0)} = \frac{2}{3}$$

$$q2 = \frac{4-0}{(4-0)+(8-0)} = \frac{1}{3}$$

$$v = \frac{4\times8-0\times0}{(4-0)+(8-0)} = \frac{8}{3}$$

Hence, the solution of the game is as follows:

The optimal strategy for Player A is ($\frac{2}{3}$, $\frac{1}{3}$), the optimal strategy for Player B is ($\frac{2}{3}$, $\frac{1}{3}$), and the value of the game is $\frac{8}{3}$.

7.5 The Principle of Dominance

In Section 7.4, the algebraic method for solving 2 × 2 two-person, zero-sum games without a saddle point was derived and applied. This approach can not, therefore, be extended directly to two-person, zero-sum games with $m \times n$ without a saddle point if either m or n or both are greater than two. But some of those games can be reduced to 2 × 2 by applying the rules of dominance, if applicable.

The dominance principle states that if one player's strategy dominates the other in all conditions then the later strategy may be ignored. A strategy dominates the other only if, under all circumstances, it is superior to other ones. The definition of dominance is particularly useful for determining two-person, zero-sum games where there is no saddle point.

LO 7. Understand the principle of dominance

Rule 1. When all of a column's elements (say ith column) are greater than or equal to any other column's corresponding elements (say jth column), the ith column is then dominated by the jth column and can be removed from the matrix.

Rule 2. When all of the elements in a row (say ith row) are less than or equal to every other row's corresponding elements (say jth row), then the ith row is dominated by the jth row and can be eliminated from the matrix.

Dominance rules (where applicable) minimize game size. Such rules can apply to any game whether or not it has a saddle point. When this rule applies to the game with a saddle point, the game size will be reduced to 1×1. And if the game size is reduced to 1×1 that means the problem has been solved. (See solved Example 2.)

Examples based on the principle of dominance

EXAMPLE 1:

For the game of payoff matrix for Player A is given in Table 7.14, obtain the (i) optimal strategy for Player A, (ii) optimal strategy for Player B, and (iii) value of the game. *(BT level 3)*

Solution: We first check whether a saddle point exists or not. Identify the minimum element of each row first, and then select the maximum among these minimum elements. This is called the value maximin, as shown in Table 7.15. Then identify each column's maximum element and choose the minimum among these maximum elements. This is called the value minimax, as shown in Table 7.15.

TABLE 7.14

Payoff Matrix for Player A

		Player B		
	Strategy	I	II	III
	I	−1	−3	7
Player A	II	6	5	2
	III	4	3	9

TABLE 7.15

Maximin Value

		Player B			Row	
	Strategy	I	II	III	minimum	Maximin
	I	−1	−3	7	−3	
Player A	II	6	5	2	2	3
	III	4	3	9	3	
Column maximum		6	5	9		
Minimax			5			

Since the maximin value (= 3) ≠ minimax value (= 5), there is no saddle point. Hence, we have to solve the game in terms of mixed strategies.

Let us first reduce the payoff matrix by applying the dominance rules.

In the payoff matrix (Table 7.14) each element of the first row is less than the corresponding element of the third row. That implies that the third row dominates the first row. So we can eliminate the first row from the given payoff matrix. Thus, we write the reduced payoff matrix as shown in Table 7.16.

In Table 7.16, each element in the first column is larger than the corresponding element in the second column. That implies the second column dominates the first column. So we can eliminate the first column and write the reduced payoff matrix as shown in Table 7.17.

In the reduced payoff matrix shown above, neither row dominates any other row, nor does any column dominate any other column. So this matrix can't be further reduced using the rules of dominance. Since the reduced payoff matrix is of order 2 x 2.

Let (p_1, p_2, p_3) and (q_1, q_2, q_3) be the mixed strategies for Players A and B, respectively, for the original payoff matrix. Therefore, from the reduced payoff matrix, (p_2, p_3) and (q_2, q_3) are the mixed strategies for the Players A and B, respectively. If v is the value of the game, then from equations (7.1–7.5), we have

$$p2 = \frac{9-3}{(5-2)+(9-3)} = \frac{2}{3}$$

$$p3 = \frac{5-2}{(5-2)+(9-3)} = \frac{1}{3}$$

$$q2 = \frac{9-2}{(5-2)+(9-3)} = \frac{7}{9}$$

TABLE 7.16

Payoff Matrix

		Player B		
	Strategy	I	II	III
Player A	II	6	5	2
	III	4	3	9

TABLE 7.17

Payoff Matrix

		Player B	
	Strategy	II	III
Player A	II	5	2
	III	3	9

$$q3 = \frac{5-3}{(5-2)+(9-3)} = \frac{2}{9}$$

$$v = \frac{5 \times 9 - 2 \times 3}{(5-2)+(9-3)} = \frac{13}{3}$$

Hence, the solution of the game is
the optimal strategy for Player A is $(0, \frac{2}{3}, \frac{1}{3})$, the optimal strategy for Player B is
$(0, \frac{7}{9}, \frac{2}{9})$, and the value of the game is $\frac{13}{3}$.

EXAMPLE 2:

For the game of payoff matrix for Player A is given in Table 7.18, obtain the (i) optimal strategy for Player A, (ii) optimal strategy for Player B, and (iii) value of the game. *(BT level 3)*

Solution: We first check whether a saddle point exists or not. . Identify the minimum element of each row first, and then select the maximum among these minimum elements. This is called the value maximin, as shown in Table 7.19. Then identify each column's maximum element and choose the minimum among these maximum elements. This is called the value minimax, as shown in Table 7.19.

TABLE 7.18

Payoff Matrix for Player A

	Strategy	Player B		
		I	II	III
	I	10	14	8
Player A	II	8	4	0
	III	12	2	6

TABLE 7.19

Minimax Value

	Strategy	Player B			Row minimum	Maximin
		I	II	III		
	I	10	14	8	8	
Player A	II	8	4	0	0	8
	III	12	2	6	2	
Column maximum		12	14	8		
Minimax			8			

Since the maximin value (= 8) = minimax value (= 8), there is a saddle point and we obtain the following results:

  i) The optimal strategy for Player A is I as it corresponds to the maximin value,

 ii) The optimal strategy for Player B is III as it corresponds to the minimax value,

iii) The value of the game is 8 [maximin value = minimax value = 8]

Now let us first reduce the payoff matrix by applying the dominance rules.

Note that each element of the first row is greater than or equal to the corresponding element of the second row. This means the second row is dominated by the first row. So we can remove the second row from the payoff matrix and write down the reduced payoff matrix as shown in Table 7.20.

Each element in the first column is greater than or equal to the corresponding element in the third column. That implies the first column is dominated by the third column. And we can remove the first column from the matrix. Following this step, the reduced payoff matrix is shown in Table 7.21.

Each element of the first row is greater than or equal to the corresponding element of the second row. This implies that the first row dominates the second row. So we can eliminate the second row from the matrix. Following this step, the reduced payoff matrix is shown in Table 7.22.

TABLE 7.20

Payoff Matrix

		Player B		
	Strategy	I	II	III
	I	10	14	8
Player A	III	12	2	6

TABLE 7.21

Payoff Matrix

		Player B	
	Strategy	II	III
	I	14	8
Player A	III	2	6

TABLE 7.22

Payoff Matrix

		Player B	
	Strategy	II	III
Player A	I	14	8

TABLE 7.23

Payoff Matrix

		Player B
	Strategy	III
Player A	I	8

The single element 14 of the first column is greater than the corresponding single element of the second column. This implies that the second column dominates the first column. So we can eliminate the first column from the reduced payoff matrix. The reduced payoff matrix after this step is shown in Table 7.23.

Thus payoff matrix reduces to order 1×1. Hence, the solution of the game is as follows:

The optimal strategy for Player A is $(1, 0, 0)$, the optimal strategy for Player B is $(0, 0, 1)$ and the value of the game is 8. This result is the same as obtained in the first part of this example.

7.6 Graphical Method for Solving Games

Many times, we have to deal either directly with $2 \times n$ and $m \times 2$ games or, after applying dominance rules, they reduce to one of those types. It is therefore important to learn how to solve $2 \times n$ and $m \times 2$ games. We address one of those approaches known as the graphical approach in this section.

LO 8. Understand the graphical method of solving $2 \times n$ and $m \times 2$ games

The concept of a $2 \times n$ zero-sum game

If the first Player A has exactly two strategies, and the second Player B has n strategies (where n is three or more), the result is a $2 \times n$ game. It is also known as a rectangular game. Since A has only two strategies, he/she cannot attempt to abandon either one of them. Because B does have several strategies, however, he/she can make some choices among them. He/she can retain some of the advantageous strategies and discard some strategies that are disadvantageous. B's goal is to offer A as little as possible for the payoff. To put it another way, B will always try to minimize the loss to itself.

The concept of an $m \times 2$ zero-sum game

If the second Player B has exactly two strategies, and the first Player A has m (where m is three or more) strategies, then $m \times 2$ is the result. It's also called a rectangular game. Since B has only two strategies, it will find it difficult to discard any of them. However, since A has more strategies, he/she will be in a position to make a choice between them. Some of the most advantageous strategies can be retained and some other strategies can be abandoned. A's motive is to get the maximum payoff possible.

TABLE 7.24

Payoff Matrix for Player A

		Player B			
	Strategy	B_1	B_2	....	B_n
	A_1	a_{11}	a_{12}	....	a_{1n}
Player A	A_2	a_{21}	a_{22}	....	a_{2n}

Consider the payoff matrix for Player A for a $2 \times n$ game as shown in Table 7.24. Assume that the game does not have a saddle point.

Let (p_1, p_2) and $(q_1, q_2, ... q_n)$ be the optimal mixed strategies for Player A and Player B, respectively.

Now, if $e_1, e_2, ... e_n$ be the expected payoffs for Player A corresponding to different strategies of Player B, then these are given by:

$$e_1 = a_{11} p_1 + a_{21} p_2 \text{ if Player B employs strategy } B_1$$

$$e_2 = a_{12} p_1 + a_{22} p_2 \text{ if Player B employs strategy } B_2$$

..

..

..

$$e_n = a_{1n} p_1 + a_{2n} p_2 \text{ if Player B employs strategy } B_n$$

But $p_1 + p_2 = 1$ so $p_2 = 1 - p_1$

So writing the above expected payoffs for Player A in terms of p_1, we get,

$$e_1 = (a_{11} - a_{21}) p_1 + a_{21} \text{ if Player B employs strategy } B_1$$

$$e_2 = (a_{12} - a_{22}) p_1 + a_{22} \text{ if Player B employs strategy } B_2$$

..

..

..

$$e_n = (a_{1n} - a_{2n}) p_1 + a_{2n} \text{ if Player B employs strategy } B_n$$

Depending on the maximum criterion for mixed strategy games, Player A should select the value of p_1 and p_2 in order to maximize the minimum expected payoffs. This can be done by plotting the expected payoff lines. Thus, the expected payoffs of row Player A corresponding to the different strategies employed by column Player B represent n straight lines.

The highest point on the lower envelope of these lines will give Player A maximum of the minimum (maximum) expected payouts as well as the maximum value of Pi. If there are only two straight lines going through the highest level, then we can identify the two strategies that correspond to those lines and exclude all other Player B strategies from the payoff matrix. That reduces the game to a two-person, zero-sum game of 2 × 2. But if there are more than two straight lines passing through the highest level, alternate solutions can come up. We select any two straight lines from among the lines passing through the highest point to obtain one solution and identify the strategies corresponding to those two lines. This reduces the game to a 2 × 2 two-person, zero-sum game again.

The $m \times 2$ games are also treated the same way where the upper envelope of the straight lines corresponding to the expected payoffs of Player B gives Player B the maximum expected payoff, and the lowest point on that gives the minimum expected payoff (minimax) and the optimum qi value.

Examples of the graphical method for solving games

EXAMPLE 1:

For the game of payoff matrix for Player A is given in Table 7.25, obtain the (i) optimal strategy for Player A, (ii) optimal strategy for Player B, and (iii) value of the game. *(BT level 3)*

Solution: We first check whether a saddle point exists or not. Identify the minimum element of each row first, and then select the maximum among these minimum elements. This is called the value maximin, as shown in Table 7.26. Then identify each column's maximum element and choose the minimum among these maximum elements. This is called the value minimax, as shown in Table 7.26.

TABLE 7.25

Payoff Matrix for Player A

		Player B		
	Strategy	I	II	III
Player A	I	2	3	8
	II	7	5	3

TABLE 7.26

Minimax Value

		Player B			Row	
	Strategy	I	II	III	minimum	Maximin
Player A	I	2	3	8	2	
	II	7	5	3	3	3
Column maximum		7	5	8		
Minimax			5			

Since the maximin value (= 3) ≠ minimax value (= 5), there is no saddle point. Then we will solve the game in terms of mixed strategies. Moreover, using dominance rules, this game cannot be reduced, since it has no inferior row or column. But this is type $2 \times n$ game, where $n = 3$.

Let (p_1, p_2) and (q_1, q_2, q_3) be the mixed strategies for Players A and B, respectively. Then the expected payoffs of Player A corresponding to the strategies I, II, and III of the Player B are given by:

$$e_1 = 2p_1 + 7p_2 = 2p_1 + 7(1-p_1) = -5p_1 + 7 \tag{i}$$

$$e_2 = 3p_1 + 5p_2 = 3p_1 + 5(1-p_1) = -2p_1 + 5 \tag{ii}$$

$$e_3 = 8p_1 + 3p_2 = 8p_1 + 3(1-p_1) = 5p_1 + 3 \tag{iii}$$

Draw two ST and UV vertical lines. Notice they are parallel to one another. Label units of equivalent scale on ST and UV. The units on the two lines ST and UV are taken as numbers of payoffs. The payoffs in the first row of the given matrix are taken along the UV line, while the payoffs in the second row go along the ST line.

By setting $p_1 = 0$, the payoff values for the second row are (7, 5, 3) and by setting $p_1 = 1$, the payoff values for the first row are (2, 3, 8).

Plot the straight lines given by (i), (ii), and (iii) by joining the payoff values of two rows with their corresponding values. Here we plot the lines by joining 7 to 2, 5 to 3, and 3 to 8 as shown in Figure 7.1. Draw bold lines along the line segments, which form the lowest boundary of expected p.

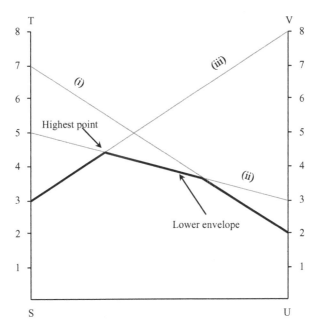

FIGURE 7.1
Graphical Solution for Example 1.

TABLE 7.27

Payoff Matrix

		Player B	
	Strategy	II	III
Player A	I	3	8
	II	5	3

In Figure 7.1, the highest point is the point of intersection of expected payoff lines given by (ii) and (iii). Thus, we write the reduced payoff matrix as shown in Table 7.27.

Since the reduced payoff matrix is of order 2 × 2. Therefore, from the reduced payoff matrix, (p_1 , p_2) and (q_2 , q_3) are the mixed strategies for the Players A and B, respectively. If v is the value of the game, then from equations (7.1–7.5), we have

$$p_1 = \frac{3-5}{(3-8)+(3-5)} = \frac{2}{7}$$

$$p_2 = \frac{3-8}{(3-8)+(3-5)} = \frac{5}{7}$$

$$q_2 = \frac{3-8}{(3-8)+(3-5)} = \frac{5}{7}$$

$$q_3 = \frac{3-5}{(3-8)+(3-5)} = \frac{2}{7}$$

$$v = \frac{3 \times 3 - 8 \times 5}{(3-8)+(3-5)} = \frac{31}{7}$$

Hence, the solution of the game is

the optimal strategy for Player A is ($\frac{2}{7}$, $\frac{5}{7}$), the optimal strategy for Player B is ($0, \frac{5}{7}$, $\frac{2}{7}$), and the value of the game is $\frac{31}{7}$.

EXAMPLE 2:

For the game of payoff matrix for Player A is given in Table 7.28, obtain the (i) optimal strategy for Player A, (ii) optimal strategy for Player B, and (iii) value of the game. *(BT level 3)*

Solution: We first check whether a saddle point exists or not. Identify the minimum element of each row first, and then select the maximum among these minimum elements. This is called the value maximin, as shown in Table 7.29. Then identify each column's maximum element and choose the minimum among these maximum elements. This is called the value minimax, as shown in Table 7.29.

TABLE 7.28

Payoff Matrix for Player A

	Strategy	Player B	
		I	II
Player A	I	4	−1
	II	−2	5
	III	−1	2

TABLE 7.29

Minimax Value

	Strategy	Player B		Row minimum	Maximin
		I	II		
Player A	I	4	−1	−1	−1
	II	−2	5	−2	
	III	−1	2	−1	
Column maximum		4	5		
Minimax		5			

Since the maximin value $(= -1) \neq$ minimax value $(= 5)$, there is no saddle point. Hence, we have to solve the game in terms of mixed strategies. Moreover, this game cannot be reduced by using dominance rules, because it has neither any inferior row nor any inferior column. But this game is of the type $m \times 2$ where $m = 3$.

Let (p_1, p_2, p_3) and (q_1, q_2) be the mixed strategies for Players A and B, respectively. Then the expected payoffs of Player B corresponding to the strategies I, II, and III of the Player A are given by:

$$e_1 = 4q_1 - q_2 = 4q_1 - (1 - q_1) = 5q_1 - 1 \tag{i}$$

$$e_2 = -2q_1 + 5q_2 = -2q_1 + 5(1 - q_1) = -7q_1 + 5 \tag{ii}$$

$$e_3 = -q_1 + 2q_2 = -q_1 + 2(1 - q_1) = -3q_1 + 2 \tag{iii}$$

Draw two ST and UV vertical lines. Notice they are parallel to one another. Label units of equivalent scale on ST and UV. The units on the two lines ST and UV are taken as numbers of payoffs. The payoffs in the first row of the given matrix are taken along the UV line, while the payoffs in the second row go along the ST line

By setting $q_1 = 0$, the payoff values for the second column are $(−1, 5, 2)$ and by setting $q_1 = 1$, the payoff values for first column are $(4, −2, −1)$.

Plot the straight lines given by (i), (ii), and (iii) by joining the payoff values of two columns with their corresponding values. Here we plot the lines by joining −1 to 4, 5 to −2, and 2 to −1 as shown in Figure 7.2. Draw bold lines along the line segments, which form the uppermost boundary of expected payoff lines given by (i), (ii), and (iii) to obtain the upper envelope (see Figure 7.2).

In Figure 7.2, the lowest point is the point of intersection of expected payoff lines given by (i) and (ii). Thus, we write the reduced payoff matrix as shown in Table 7.30.

Since the reduced payoff matrix is of order 2 × 2. Therefore, from the reduced payoff matrix, (p_1, p_2) and (q_1, q_2) are the mixed strategies for the Players A and B, respectively. If v is the value of the game, then from equations (7.1–7.5), we have

$$p_1 = \frac{5-(-2)}{\left(4-(-1)\right)+\left(5-(-2)\right)} = \frac{7}{12},$$

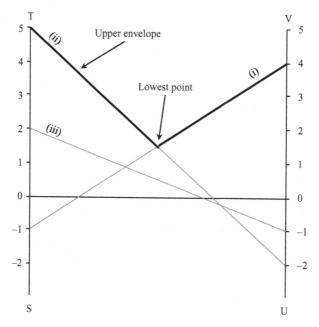

FIGURE 7.2
Graphical Solution for Example 2.

TABLE 7.30

Payoff Matrix

		Player B	
	Strategy	I	II
Player A	I	4	−1
	II	−2	5

$$p_2 = \frac{4-(-1)}{\left(4-(-1)\right)+(5-(-2))} = \frac{5}{12},$$

$$q_1 = \frac{5-(-1)}{\left(4-(-1)\right)+(5-(-2))} = \frac{6}{12},$$

$$q_2 = \frac{4-(-2)}{\left(4-(-1)\right)+(5-(-2))} = \frac{6}{12},$$

$$v = \frac{4 \times 5 - (-1) \times (-2)}{\left(4-(-1)\right)+(5-(-2))} = \frac{18}{12}.$$

Hence, the solution of the game is
the optimal strategy for Player A is ($\frac{7}{12}$, $\frac{5}{12}$, 0), the optimal strategy for Player B is
($\frac{6}{12}$, $\frac{6}{12}$), and the value of the game is $\frac{18}{12}$.

CHAPTER SUMMARY

- Game theory deals with player decision-making processes in competing and competitive situations where a player's strategy (or action or move) is based on the opponent's move.
- A strategy is an action plan by which a player can agree on every move under all possible circumstances in a game.
- If the game uses only one alternative, this is considered a pure strategy.
- When several alternatives are used to play the game with various strategies, this is called a game with mixed strategies.
- A game can be a two-person game or a multi-person game.
- Each player selects one strategy every time they play the game.
- Saddle point is an element in the payoff matrix, which is the minimum in one direction and the maximum in the other.
- The dominance principle reduces the size of the payoff matrix and thus encourages a fast solution to the problem.

Using the graphical form, games of the order $2 \times n$ and $m \times 2$ can be reduced to 2×2.

Questions

1 Explain the concept of a game and state competitive game assumptions. *(BT level 1 and 2)*

2 Explain the following terms with reference to game theory. *(BT level 2)*

a) Strategy (b) payoff matrix (c) saddle point (d) pure strategy and mixed strategy e) Two-person, zero-sum game (f) value of a game (g) $2 \times n$ game and $m \times 2$ game

3 Explain minimax and maximin principles. *(BT level 2)*

4 Give an example of how to reduce the size of rectangular gaming by using dominance properties? *(BT level 4)*

5 How are you going to interpret the payoff matrix results for a two-person, zero-sum game? Explain. *(BT level 2)*

6 Describe the solution method for a $2 \times n$ zero-sum game. *(BT level 2)*

7 Describe the available solution methods for games with a saddle point and without a saddle point? *(BT level 2)*

Practice Problem

1. Two Players A and B face each other and show their hands simultaneously in one of the following three shapes: a fist denoting a rock (R), the forefinger and middle finger extending and spreading denoting scissors (S), and a downward-faced palm denoting a sheet of paper (P). Rock beats scissors, scissors beats (cuts) paper, and paper beats (covers) rock. The winner receives a penny from the opposing player and in case of a tie, no money is exchanged. Create the payoff matrix for Player A. *(BT level 6)*

2. Beside each other, 'A' and 'B' own stores. Every day a 10% or 20% discount is revealed for sale. 'A' gets 70% of the customers if they both give 10% off. 'A' gets 30% of customers if 'A' goes 10% and 'B' 20% sale. If 'A' announces a sale of 20% and 'B' a sale of 10%, 'A' will gain 90% of customers. If both go 20%, 'A' will get 50% of the customers. Among them 'A' and 'B' get all the customers every day and every customer patronizes only one of the shops every day. Create the payoff matrix for Player A. *(BT level 6)*

3. Solve the two-person, zero-sum game that has the following payoff matrix for Player A (Table 7.31). *(BT level 3)*

TABLE 7.31

Payoff Matrix for Player A

	Strategy	Player B I	Player B II
Player A	I	−3	−2
	II	3	−4

4. The payoff for Player A is in the game given below (Table 7.32). Determine the M and N values which make a saddle point (A2, B2). *(BT level 3)*

TABLE 7.32

Payoff for Player A

		Player B		
	Strategy	**I (B1)**	**II (B2)**	**III (B3)**
	I (A1)	2	N	7
Player A	II (A2)	M	6	11
	III (A3)	7	3	4

5. Two players, A and B, each call one of the numbers 1 and 2 at a time. If they both call 1, there will be no charge. If both calls 2, B pays A Rs. 1. Where A calls 1 and B 2, B pays A Rs. 1. Where A calls 2 and B 1, A pays B Rs. 1. What's the payoff matrix for this game? Is the game fair for both players? *(BT level 6)*

TABLE 7.33

Payoff Matrix

		Player B	
	Strategy	Sale on all	Sale on popular
	Sale on all	3	−2
Player A	Sale on popular	−2	1

6. Two competitive ice-cream chains, A and B, plan to retain sales of ice cream to gain any additional business from one another. We each have the choice of selling their most famous flavours or all of their ice cream. For each of their decisions, they consider the following payoff matrix (Table 7.33) to demonstrate how the business is turning. In the matrix, the number of B to A customers is given based on a hundred people. A negative number indicates they shift from A to B. Identify the best strategy for each of them. *(BT level 3)*

TABLE 7.34

Payoff Matrix

		Player B		
	Strategy	**I**	**II**	**III**
	I	3	−2	9
Player A	II	4	4	−5

7. Solve the games given below graphically. The payoff matrix is for Player A is given in Table 7.34 and Table 7.35. *(BT level 3)*

a)

b)

TABLE 7.35

Payoff Matrix

| | Strategy | Player B | |
		I	II
	I	4	7
Player A	II	5	4
	III	4	6

8

Queuing Theory

Learning Outcomes: After studying this chapter, the reader should be able to

LO 1. *Understand the queuing system. (BT level 2)*

LO 2. *Understand elements of the queuing system. (BT level 2)*

LO 3. *Understand the assumptions of the common model. (BT level 2)*

LO 4. *Understand and calculate the characteristics of a single-server queue. (BT level 2 and 3)*

LO 5. *Understand and calculate the characteristics of a multi-server queue. (BT level 2 and 3)*

LO 6. *Analyse a variety of operating characteristics of queuing models. (BT level 4)*

8.1 Introduction

LO 1. Understand the queuing system

Queues (waiting lines) are one of the most prevalent occurrences in the life of all. In post offices, banks, restaurants, railways, and airline booking counters, we wait in queues. Not only do people spend a large amount of their time standing in queues but also goods queue up in manufacturing plants, devices waiting to be serviced in line, planes waiting to take off and land, etc. Because time is a valuable resource, a significant topic of study is the reduction of waiting time. We want to find ways to reduce the time the customer has spent waiting for the service provider and at the same time minimize the costs. This is where the theory of queuing, also known as the theory of the waiting line, helps us. It was designed in 1909 when A. K. Erlang, a Danish engineer, made an attempt to examine the congestion of telephone traffic. After then, queuing theory has become even more complex with implementations of different waiting lines.

The aim of the queuing study is to provide information to assess an appropriate level of service and service efficiency because it is expensive (because of idle staff or equipment) to provide too much service efficiency. Yet it's also expensive to have too little service capacity (because of waiting members in the queue). Queuing is a probabilistic, not deterministic, method of analysis. The results of queuing analysis, called performance indicators, are thus probabilistic. Such operating statistics (such as the average time a person has to wait in line to be served) are used by the process manager for making decisions. Queuing issues are extremely complicated, so analytical methods are only available for relatively small

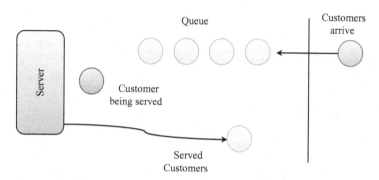

FIGURE 8.1
Basic Queuing System.

problems. Simulation (not mentioned in this book) offers a more robust way of addressing bigger and more complicated issues.

Queues form if customers want some kind of service but find the server is already busy. Then they have the option either to enter a queue and wait to be served or to leave and probably go to a competitor with a non-occupied server. Here we take the situation in which they agree to enter the queue. By convention, "the customer" is someone or anything that wants a service, and the "server" is the person or thing that provides that service. Whatever parties involved all queues have similar features-a simple queuing system is shown in Figure 8.1.

A movement of customers from the finite or infinite population to the service facility forms a queue (waiting line) an account of a lack of ability to serve them all at a time. If the service facilities and the customers are not in complete harmony, the waiting time is required either for the customer's arrival or for the services. The queuing system typically consists of one or more queues and one or more servers and operates within a set of procedures. Depending on the status of the server, the incoming customer either waits for the queue or gets the turn to be served. If the server is free when a customer arrives, the customer can enter the counter directly to get service and then exit the system. During this process, the system can experience "customer waiting" and/or "server idle time" over a period of time.

8.1.1 Elements of the Queuing System

LO 2. Understand elements of the queuing system

The main elements of a queue system are the call population, arrival rates, servers, queue or waiting line, and a queue discipline.

Calling population (source of input) is the source of customer supply. Customers are not necessarily human in the queue system. For example, "customers" may be trucks entering a weighbridge, orders being processed, and machines waiting for maintenance, etc. The customer source for the queuing system can be either finite or infinite. There is a specified number of customers on a finite source. For instance, if a maintenance person serves four assembly lines, then the number of maintenance personnel's customers is known, that is to say, four. The probability of the arrival of a customer depends on the number of customers already served. In comparison, infinite customer sources believe that there are a vast number of potential customers such that another customer will still arrive irrespective of how many customers are served. For most queuing systems dealing with external markets, customer sources are infinite or "close to infinite."

Arrival rate is the rate at which customers that need to be served arrive at the server. Customers rarely arrive at a steady and predictable rate. Usually, there is variability in the rate of arrival. As a result, it is necessary to describe the rate of arrival in terms of probability distributions. The key problem here is that it is common in queuing systems that, at times, no customers arrive and at other times, many arrive fairly close together. The arrivals could be represented with many distributions, but the number of arrivals per unit of time at a service facility could often be defined by the *Poisson distribution* (through years of study and practical experience in the field of queuing). In a queuing, λ indicates the average rate of arrival or how many customers arrive over time. For the models described in this chapter, we presume that the waiting-line method does not involve balking (refusing to join a line), reneging (leaving a line), or jockeying (changing lines) by customers.

The arrivals occur randomly and independently of other arrivals for many waiting-line situations, and we cannot anticipate when an arrival will occur. For such cases, analysts find that a clear explanation of the arrival pattern is given by the Poisson probability distribution.

The Poisson probability function provides the probability of x arrivals in a specific time period. The probability function is as follows:

$$P(x) = \frac{\lambda^x e^{-\lambda}}{x!} \text{ for } x = 0, 1, 2, \ldots, \tag{8.1}$$

where

x = the number of arrivals in the time period,

λ = the *mean* number of arrivals per time period,

e = 2.71828, and

$x!$ = the factorial of a value x.

Server is the place where customers are served in the queue. Any number of servers configured in different ways can be designed in any queuing system. Servers are parallel configured, but some may have servers in a series system. Obviously, many queue systems are dynamic series and parallel connections arrangements. The time it takes to service each customer is also likely to differ. Even if customers don't have different requirements, the time they have to perform repetitive service tasks is different for human servers.

The time spent on a servicing facility from beginning to end is referred to as the service time for the customer. The probability distribution of service times for each server (and likely for different customers) must be described by a model of a particular queuing system, though for every server the same distribution is common. The *exponential* distribution is the service-time distribution most widely assumed in practice (especially because of its far more tractable format than any other).

If an exponential probability distribution for the service time is believed to be followed, formulas for providing useful details of the operation of the waiting line are available. Using an exponential probability distribution, the probability that the service time will be less than or equal to a time of length t is

$$P(\text{service time} \leq t) = 1 - e^{-\mu t}, \tag{8.2}$$

where

μ = the *mean* number of units that can be served per time period, and

e = 2.71828.

Queue is where the clients are waiting to be served. The queue is defined by the maximum number of customers it can contain. Depending on whether the number is infinite or finite, the queues are called infinite or finite. For most queuing models, the concept of an infinite queue is the usual one, even in cases where the allowable number of customers actually has a (relatively large) finite upper bound, as dealing with such an upper bound would be a complicating factor in the analysis. However, in the case of queuing systems where this upper bound is small enough that it would actually be reached with some frequency, it is necessary to assume a finite queue.

Queue discipline is the order in which waiting customers are served. Fast Shop Market customers are served on the basis of "first come, first served." The first user who is in line is first served at the checkout counter. It is the most common form of a queue. Nevertheless, there are other disciplines. A machine operator that, for example, stack in-process pieces on a machine so that the last piece is above the stack and is chosen first. It's called "last-in, first-out" in this queue discipline. Or the machine operator might simply reach into a box full of parts and randomly pick one. In this case, the order of the queue is random. Customers are also scheduled to work on a predetermined appointment, such as patients in a doctor's office or dentist's office or diners in a restaurant where reservations are necessary. In this situation, consumers are taken on a prearranged timetable, regardless of when they arrive at the facilities. A final example of the various types of queue categories is where customers are sorted alphabetically by their last names, such as registration at school or work interviews.

8.1.2 Kendall Notation

D. G. Kendall (1951) developed a notation commonly accepted for defining arrival patterns, service time distribution, and server numbers in a queuing model. This notation is often seen in software for queuing models. A queuing system is usually described by five symbols and denoted as 1/2/3: 4/5 or separated by slashes as 1/2/3/4/5. The first '1' mark defines the process of arrival. The second '2' symbol defines the time distribution of the service. The third '3' symbol is the number of servers. The symbols '4' and '5,' respectively, represent the system capacity and the queue discipline. If the system has an infinite capacity with the FCFS queue discipline, the queuing system can be represented as 1/2/3: ∞/FCFS or simply 1/2/3. When the arrivals obey the Poisson distribution, the M symbol is used instead of '1.' If the departures are exponential, the symbol M is used instead of '2.'

Thus a single-server model with Poisson arrivals and exponential service times would be represented by

$$M/M/1.$$

When a second server is added, we would have

$$M/M/2.$$

If there are m distinct servers in the queuing system with Poisson arrivals and exponential service times, the Kendall notation would be

$$M / M / m.$$

A three-server system with Poisson arrivals and constant service time would be identified as

$$M / D / 3.$$

A three-server system with Poisson arrivals and service times that are normally distributed would be identified as

$$M / G / 3.$$

8.2 Single-Server Queuing Model with Poisson Arrivals and Exponential Service Times (*M/M/1*)

LO 3. Understand the assumptions of the common model

As the symbol indicates, the M/M/1 queuing model deals with a queuing system that has a single Poisson input server, an exponential distribution for services, and no limit to the capacity of the system while the customers are served on the "first come, first served" basis. It is one of the simplest and most commonly used queuing models. Figure 8.1 provides a representation of a single-server line of waiting. This involves the following assumptions.

1. Arrivals are served on the basis of first come, first served.
2. No matter the length of the queue, every arrival is waiting for it to be served – that is, no reneging or balking.
3. Arrivals do not change with the preceding arrivals, but the average arrival rate does not change with time.
4. Arrivals are described by a distribution of Poisson probabilities and come from a vast or infinite population.
5. Service times also differ and are different from one customer to the next, but its average rate is known.
6. Service times are based on the negative exponential probability distribution.
7. The average rate of service is higher than the average rate of arrival.

Such assumptions were used to build a single-server queuing system model. The mathematical approach used to derive the formulas for the queuing system's operating characteristics is however quite complex. In this chapter, however, we are not aiming at the theoretical development of the models but at showing how the formulas that are created can inform about the operational characteristics of the system. Specialized texts can be consulted by readers interested in the mathematical development of formulas.

Given that

λ = the mean number of arrivals per time period (the arrival rate),

μ = the mean number of services per time period (the service rate).

The same time period shall be used when determining the arrival rate and service rate. For example, if there is the average number of arrivals per hour, then the average number that could be served per hour must be noted.

Imagine that the average arrival rate is two customers an hour and the average operation rate is four customers an hour. On average, every 30 minutes a customer arrives, and it takes about 15 minutes to serve. In extending this logic, we may say that a server is generally busy for a ratio of λ/μ time, and this is defined as the factor of utilization. This is also the time a customer is served or waiting, which is technically defined as "in the system." Looking at it differently, it is also the probability that an incoming customer may have to wait, which is the average number of customers served at any time.

The probability that there is no customer in the system (it is also the probability that the server is idle) is

$$P0 = 1 - \frac{\lambda}{\mu}. \tag{8.3}$$

The probability that n customers are in the queuing system is

$$Pn = P0 \left(\frac{\lambda}{\mu} \right)^n. \tag{8.4}$$

The average number of customers in the queuing system (i.e., the customers being serviced and in the waiting line) is

$$L = \frac{\lambda}{\mu - \lambda}. \tag{8.5}$$

<table>
<tr><td>

LO 4. Understand and calculate the characteristics of a single-server queue

</td><td>

The average number of customers in the queue is the average number in the system minus the average number being served:

</td></tr>
</table>

$$Lq = \frac{\lambda^2}{\mu(\mu - \lambda)}. \tag{8.6}$$

The average time a customer spends in the total queuing system (i.e., waiting and being served) is

$$W = \frac{1}{\mu - \lambda} = \frac{L}{\lambda}. \tag{8.7}$$

The average time a customer spends waiting in the queue to be served is the average time in the system minus the average service time:

$$Wq = W - \frac{1}{\mu} = \frac{\lambda}{\mu(\mu - \lambda)}. \tag{8.8}$$

Remember that these operating features are averages over a period of time and are not absolute. In other words, clients arriving on the list, such as 2.2 customers, will not be placed in line. No customers or 1, 2, 3, or 4 can be found. Value 2.2, like other operating characteristics, is simply an average over time.

Examples based on the single-server queuing model

EXAMPLE 1:

The average rate of arrival at the petrol pump is 30 customers an hour, and the average service time is 48 an hour. Determine (a) the average number of customers in the system, (b) the average number of customers in the queue, (c) the average time of the customer in the system, and (d) the average time the customer waits in the queue. *(BT level 3)*

Solution: It is given that

Arrival rate (λ) = 30 per hour,

Service rate (μ) = 48 per hour.

a) The average number of customers in the system is

$$L = \frac{\lambda}{\mu - \lambda} = \frac{30}{48 - 30} = 1.66 \text{ customers.}$$

b) The average number of customers in the queue is

$$Lq = \frac{\lambda^2}{\mu(\mu - \lambda)} = \frac{(30)^2}{48(48 - 30)} = 1.04 \text{ customers.}$$

c) The average time of customer in the system is

$$W = \frac{1}{\mu - \lambda} = \frac{1}{48 - 30} = 0.055 \text{ hour.}$$

d) The average time a customer waits in queue is

$$Wq = \frac{\lambda}{\mu(\mu - \lambda)} = \frac{30}{48(48 - 30)} = 0.034 \text{ hour.}$$

EXAMPLE 2:

Customers arrive at the hospital at a rate of four per hour (Poisson arrival) and the doctor will serve at a rate of five per hour (exponential). (a) What is the probability of a customer not joining the queue? (b) What is the probability of six customers in the system? (c) What is the average number of customers in the system? (d) What is the average time a customer is waiting in a queue in minutes? *(BT level 3)*

Solution: It is given that

Arrival rate $(\lambda) = 4$ per hour
Service rate $(\mu) = 5$ per hour

a) A customer does not join the queue is when there is no one in the system. The probability is

$$P0 = 1 - \frac{\lambda}{\mu} = 1 - \frac{4}{5} = 0.2.$$

b) The probability that there are six customers in the system is

$$P6 = P0 \left(\frac{\lambda}{\mu} \right)^n = 0.2 \times \left(\frac{4}{5} \right)^6 = 0.0524.$$

c) The average number of customers in the system is

$$W = \frac{\lambda}{\mu - \lambda} = \frac{4}{5 - 4} = 4.$$

d) The average time a customer waits in queue is

$$Wq = \frac{\lambda}{\mu(\mu - \lambda)} = \frac{4}{5(5 - 4)} = 0.8 \text{ hour} = 48 \text{ minutes.}$$

EXAMPLE 3:

At the car wash centre, the worker finds that the time spent at his job has an exponential distribution of 30 minutes. If he washes in the order in which the cars entered, and if the arrival of the sets is approximately Poisson with an average rate of 12 per eight-hour day, what is the expected idle time of the worker each day? How many average numbers of jobs that are ahead of the car just brought in? *(BT level 3)*

Solution: It is given that

Arrival rate $(\lambda) = \dfrac{12}{8}$ arrivals per hour,
Service rate $(\mu) = 2$ cars per hour.

The worker's expected idle time is

$$P0 = 1 - \frac{\lambda}{\mu} = 1 - \frac{\frac{12}{8}}{2} = 1 - \frac{6}{8} = 0.25.$$

Hence, the idle time in the eight-hour day = 8 × 0.25 = 2 hours.

Now, the average number of jobs that are ahead of the car just brought in = Average number of units in the system,

$$W = \frac{\lambda}{\mu - \lambda} = \frac{\frac{12}{8}}{2 - \frac{12}{8}} = 3.$$

EXAMPLE 4:

Customers arrive at a bank teller randomly at an average rate of 35 an hour. If the teller takes each customer an average of one minute to serve, what is the average number of customers in the queue, and how long are they waiting to be served? What if the average service time falls to 1.5 or 2 minutes? *(BT level 3)*

Solution: It is given that

Arrival rate (λ) = 35 arrivals per hour,
Service rate (μ) = 60 customers per hour.
a) The average number of customers in the queue is

$$Lq = \frac{\lambda^2}{\mu(\mu - \lambda)} = \frac{(35)^2}{60(60 - 35)} = 0.816 \text{ customers.}$$

b) The average time a customer waits in queue is

$$Wq = \frac{\lambda}{\mu(\mu - \lambda)} = \frac{35}{60(60 - 35)} = 0.0233 \text{ hour} = 1.4 \text{ minute.}$$

c) With an average service time of 1.5 minute, service rate (μ) = 40 customers per hour.
So the average number of customers in the queue is

$$Lq = \frac{\lambda^2}{\mu(\mu - \lambda)} = \frac{(35)^2}{40(40 - 35)} = 6.12 \text{ customers.}$$

And the average time a customer waits in a queue is

$$Wq = \frac{\lambda}{\mu(\mu - \lambda)} = \frac{35}{40(40 - 35)} = 0.175 \text{ hour} = 10.5 \text{ minutes.}$$

d) The service rate is $\mu = 30$ if the average travel time rises to two minutes. This doesn't satisfy the condition that $\mu > \lambda$, so the system won't settle down to a steady-state, and the queue will continue to grow forever.

EXAMPLE 5:

Vehicles travel through a toll gate at a rate of 75 an hour. The average passage time to the gate is 40 seconds. When the toll gate idle time is less than 10% and the average queue length at the gate is more than six vehicles, are the authorities able to add one more gate to minimize the average time to travel through the toll gate to 30 seconds, test if the installation of the second gate is justified? *(BT level 3 and 5)*
Solution: It is given that

$$\text{Arrival rate}(\lambda) = 75 \text{ arrivals per hour,}$$

$$\text{Service rate}(\mu) = \frac{3600}{40} = 90 \text{ vehicles per hour.}$$

a) The average number of vehicles in the queue is

$$Lq = \frac{\lambda^2}{\mu(\mu - \lambda)} = \frac{(75)^2}{90(90 - 75)} = 4.16 \text{ vehicles.}$$

b) Now, the revised time taken to pass through the gate = 30 seconds,

$$\therefore \text{New service rate}(\mu) = \frac{3600}{30} = 120 \text{ vehicles per hour.}$$

So the average number of vehicles in the queue is

$$Lq = \frac{\lambda^2}{\mu(\mu - \lambda)} = \frac{(75)^2}{120(120 - 75)} = 1.04 \text{ vehicles.}$$

And the percentage of the idle time of the toll gate is:

$$P0 = 1 - \frac{\lambda}{\mu} = 1 - \frac{75}{120} = 0.375 = 37.5\%.$$

This idle time is not under 10%. The installation of the second gate is therefore not justified, as the total number of vehicles waiting in the queue is not more than six and the idle time is not less than 10%.

8.3 Multiple-Server Queuing Model with Poisson Arrivals and Exponential Service Times (*M/M/m*)

A multi-server queuing model consists of two or more servers, which are meant to be similar in terms of service capacity. There are two common queuing options for multiple-server systems: (1) arriving customers wait in a single waiting line (called a "pooled" or "shared" queue) and then move to the first server available for processing, or (2) each server has a "dedicated" queue and an arriving customer chooses one of those lines to join (and usually is not permitted to switch lines). We focus here on the system design for all servers with a single shared waiting line. The operating characteristics of a multiple-server system are generally better when a single shared queue is used rather than multiple dedicated queues. Examples of this type of waiting line include an airline ticket and check-in desk, where passengers line up in a single line, waiting for one of several service agents, and a post office line, where customers in a single line are waiting for service from a number of postal clerks. Figure 8.2 shows a two-server waiting line model.

In this section, we present formulas that can be used to evaluate the operating characteristics of a multiple-server waiting line for the steady-state. These formulas are valid, assuming the following:

1. Arrivals follow Poisson's probability distribution.
2. For each server, the service time follows an exponential distribution.
3. The service rate μ for each server is the same.
4. The arrivals wait in a single waiting line and then move for service to the first available server.

The parameters of the multiple-server model are as follows:

λ = the arrival rate (average number of arrivals per time period),

μ = the service rate (average number served per time period) per server,

m = the number of servers.

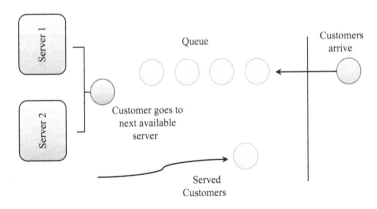

FIGURE 8.2
Two-Server Waiting Line Model.

Since μ is the server service rate, $m\mu$ is the multi-server system service rate. The formulas for the operating characteristics of multiple-server waiting lines can be used, as in the case of the one-server waiting-line model, only in cases where the system's service rate exceeds the system's arrival rate; in other words, it exists only when $m\mu$ is more than λ.

LO 5 Understand and calculate the characteristics of a multi-server queue

The probability that there are no customers in the system (all servers are idle) is

$$PO = \frac{1}{\left[\sum_{n=0}^{n=m-1} \frac{1}{n!}\left(\frac{\lambda}{\mu}\right)^n\right] + \frac{1}{m!}\left(\frac{\lambda}{\mu}\right)^m \left(\frac{m\mu}{m\mu-\lambda}\right)}. \tag{8.9}$$

The average number of customers or units in the system:

$$L = \frac{\lambda\mu\left(\frac{\lambda}{\mu}\right)^m}{(m-1)!(m\mu-\lambda)^2} P_0 + \frac{\lambda}{\mu}. \tag{8.10}$$

The average number of customers in the queue is

$$Lq = L - \frac{\lambda}{\mu}. \tag{8.11}$$

The average time a customer spends in the queuing system (waiting and being served) is

$$W = \frac{L}{\lambda}. \tag{8.12}$$

The average time a customer spends in the queue waiting to be served is

$$Wq = W - \frac{1}{\mu} = \frac{L_q}{\lambda}. \tag{8.13}$$

Obviously, these equations are more complicated than those used in the single-server model, but they are used in almost the same way to provide the same information type as the simpler model.

Examples based on the multiple-server queuing model

EXAMPLE 1:

Customers arrive randomly at a service centre at an average rate of 25 an hour. If there are three servers with an average service rate of each server is ten per hour, find the probability that there are no customers in the system. *(BT level 3)*

Solution: It is given that

Arrival rate (λ) = 25 arrivals per hour,

Service rate of each server (μ) =10 customers per hour,

Number of servers $m = 3$.

The probability that there are no customers in the system (all servers are idle) is

$$PO = \frac{1}{\left[\sum_{n=0}^{n=m-1} \frac{1}{n!}\left(\frac{\lambda}{\mu}\right)^n\right] + \frac{1}{m!}\left(\frac{\lambda}{\mu}\right)^m\left(\frac{m\mu}{m\mu - \lambda}\right)}$$

$$PO = \frac{1}{\left[\frac{1}{0!}\left(\frac{25}{10}\right)^0 + \frac{1}{1!}\left(\frac{25}{10}\right)^1 + \frac{1}{2!}\left(\frac{25}{10}\right)^2\right] + \frac{1}{3!}\left(\frac{25}{10}\right)^3\left(\frac{3\times10}{3\times10 - 25}\right)} = 0.0449.$$

The system will be completely idle for about 4.5% of the time.

EXAMPLE 2:

According to a Poisson distribution, customers wishing to open new accounts arrive at the bank at the rate of five an hour and each customer is spent an average of ten minutes by the account officer setting up a new account. There are two account officers. Determine (a) the probability that there are no customers in the system, (b) the average number of customers in the system, (c) the average number of customers in the queue, (d) the average time spent by the customer in the queue system, and (e) the average time spent by the customer in the queue. *(BT level 3)*

Solution: It is given that

Arrival rate (λ) = 5 arrivals per hour,

Service rate of each server (μ) = 6 customers per hour,

Number of servers $m = 2$.

a) The probability that there are no customers in the system (all servers are idle) is

$$PO = \frac{1}{\left[\sum_{n=0}^{n=m-1} \frac{1}{n!}\left(\frac{\lambda}{\mu}\right)^n\right] + \frac{1}{m!}\left(\frac{\lambda}{\mu}\right)^m\left(\frac{m\mu}{m\mu - \lambda}\right)}$$

$$PO = \frac{1}{\left[\frac{1}{0!}\left(\frac{5}{6}\right)^0 + \frac{1}{1!}\left(\frac{5}{6}\right)^1\right] + \frac{1}{2!}\left(\frac{5}{6}\right)^2\left(\frac{2\times6}{2\times6 - 5}\right)} = 0.4118.$$

b) The average number of customers in the system:

$$L = \frac{\lambda \mu \left(\dfrac{\lambda}{\mu}\right)^m}{(m-1)!\,(m\mu - \lambda)^2} P_0 + \frac{\lambda}{\mu}$$

$$= \frac{5 \times 6 \left(\dfrac{5}{6}\right)^2}{(2-1)!\,(2 \times 6 - 5)^2} \times 0.4118 + \frac{5}{6} = 1.00.$$

c) The average number of customers in the queue is

$$Lq = L - \frac{\lambda}{\mu}$$

$$= 1 - \frac{5}{6} = 0.17.$$

d) The average time a customer spends in the queuing system (waiting and being served) is

$$W = \frac{L}{\lambda}$$

$$= \frac{1}{5} = 0.20\,\text{hour}.$$

e) The average time a customer spends in the queue waiting to be served is

$$Wq = \frac{L_q}{\lambda} = \frac{0.17}{5} = 0.034\,\text{hour}.$$

EXAMPLE 3:

Customers arrive at a game-playing centre at the rate of eight per hour with Poisson arrival. There are two game-playing stations and the average time taken by a person is ten minutes at each station with exponentially distributed. Find (a) the probability that both game-playing stations are free when a customer arrives, (b) the probability that a customer can use a game-playing station immediately on arrival, (c) the probability that there is no queue on arrival, (d) the average number of customers in the system, and (e) average waiting time in the queue. **(BT level 3)**

Solution: It is given that

Arrival rate $(\lambda) = 8$ arrivals per hour,

Service rate of each server (μ) = 6 customers per hour,
Number of servers $m = 2$.

a) The probability that both game-playing stations are free when a customer arrives:

$$PO = \cfrac{1}{\left[\displaystyle\sum_{n=0}^{n=m-1} \frac{1}{n!}\left(\frac{\lambda}{\mu}\right)^{n}\right] + \frac{1}{m!}\left(\frac{\lambda}{\mu}\right)^{m}\left(\frac{m\mu}{m\mu - \lambda}\right)}$$

$$PO = \cfrac{1}{\left[\dfrac{1}{0!}\left(\dfrac{8}{6}\right)^{0} + \dfrac{1}{1!}\left(\dfrac{8}{6}\right)^{1}\right] + \dfrac{1}{2!}\left(\dfrac{8}{6}\right)^{2}\left(\dfrac{2\times 6}{2\times 6 - 8}\right)} = 0.2.$$

b) A game-playing station is free when there is no customer in the system or when there is one customer in the system.

∴ The probability that the customer can use a game-playing station immediately on arrival is:

$$= PO + P1 = PO + PO\left(\frac{\lambda}{\mu}\right)$$

$$= 0.2 + 0.2\left(\frac{8}{6}\right) = 0.467.$$

c) The probability that there is no queue on arrival is the probability of having no customer or one customer or two customers in the system.

$$= PO + P1 + P2 = PO + PO\left(\frac{\lambda}{\mu}\right) + PO\frac{\left(\frac{\lambda}{\mu}\right)^{2}}{2}$$

$$= 0.2 + 0.2\left(\frac{8}{6}\right) + 0.2\left(\frac{\left(\frac{8}{6}\right)^{2}}{2}\right) = 0.644$$

d) The average number of customers in the system is

$$L = \frac{\lambda\mu\left(\frac{\lambda}{\mu}\right)^{m}}{(m-1)!\left(m\mu-\lambda\right)^{2}}P_0 + \frac{\lambda}{\mu}$$

$$= \frac{8\times6\left(\frac{8}{6}\right)^{2}}{(2-1)!\left(2\times6-8\right)^{2}}\times0.2 + \frac{8}{6} = 2.4.$$

e) Average waiting time in the queue is

$$Wq = \frac{L_q}{\lambda},$$

$$Lq = L - \frac{\lambda}{\mu} = 2.4 - \frac{8}{6} = 1.06,$$

$$\therefore Wq = \frac{1.06}{8} = 0.133 \text{ hours.}$$

EXAMPLE 4:

In a single-server queuing system, customers arrive at a rate of ten per hour and the average service time is 16 per hour. Customers arrive at the rate of 20 per hour in a two-server queuing system and the mean service time is 16 per hour. Compare the performance of both systems. *(BT level 3 and 4)*

Solution: For single-server queuing system it is given that

Arrival rate (λ) = 10 arrivals per hour,
Service rate of each server (μ) = 16 customers per hour.
a) The average number of customers in the system is

$$L = \frac{\lambda}{\mu-\lambda} = \frac{10}{16-10} = 1.66 \text{ customers.}$$

b) The average number of customers in the queue is

$$Lq = \frac{\lambda^{2}}{\mu(\mu-\lambda)} = \frac{(10)^{2}}{16(16-10)} = 1.04 \text{ customers.}$$

c) The average time of customer in the system is

$$W = \frac{1}{\mu - \lambda} = \frac{1}{16 - 10} = 0.166 \text{ hour.}$$

d) The average time a customer waits in queue is

$$Wq = \frac{\lambda}{\mu(\mu - \lambda)} = \frac{10}{16(16 - 10)} = 0.104 \text{ hour.}$$

For two-server queuing system it is given that
Arrival rate (λ) = 20 arrivals per hour,
Service rate of each server (μ) = 16 customers per hour,
Number of servers $m = 2$.
The probability that there are no customers in the system (all servers are idle) is

$$PO = \frac{1}{\left[\sum_{n=0}^{n=m-1} \frac{1}{n!} \left(\frac{\lambda}{\mu} \right)^n \right] + \frac{1}{m!} \left(\frac{\lambda}{\mu} \right)^m \left(\frac{m\mu}{m\mu - \lambda} \right)}$$

$$PO = \frac{1}{\left[\frac{1}{0!} \left(\frac{20}{16} \right)^0 + \frac{1}{1!} \left(\frac{20}{16} \right)^1 \right] + \frac{1}{2!} \left(\frac{20}{16} \right)^2 \left(\frac{2 \times 16}{2 \times 16 - 20} \right)} = 0.2308.$$

a) The average number of customers in the system:

$$L = \frac{\lambda \mu \left(\frac{\lambda}{\mu} \right)^m}{(m-1)!(m\mu - \lambda)^2} P_0 + \frac{\lambda}{\mu}$$

$$= \frac{20 \times 16 \left(\frac{20}{16} \right)^2}{(2-1)!(2 \times 16 - 20)^2} \times 0.2308 + \frac{20}{16} = 2.051.$$

b) The average number of customers in the queue is

$$Lq = L - \frac{\lambda}{\mu}$$

$$= 2.051 - \frac{20}{16} = 0.801.$$

c) The average time a customer spends in the queuing system (waiting and being served) is

$$W = \frac{L}{\lambda}$$

$$= \frac{2.051}{20} = 0.103 \, \text{hour}.$$

d) The average time a customer spends in the queue waiting to be served is

$$Wq = \frac{L_q}{\lambda} = \frac{0.801}{20} = 0.04 \, \text{hour}.$$

The comparison of the two models is given in Table 8.1.

TABLE 8.1

Comparison of the Two Models

Performance Measure	Single-Server Model	Two-Server Model
Average number of customers in the system	1.66	2.05
Average number of customers in the queue	1.04	0.801
Average time a customer spends in the system	0.166 hour	0.103 hour
Average time a customer spends in the queue	0.104 hour	0.04 hour

From Table 8.1 by providing two servers the queue length reduced from 1.04 to 0.801. The number of customers in the system is higher in two-server queuing system. The time in the second system is lesser because of two servers.

8.4 Little's Relationships for Queuing Models

For single-server and multi-server queue models, we presented formulas for calculating operative characteristics with Poisson arrivals and exponential service times in Sections 8.2 and 8.3. For any queuing system in a steady-state, there are certain relationships among different operating characteristics. A steady-state occurs when a queuing system is in its normal stabilized operating condition, usually after an initial or transient state (for example, if a business begins in the morning, customers wait at the door). In this state, both the rate of arrival and the rate of service should be constant.

John D. C. Little showed that there are many relationships between different characteristics and that they extend to various queuing systems. Two of the relationships, called Little's equations of flow, are

$$L = \lambda W, \tag{8.14}$$

$$L_q = \lambda W_q. \tag{8.15}$$

Equation (8.14) shows that the average number of units in the system, L, can be found by multiplying the arrival rate, λ, by the average time a unit spends in the system, W. Equation (8.15) shows that the same relationship holds between the average number of units in the waiting line, L_q, and the average time a unit spends in the waiting line, W_q.

Another general expression that applies to waiting-line models is that the average time in the system, W, is equal to the average time in the waiting line, W_q, plus the average service time.

$$W = Wq + \frac{1}{\mu} \tag{8.16}$$

The advantage of these formulas is that once one of these four characteristics is known, it is easy to find the other characteristics. This is important because one of these may be much easier to determine for certain queuing models than the other. These are applicable to all queuing systems, except the finite population model.

CHAPTER SUMMARY

- Queuing systems are important parts of the business world.

- If the customer enters a server and finds that the server is already busy, a queue is generated.

- A single-server with random arrival and service times is the simplest queue.

- Key operating characteristics for a system are shown to be (1) utilization rate, (2) percent idle time, (3) average time spent waiting in the system and the queue, (4) an average number of customers in the system and the queue, and (5) probabilities of various numbers of customers in the system.

Questions

1. What is the queuing system? What are the components in a queuing system? *(BT level 2)*
2. What are the assumptions behind common queuing models? *(BT level 2)*
3. Provide an example of where it would not be acceptable to provide a first-in, first-out rule for queue discipline. *(BT level 2 and 3)*
4. Describe the important operating characteristics of a queuing system. *(BT level 2)*

5. What is Kendall's notation? Classify the queuing system based on Kendall notation. *(BT level 2)*

6. "Basic problem in queuing theory is to determine an optimal service level." Elucidate this statement. *(BT level 4)*

7. For each of the following queuing systems, indicate if it is a single- or multiple-server model, the queue discipline, or its calling population is infinite or finite. *(BT level 4)*
   a. Hair salon b. Bank c. Laundromat d. Doctor's office

Practice Problem

1. The bank estimates that the arrival rate during this period will be Poisson distributed with an average of four customers per hour. The service time for opening a new account is spread exponentially with an average of 12 minutes per customer. The bank needs to determine the operating characteristics of this system and decide if the current person is adequate and manage the increased traffic. *(BT level 3)*

2. The bank needs to measure the operational characteristics in practice question 1 if an additional employee has been hired to assist with new account enrolments. *(BT level 4)*

3. Requests for assistance shall be received from the reference desk of the university library. Assume that the Poisson probability distribution with an arrival rate of five requests per hour can be used to describe the arrival pattern and that the service times follow the exponential probability distribution with a service rate of six requests per hour. *(BT level 3)*
   a.   What is the probability there will be no requests for assistance in the system?
   b.   What is the average number of requests that await service?
   c.   What is the average time to wait in minutes before the service starts?
   d.   What is the average time at the reference desk in minutes (waiting time plus service time)?
   e.   What is the probability that a new arrival will have to wait for the service?

4. A single-server automotive oil change and lubrication service are offered by ABC Oil. Customers are provided with an arrival rate of five cars per hour. The service rate is ten cars per hour. Assume that arrivals follow a distribution of Poisson probability and that service times follow an exponential distribution of probability. *(BT level 3)*
   a.   What is the mean number of cars in the system?
   b.   What is the average time a car is waiting for the oil and lubrication service to start?
   c.   What is the average amount of time a car spends in the system?
   d.   What is the probability an arrival would have to wait for service?

5. The clinic has two general practitioners who see patients every day. Averages of seven patients arrive at the clinic per hour (Poisson distributed). Each physician spends an average of 15 minutes (exponentially distributed) with the patient. In a waiting area, the patients wait until one of the two doctors can see them. However, since patients usually don't feel well when they come to the clinic, doctors don't think it is a good practice for a patient to wait longer than 20 minutes on average. Is this clinic supposed to add a third doctor, and if so, is this going to ease the waiting problem? *(BT level 5)*

6. A small local supermarket with a single checkout counter. Assume shoppers arrive at the checkout lane according to a Poisson distribution of probability, with an arrival rate of 20 customers an hour. Checkout times are followed by an exponential probability distribution, with a service rate of 25 customers per hour. *(BT level 4)*

   a. Calculate the operating characteristics of the waiting line.

   b. If the service goal of the manager is to reduce the waiting period until the checkout process starts to no more than five minutes, what suggestions will you make with respect to the existing checkout system?

7. After reviewing the problem six waiting-line analysis, the owner of the store would like to consider one of the following alternatives for improving the service. Which alternative would you suggest? Justify your suggestion. *(BT level 5)*

   c. Employ a second person to bag the groceries while the operator of the cash register enters the cost data and collects money from the customer. This increased single-server operation could increase the service rate to 35 customers an hour.

   d. Employ a second person to run a second counter for the checkout. The two-server system will give each served a service rate of 25 customers an hour.

9

Dynamic and Integer Programming

Learning Outcomes: *After studying this chapter, the reader should be able to*

LO 1. *Explain how dynamic programming is important in decision making. (BT level 2)*

LO 2. *Understand the terms used in dynamic programming. (BT level 2)*

LO 3. *Solve problems using dynamic programming. (BT level 3)*

LO 4. *Understand integer programming and differentiate between linear and integer programming. (BT level 2)*

LO 5. *Formulate the integer programming problem. (BT level 6)*

LO 6. *Understand and apply the branch-and-bound method for solving the integer programming problem. (BT level 2 and 3)*

LO 7. *Understand and apply the cutting plane method for solving the integer programming problem. (BT level 2 and 3)*

9.1 Introduction to Dynamic Programming

LO 1. Explain how dynamic programming is important in decision making

Dynamic programming is a technique for formulating problems where decisions are to be taken in stages – a problem of multi-stage decision making. A multi-stage decision problem is a problem in which the decision has to be taken in successive stages. In this case, the problem solver will make a decision at all stages, so that the overall efficiency of all stages is optimal. Here the original problem is broken down or decomposed into small problems, known as sub-problems or stages that are very convenient for handling and finding the optimal stage. It is a systemic procedure in which the optimal combination of decisions is determined. The dynamic programming technique was developed by Richard Bellman in the early 1950s.

Unlike linear programming, there is no standard mathematical formulation of the dynamic programming problem. Instead, dynamic programming is a general way of solving problems, and specific equations should be developed to fit every situation. Mathematically, a dynamic programming problem is a decision-making problem in n variables, the problem is subdivided into n sub-problems, and each problem is a decision-making problem in one variable only. The solution to the dynamic programming problem is achieved sequentially from one (initial) stage to the next until the final stage is reached. Dynamic programming can be referred to as recursive optimization.

9.2 Terms Used in Dynamic Programming

LO 2. Understand the terms used in dynamic programming

The following are some terms that we very often come across in dynamic programming.

Stage: Stage means a part of the total problem for which a decision can be taken. At each stage, there are a number of options for decision making, the best of which is called stage decision making, which may not be optimal for the stage but which contributes to the accomplishment of an optimal decision-making policy.

State: The condition of the decision-making process at a stage is referred to as its state. The variables that specify the condition of the decision process (i.e., the status of the system at a particular stage) are called state variables. The number of state variables should be as small as possible as the decision process is more complicated if there is a greater number of state variables.

Principle of Optimality: Bellman's principle of optimality states, "An optimal policy (a sequence of decisions) has the property that whatever the initial state and decision are, the remaining decisions must constitute an optimal policy with regard to the state resulting from the first decision."

This implies that the optimal policy for subsequent stages does not depend on the policy adopted at the previous stages given the initial state of the system.

The steps for getting the solution for the dynamic programming problem are as follows:

LO 3. Solve problems using dynamic programming

- The mathematical formulation of the problem and the writing of the recursive equation (recursive relationship linking the optimal decision-making function for the 'n' stage problem to the optimal decision-making function for the $(n - 1)$ stage sub-problem).

- To write and solve the relationship giving the optimal decision function for a single-stage sub-problem.

- To solve the optimal decision-making function for 2-stage, 3-stage .... $(n - 1)$ stage problem and then an n-stage problem.

In general, the recursive relationship can be formulated and stated as follows. Let

N = number of stages,

n = current stage (n = 1, 2, ..., N),

$n + 1$ = previous stage,

S_n = state of the system in the current stage for which the recursive relationship holds,

S_{n+1} = state in the previous stage,

$f_n(S_n)$ = total payoff for each alternative,

$f^*_n(S_n)$ = optimal total payoff,

$f^*_{n+1}(S_{n+1})$ = optimal total payoff obtained in the previous stage,

x_n= decision among alternatives made at stage n in the state under consideration, and

$C_n(S_n, x_n)$ = immediate payoff in stage n when decision x_n is made for a specific value S_n of the state variable.

The recursive relationship for minimization for state S_n at stage n is given as follow:

$$f^*_n(Sn) = \min_{x_n} \left\{ C_n(S_n, x_n) + f^*_{n+1}(S_{n+1}) \right\}. \tag{9.1}$$

The recursive relationship for maximization for state S_n at stage n is given as follow:

$$f^*_n(Sn) = \max_{x_n} \left\{ C_n(S_n, x_n) + f^*_{n+1}(S_{n+1}) \right\}. \tag{9.2}$$

9.3 Characteristics of Dynamic Programming

The important characteristics of dynamic programming which differentiate it from other quantitative decision-making techniques can be summarized as follows.

1. The problem can be broken down into stages where a decision is needed at each stage.
2. There are a number of associated states in each stage. By a state we mean the information needed to make an optimal decision at any stage.
3. The decision chosen at any stage describes how the state at the present stage is transformed into a state at the next stage.
4. In view of the current state, the optimal decision for each of the remaining stages must not depend on previously reached states or previously chosen decisions. This idea is referred to as the principle of optimality.
5. The solution procedure begins with the finding of an optimal policy for the last stage.
6. To identify the optimal policy for each state of the system, a recursive relationship is formulated with the remaining n stages, given the optimal policy for each state with $(n - 1)$ stages left.
7. Using this recursive relationship, the solution procedure starts at the end and moves stage by stage backward—each time finding the optimal policy for that stage—until the optimal policy starts at the initial stage.

Examples based on dynamic programming

EXAMPLE 1:

The manufacturing company has three sections producing three components: $P1$, $P2$, and $P3$. The management allocated Rs. 40,000 to the expansion of the production facilities. For $P1$ and $P2$, production can be increased either by adding new machines or by replacing old, inefficient machines with automatic machines. As the production of $P3$ started only a few years ago, the additional amount can only be invested by adding new machines. Costs for the addition and replacement of the machines, along with the associated expected returns for the production of the components set out in Table 9.1. Select a set of expansion plans that may yield the maximum return. *(BT level 3)*

TABLE 9.1

Data for Example 1

Alternative	Section Producing $P1$		Section Producing $P2$		Section Producing $P3$	
	Cost	Return	Cost	Return	Cost	Return
1. No expansion	0	0	0	0	0	0
2. Add new machine	5,000	10,000	10,000	15,000	5,000	10,000
3. Replace old machine	10,000	15,000	15,000	20,000	–	–

Solution:
Each section of the company is a stage here. The number of stages is therefore equal to three. There are a number of alternatives for expansion at each stage. Let capital be the state variable.

Stage 1. The following is provided the recursive function for the combination of the first stage state variable, $S1$, and alternative $x1$. The corresponding returns are summarized in Table 9.2. Optimal return and the corresponding alternative are shown in the last two columns.

Stage 2. Now let's move on to stage 2. Here, again, there are three alternatives available. The recursive function for a given combination of state variables, $S2$, and alternative x_2 in the second stage is presented next. The calculations are made in Table 9.3.

Here, state $S2$ represents the total amount allocated to the current stage and the preceding stage. Likewise, the return is also the sum of the current stage and the preceding stage.

Stage 3. Now let's move on to stage 3. There are two alternatives available. The recursive function for the combination of the state variable $S3$ and the alternative x_3 in the third stage is presented below. The calculations are made in Table 9.4.

For $S_3 = 40,000$, the optimal decision for stage 3 is alternative 2, which gives a total return of Rs. 45,000. This involves a cost of Rs. 5,000 and leaves Rs. 35,000 to be allotted for stage 2 and 1 combined. From Table 9.3, for allocation of Rs. 35,000, alternative 3 is to be

TABLE 9.2

Optimal Return

State Variable S_1	1. No Expansion Cost = 0 Return	2. Add New Machine Cost = 5,000 Return	3. Replace Old Machine Cost = 10,000 Return	Optimal Return	Decision
0	0	–	–	0	1
5,000	0	10,000	–	10,000	2
10,000	0	10,000	15,000	15,000	3
15,000	0	10,000	15,000	15,000	3
20,000	0	10,000	15,000	15,000	3
25,000	0	10,000	15,000	15,000	3
30,000	0	10,000	15,000	15,000	3
35,000	0	10,000	15,000	15,000	3
40,000	0	10,000	15,000	15,000	3

TABLE 9.3

Optimal Return

State variable S_2	1. No Expansion Cost = 0 Return	2. Add New Machine Cost = 10,000 Return	3. Replace Old Machine Cost = 15,000 Return	Optimal Return	Decision
0	0 + 0	–	–	0	1
5,000	0 + 10,000 = 10,000	–	–	10,000	1
10,000	0 + 15,000 = 15,000	15,000 + 0 = 15,000	–	15,000	1 and 2
15,000	0 + 15,000 = 15,000	15,000 + 10,000 = 25,000	20,000 + 0 = 20,000	25,000	2
20,000	0 + 15,000 = 15,000	15,000 + 15,000 = 30, 000	20,000 + 10,000 = 30,000	30,000	2 and 3
25,000	0 + 15,000 = 15,000	15,000 + 15,000 = 30, 000	20,000 + 15,000 = 35,000	35,000	3
30,000	0 + 15,000 = 15,000	15,000 + 15,000 = 30, 000	20,000 + 15,000 = 35,000	35,000	3
35,000	0 + 15,000 = 15,000	15,000 + 15,000 = 30, 000	20,000 + 15,000 = 35,000	35,000	3
40,000	0 + 15,000 = 15,000	15,000 + 15,000 = 30, 000	20,000 + 15,000 = 35,000	35,000	3

chosen which costs Rs. 15,000. For remaining Rs. 20,000, from Table 9.2, decision alternative 3 is to be selected. The optimum expansion of the production facility is, therefore, 3-3-2, which can be elaborated as follows: replace the old machine with automatics in the $P1$ component production section, replace the old machine with automatics in the $P2$ component production section, and add a new machine to the $P3$ component production section. This policy gives an optimum return of Rs. 45.000.

TABLE 9.4

Optimal Return

State variable S_3	1. No Expansion	2. Add New Machine	Optimal Return	Decision
	Cost = 0 Return	Cost = 5,000 Return		
0	0 + 0	–	0	1
5,000	0 + 10,000 = 10,000	10, 000 + 0 = 10,000	10,000	1 and 2
10,000	0 + 15,000 = 15,000	10,000 + 10,000 = 20,000	20,000	2
15,000	0 + 25,000 = 25,000	10,000 + 15,000 = 25,000	25,000	1 and 2
20,000	0 + 30,000 = 30,000	10,000 + 25,000 = 35,000	35,000	2
25,000	0 + 35,000 = 35,000	10,000 + 30,000 = 40,000	40,000	2
30,000	0 + 35,000 = 35,000	10,000 + 35,000 = 45,000	45,000	2
35,000	0 + 35,000 = 35,000	10,000 + 35,000 = 45,000	45,000	2
40,000	0 + 35,000 = 35,000	10,000 + 35,000 = 45,000	45,000	2

EXAMPLE 2:

The Company has nine salesmen who have to be assigned to three marketing zones. The return of profits from each zone depends on the number of salesmen working in that zone. The expected returns for the different number of salesmen in different zones, as estimated from past records, are shown in Table 9.5. Determine an optimum allocation policy. *(BT level 3)*

TABLE 9.5

Data for Example 2

No. of Salesmen	Zone 1	Zone 2	Zone 3
0	20,000	25,000	32,000
1	35,000	35,000	44,000
2	50,000	42,000	50,000
3	60,000	54,000	60,000
4	69,000	62,000	72,000
5	80,000	72,000	85,000
6	88,000	83,000	92,000
7	95,000	88,000	100,000
8	90,000	90,000	100,000
9	80,000	90,000	100,000

Solution:

In this problem, each zone can be considered as a stage, the number of salesmen at each stage as decision variables. The status variable of the problem is the number of salesmen who can be assigned at a stage.

Stage 1. The return of profit corresponding to a different number of salesmen allocated to zone 1 are given in Table 9.6.

TABLE 9.6

Profit in Zone 1

No. of Salesmen	Zone 1
0	20,000
1	35,000
2	50,000
3	60,000
4	69,000
5	80,000
6	88,000
7	95,000
8	90,000
9	80,000

Stage 2. Consider zone 1 and zone 2. Nine salesmen can be divided between two zones in ten different ways, such as 9 in zone 1 and 0 in zone 2, 8 in zone 1 and 1 in zone 2, 7 in zone 1 and 2, and so on. Returns for all salesmen are shown in Table 9.7. The profit for all possible combinations can be read along the diagonal for a particular number of salesmen. Maximum profits are marked with *.

The outcomes from Table 9.7 are shown in Table 9.8.

TABLE 9.7

Return for All Salesmen

Zone 1		0	1	2	3	4	5	6	7	8	9
Return (1,000 of Rs.)		20	35	50	60	69	80	88	95	90	80
Zone 2	Return (1,000 of Rs.)										
0	25	45*	60*	75*	85*	94	105*	113	120	115	105
1	35	55	70	85*	95*	104	115*	123*	130	125	
2	42	62	77	92	102	111	122	130	137		
3	54	74	89	104	114	123*	134*	142			
4	62	82	97	112	122	131	142				
5	72	92	107	122	132	141					
6	83	103	118	133	143*						
7	88	108	123	138							
8	90	110	125								
9	90	110									

TABLE 9.8

Optimal Outcome

No. of Salesmen	Optimal Outcome in 1,000 of Rupees	No. of Salesmen in Zone 1	No. of Salesmen in Zone 2
0	45	0	0
1	60	1	0
2	75	2	0
3	85	3	0
		2	1
4	95	3	1
5	105	5	0
6	115	5	1
7	123	6	1
		4	3
8	134	5	3
9	143	3	6

Stage 3. Consider now the distribution of nine salesmen across three zones 1, 2, and 3. The decision at this stage will result in the allocation of a certain number of salesmen to zone 3 and the remaining to zone 2 and 1 combined. Returns for all salesmen are shown in Table 9.9. The profit for all possible combinations can be read along the diagonal for a particular number of salesmen. Maximum profits are marked with *.

TABLE 9.9

Returns for All Salesmen

Salesmen		0	1	2	3	4	5	6	7	8	9
Zone 1 + Zone 2 (From Table 9.8)		0+0	1+0	2+0	3+0 2+1	3+1	5+0	5+1	6+1 4+3	5+3	5+6
Total Return (1,000 of Rs.)		45z	60	75	85	95	105	115	123	134	143
Zone 3	Return (1,000 of Rs.)										
0	32	77*	92*	107*	117	127	137	147	155	166	175
1	44	89	104	119*	129*	139*	149*	159	167	178	
2	50	95	110	125	135	145	155	165	173		
3	60	105	120	135	145	155	165	175			
4	72	117	132	147	157	167	177				
5	85	130	145	160*	170*	180*					
6	92	137	152	167	177						
7	100	145	160	175							
8	100	145	160								
9	100	145									

The outcomes from Table 9.9 are shown in Table 9.10.

TABLE 9.10

Optimal Outcome

No. of Salesmen	Optimal Outcome in 1,000 of Rupees	No. of Salesmen in Zone 1	No. of Salesmen in Zone 2	No. of Salesmen in Zone 3
0	77	0	0	0
1	92	1	0	0
2	107	2	0	0
3	119	2	0	1
4	129	3	0	1
		2	1	1
5	139	3	1	1
6	149	5	0	1
7	160	2	0	5
8	170	3	0	5
		2	1	5
9	180	3	1	5

From Table 9.10, the maximum profit for nine salesmen is Rs. 1,80,000 if five salesmen are allotted to zone 3, and from the remaining 4, 1 is allotted to zone 2 and 3 to zone 1.

EXAMPLE 3:

Three types of fruit cartons, mango cartons, grapes cartons, and strawberry cartons, are to be loaded into a tri-wheeler. The weight of each mango carton is 100 kg and the owner of the truck will be paid Rs.20 per mango carton for transport. The weight of each carton of grapes is 200 kg and the owner of the cart will be paid Rs.50 for each carton of grapes. The weight of each strawberry carton is 400 kg and the owner of the truck is paid Rs. 80 for each pineapple carton. The truck can carry a maximum weight of 500 kg. How many cartons of mangoes, grapes, and strawberries should the owner of the tri-wheeler be loaded to make the maximum profit? *(BT level 5)*

TABLE 9.11

Summary for Example 3

Fruit	Weight of Carton (kg)	Charge/Carton (Rs.)
Mangos	100	20
Grapes	200	50
Strawberry	400	80

Solution: The information is summarized in Table 9.11.

TABLE 9.12

Optimal Return

State Variable S_1	0	1	2	3	4	5	Optimal Return	Decision
0	0						0	0
100	0	20					20	1
200	0	20	40				40	2
300	0	20	40	60			60	3
400	0	20	40	60	80		80	4
500	0	20	40	60	80	100	100	5

It's a three-stage problem. Let x_1, x_2, and x_3 be the fruit carton loaded then we have to maximize the sum of $x_i v_i$. Where $x_i v_i$ denotes the obtained excepted value. First, we load the mangoes, then the grapes, followed by the strawberry. After we finish the formulation, you will see that all possible combinations for loading the three fruits are taken into consideration. So it is not important to load the fruits in order.

Stage 1   The largest value of $x1$ is $\dfrac{Maximum\ capacity\ of\ the\ truck}{Weight\ of\ carton\ of\ mangos} = \dfrac{500}{100} = 5$

The recursive function for the combination of the state variable, $S1$, and alternative x_1 in the first stage is presented below. The corresponding profits are summarized in Table 9.12. Optimal profits and the corresponding alternative are shown in the last two columns.

Stage 2

The largest value of $x2$ is $\dfrac{Maximum\ capacity\ of\ the\ truck}{Weight\ of\ carton\ of\ grapes} = \dfrac{500}{200} = 2.5 = 2\,(\text{integral value})$

The recursive function for the combination of the state variable, $S2$, and alternative x_2 in the first stage is presented below. The corresponding profits are summarized in Table 9.13. Optimal profits and the corresponding alternative are shown in the last two columns.

Stage 3

The largest value of $x2$ is $\dfrac{Maximum\ capacity\ of\ the\ truck}{Weight\ of\ carton\ of\ strawberry} = \dfrac{500}{400} = 1.25 = 1\,(\text{integral value})$

The recursive function for the combination of the state variable, $S3$, and alternative x_3 in the first stage is presented next. The corresponding profits are summarized in Table 9.14. Optimal profits and the corresponding alternative are shown in the last two columns.

For $S_3 = 500$ kg, the optimal decision for stage 3 is alternative 0, which gives a total profit of Rs. 120. This involves 0 cartons of strawberry and leaves 500 kg to be considered for stage 2 and 1 combined. From Table 9.13, the optimal decision for stage 2 is alternative 2 which gives a total profit of Rs. 120. This involves 2 cartons of grapes which are 400 kg loading and leaves 100 kg to be considered for stage 1. For the remaining 100 kg, from

TABLE 9.13

Optimal Return

State Variable S_2	0	1	2	Optimal Return	Decision
0	0+0 = 0			0	0
100	0+20 = 20			20	0
200	0+40 = 40	50+0 = 50		50	1
300	0+60 = 60	50+20 = 70		70	1
400	0+80 = 80	50+40 = 90	100+0 = 100	100	2
500	0+100 = 100	50+60 = 110	100+20 = 120	120	2

TABLE 9.14

Optimal Return

State Variable S_3	0	1	Optimal Return	Decision
0	0+0 = 0		0	0
100	0+20 = 20		20	0
200	0+50 = 50		50	0
300	0+70 = 70		70	0 and 1
400	0+100 = 100	80+0 = 80	100	1
500	0+120 = 120	80+20 = 100	120	0

Table 9.12 decision alternative 1 is to be selected. Thus, the optimal policy of loading the fruit carton in the tri-wheeler is 0 – 2 – 1, which can be elaborated as: no strawberry cartons, two grape cartons, and one mango carton. This policy provides optimal profits for Rs. 120.

EXAMPLE 4:

A manufacturer has entered into a contract for the supply of the following number of units (Table 9.15) of a product at the end of each month.

Stage 5: Fifth month: To save the carrying cost, nothing should have been left at the end of the fourth month, and nothing should be left at the end of the fifth month, as this is the last month.

Produce 600 units for which the set-up cost is Rs. 4,000/- and no inventory carrying cost. Hence the total cost is Rs. 4,000/-.

Stage 4: Fourth month: There are two alternatives.

First: Produce 900 (300 + 600) units to satisfy demand of fourth and fifth month. Hence the total cost = Set-up cost + inventory carrying cost for one month = Rs. 4,000 + Rs. 2,400/- = Rs. 6,400/-.

Second: Produce 300 units in the fourth month and 600 units in the fifth month so that there will be no inventory carrying cost. We have only two set-up costs *i.e.* Rs. 4,000 + Rs. 4,000 = Rs. 8,000/-.

TABLE 9.15

Data for Example 4

Month	No. of Units
1	1,000
2	500
3	2,000
4	300
5	600

The cost of the set-up is Rs. 4,000/-, while the cost of carrying inventories is Re. 4/- per month per piece. In which month should the batches be produced and of what size, so as to minimize the total set-up and inventory loading costs? *(BT level 5)*
Solution:
Here, the five months represent the five stages, and the number of units to be manufactured is the state variable. We can start with the last month.

Here the optimum decision is to produce 900 units in the fourth month and no units in the fifth month.

Stage 3: Third month: There are three alternatives

First: Produce 2,900 (2,000 + 300 + 600) units to satisfy demand of third, fourth, and fifth month. Hence the total cost = Set-up cost + inventory carrying cost of 300 units for one month + inventory carrying cost of 600 units for two month = Rs. 4,000 + Rs. 1,200/- + Rs. 4,800/- = Rs. 10,000/-.

Second: Produce 2,300 (2,000 + 300) units in the third month and 600 units in the fifth month. Hence the total cost = Set-up cost + inventory carrying cost of 300 units for one month + Set-up cost = Rs. 4,000 + Rs. 1,200/- + Rs. 4,000/- = Rs. 9,200/-.

Third: Produce 2,000 units in the third month and 900 (300 + 600) units in the fourth month. Hence the total cost = Set-up cost + Set-up cost + inventory carrying cost of 600 units for one month = Rs. 4,000 + Rs. 4,000/- + Rs. 2,400/- = Rs. 10,400/-.

Here the optimum decision is to produce 2,300 (2,000 + 300) units in the third month and 600 units in the fifth month

Stage 2: Second month: There are four alternatives

First: Produce 3,400 (500 + 2,000 + 300 + 600) units to satisfy demand of second, third, fourth, and fifth month. Hence the total cost = Set-up cost + inventory carrying cost of 2,000 units for one month + inventory carrying cost of 300 units for two months + inventory carrying cost of 600 units for three months = Rs. 4,000 + Rs. 8,000/- + Rs. 2,400/- + Rs. 7,200/- = Rs. 21,600/- Second: Produce 2,800 (500 + 2,000 + 300) units in the second month and 600 units in the fifth month. Hence the total cost = Set-up cost + inventory carrying cost of 2,000 units for one month + inventory carrying cost of 300 units for two months + Set-up cost = Rs. 4,000 + Rs. 8,000/- + Rs. 2,400/- + Rs. 4,000/- = Rs.18,400/-.

Third: Produce 2,500 (500 + 2,000) units in the second month and 900 (300 + 600) units in the fourth month. Hence the total cost = Set-up cost + inventory carrying cost of 2,000 units for one month + Set-up cost + inventory carrying cost of 600 units for one month = Rs. 4,000 + Rs. 8,000/- + Rs. 4,000 + Rs. 2,400/- = Rs. 18,400/-.

Fourth: Produce 500 units in the second month and 2,900 (2,000 + 300 + 600) units in the third month. Hence the total cost = Set-up cost + Set-up cost + inventory carrying cost of 300 units for one month + inventory carrying cost of 600 units for two months = Rs. 4,000 + Rs. 4,000/- + Rs. 1,200 + Rs. 4,800/- = Rs. 14,000/-.

Here the optimum decision is to produce 500 units in the second month and 2,900 (2,000 + 300 + 600) units in the third month.

Stage 1: First month: There are five alternatives.

First: Produce 4,400 (1,000 + 500 + 2,000 + 300 + 600) units to satisfy demand in the first, second, third, fourth, and fifth months. Hence the total cost = Set-up cost + inventory carrying cost of 500 units for one month + inventory carrying cost of 2,000 units for two months + inventory carrying cost of 300 units for three months + inventory carrying cost of 600 units for four months = Rs. 4,000 + Rs. 2,000 + Rs. 16,000/- + Rs. 3,600/- + Rs. 9,600/- = Rs. 35,200/-.

Second: Produce 3,800 (1,000 + 500 + 2,000 + 300) units in the first month and 600 units in the fifth month. Hence the total cost = Set-up cost + inventory carrying cost of 500 units for one month + inventory carrying cost of 2,000 units for two months + inventory carrying cost of 300 units for three months + Set-up cost = Rs. 4,000 + Rs. 2,000 + Rs. 16,000/- + Rs. 3,600/- + Rs. 4,000/- = Rs. 29,600/-.

Third: Produce 3,800 (1,000 + 500 + 2,000) units in the first month and 900 (300 + 600) units in the fourth month. Hence the total cost = Set-up cost + inventory carrying cost of 500 units for one month + inventory carrying cost of 2,000 units for two months + Set-up cost + inventory carrying cost of 600 units for one month = Rs. 4,000 + Rs. 2,000 + Rs. 16,000/- + Rs. 4,000/- + Rs. 2,400/- = Rs. 28,400/-.

Fourth: Produce 1,500 (1,000 + 500) units in the first month and 2,900 (2,000 + 300 + 600) units in the third month. Hence the total cost = Set-up cost + inventory carrying cost of 500 units for one month + Set-up cost + inventory carrying cost of 300 units for one month + inventory carrying cost of 600 units for two months = Rs. 4,000 + Rs. 2,000 + Rs. 4,000/- + Rs. 1,200/- + Rs. 4,800/- = Rs. 16,000/-.

Fifth: Produce 1,000 units in the first month and 3,400 (500 + 2,000 + 300 + 600) units in the second month. Hence the total cost = Set-up cost + Set-up cost + inventory carrying cost of 2,000 units for one month + inventory carrying cost of 300 units for two months + inventory carrying cost of 600 units for three months = Rs. 4,000 + Rs. 4,000 + Rs. 8,000/- + Rs. 2,400/- + Rs. 2,400/- = Rs. 20,800/-.

Here the optimum decision is to produce 1,500 (1,000 + 500) units in the first month and 2,900 (2,000 + 300 + 600) units in the third month.

Hence the minimum cost policy is to produce a batch of 1,500 units in the first month 2,900 units in the third month, which gives a minimum of set-up and inventory carrying cost of Rs. 16,000/-.

9.4 Introduction to Integer Programming

LO 4. Understand integer programming and differentiate between linear and integer programming

The implicit assumption in the linear programming models formulated and solved in chapter second was that solutions could be either fractional or real numbers (i.e., non-integer). Non-integer solutions are however not always practical. However, some or all of the decision variables in many important applications have to be restricted to integer values. These models are called integer programming models. The main difference between linear programming and integer programming models is that linear programming models allow fractional values, such as 0.216 and 4.2155, for decision variable cells, while integer programming models allow only integer values in integer-restricted decision variable cells. There are three types of integer programming problems:

1. Pure integer programming problems are cases where all variables are required to have integer values. For example,

$$Max z = 6x_1 + 4x_2$$

$$\text{Subjected to } 2x_1 + 2x_2 \leq 10$$

$$x_1, x_2 \geq 0,$$

$$x_1, x_2 \text{ integer.}$$

2. Mixed-integer programming problems are cases in which some, but not all of the decision variables are required to have integer values. For example,

$$Max z = 6x_1 + 4x_2.$$

$$\text{Subjected to } 2x_1 + x_2 \leq 10$$

$$3x_1 - 2x_2 \leq 6$$

$$x_1, x_2 \geq 0,$$

$$x_1 \text{ is integer.}$$

3. Zero – some integer programming problems are special cases where all decision variables must have an integer solution value of 0 or 1. For example,

$$Max z = 2x_1 - x_2.$$

$$\text{Subjected to } 2x_1 + 3x_2 \leq 12$$

$$x_1 - x_2 \leq 4,$$

$$x_1, x_2 = 0 \text{ or } 1.$$

9.5 Formulating Integer Programming Problems

LO 5. Formulate the integer programming problem
It often happens that one cannot formulate an integer programming model simply by formulating a linear programming model and then adding the requirement that the decision variables have integer values; instead, one must proceed by introducing special integer-valued variables into the linear model. However, there are some situations in which a linear programming model can be immediately converted to an integer programming model. For instance, it can happen that the rounding off of fraction values of the x_j would not be adequately accurate if the x_j represents the production level of very expensive items. Requiring that the x_j has an integer value leads immediately to an integer programming model.

In a manufacturing plant, the production of items is scheduled in terms of lots or batches, and therefore the only viable option is to have integer programming. Likewise, integer programming is used in many situations, such as the shipment of goods involving a discreet number of trucks, capital budgeting, problems with travelling salesmen, and a list of many. Integer programming is used in a wide variety of management decision-making situations. The situations discussed with examples will illustrate both the variety of integer programming model applications and the variety of ideas used to formulate these models.

Examples based on formulating the integer programming problem

EXAMPLE 1:

The corporation is considering four potential investment opportunities. The expected return and the capital required for the project are given in Table 9.16.
The company allocated Rs. 600,000 for first-year investments and Rs. 400,000 for second-year investments. Formulate an integer programming model to determine which projects from the accepted project should be accepted to maximize the present value. *(BT level 6)*

TABLE 9.16

Data for Example 1

Project	Capital Required by Project (Rs. Thousand)		Expected Return (Rs. Thousand)
	First Year	Second Year	
1	350	300	350
2	400	300	400
3	150	80	125
4	200	100	150

Solution:

$$\text{Decision variable } xj = \begin{cases} 1, \text{if project } j \text{ is accepted} \\ 0, \text{if project } j \text{ is rejected} \end{cases}$$

Objective function: maximize $z = 350x_1 + 400x_2 + 125x_3 + 150x_4$
The constraints are on the availability of the funds.

$$350x_1 + 400x_2 + 150x_3 + 200x_4 \leq 600$$

$$300x_1 + 300x_2 + 80x_3 + 100x_4 \leq 400$$

$$x_i = 0 \text{ or } 1$$

EXAMPLE 2:

The WorkCentre produces two products A and *B* on a weekly basis. Each item requires 10 kg of raw material and 150 kg of raw material available on a weekly basis. Product A requires five hours of work, Product *B* requires two hours of work, and the WorkCentre operates 35 hours a week. Product A profits Rs. 150, and product *B* profits Rs. 110. The supervisor intends to determine the total number of products A and *B* to be produced in order to maximize profits. Formulate an integer programming model for this problem. *(BT level 6).*

Solution:

Decision variable : x_1 = number of product A produce

x_2 = number of product B produce

Objective function : maximize $z = 150x_1 + 110x_2$

Subject to constraints

$$10x_1 + 10x_2 \leq 150$$

$$5x_1 + 2x_2 \leq 35$$

$$x_1, x_2 \geq 0 \text{ and integer}$$

EXAMPLE 3:

XYZ Manufacturing plans to build at least one new plant and are considering three cities: City 1, City 2, and City 3. Once the plant or plants has been built, the company wishes to have enough capacity to produce at least 50,000 units per year. The costs associated with possible locations are shown in Table 9.17. Formulate an integer programming model for this problem. *(BT level 6)*

TABLE 9.17

Data for Example 3

City	Fixed Cost (in Crore of Rs. Annually)	Variable Cost (in Rs. per Unit)	Annual Capacity
City 1	24	42	35,000
City 2	17	45	25,000
City 3	19	40	25,000

Solution:

$$\text{Decision variable}: x1 = \begin{cases} 1, \text{if plant is build in City 1} \\ 0, \text{otherwise} \end{cases}$$

$$x2 = \begin{cases} 1, \text{if plant is build in City 2} \\ 0, \text{otherwise} \end{cases}$$

$$x3 = \begin{cases} 1, \text{if plant is build in City 3} \\ 0, \text{otherwise} \end{cases}$$

$$x_4 = \text{number of units produced at City 1 plant}$$

$$x_5 = \text{number of units produced at City 2 plant}$$

$$x_6 = \text{number of units produced at City 3 plant}$$

Objective function: minimize $z = 24\,x_1 + 17\,x_2 + 19\,x_3 + 42\,x_4 + 45\,x_5 + 40\,x_6$
Subject to constraints

$$x_4 + x_5 + x_6 \geq 50,000$$

$$x_4 \leq 35,000x_1$$
$$x_5 \leq 25,000x_2$$
$$x_6 \leq 25,000x_3$$
$$x_1, x_2, x_3 = 0 \, or \, 1 \, and \, x_4, x_5, x_6 \geq 0 \, and \, integer$$

EXAMPLE 4:

An aluminium channel supplier for a window frame that maintains a channel length stock of 35 ft. It has been determined that the cost of storing the channel is Rs. 1,000 per 35 ft. where the purchase, shipping, and storage costs are included but should not include a proration of any fixed overhead. Suppose the customer orders 300 channels in the length of 13 ft, 250 channels in the length of 15 ft, and 200 channels in the length of 20 ft. The supplier will complete this order by cutting the stock channels to the specified lengths and discarding any waste. The problem is how to cut the channels requested by the customer to minimize the cost of the channels taken out of stock. **(BT level 6)**

Solution: First, we're going to determine the possible ways to cut the aluminium channel. From a 35-ft stock channel, the supplier can cut two channels with a length of 13 ft or 15 ft or one channel with a length of 15 ft and 20 ft. Such possible ways of cutting a channel are called patterns. All patterns are shown in Table 9.18.

TABLE 9.18

A Pattern for Different Lengths

	Pattern					Demand
	1	2	3	4	5	
13 ft. order length	2	0	0	1	1	300
15 ft. order length	0	2	1	1	0	250
20 ft. order length	0	0	1	0	1	200
Cut length	26	30	35	28	33	
Cost per channel	1,000	1,000	1,000	1,000	1,000	

Decision variable: x_j = number of jth pattern, $j = 1, 2, 3, 4, 5$

Objective function: minimize z = $1,000x_1 + 1,000x_2 + 1,000x_3 + 1,000x_4 + 1,000x_5$

Subject to constraints

$$2x_1 + x_4 + x_5 \geq 300$$
$$2x_2 + x_3 + x_4 \geq 250$$
$$x_3 + x_5 \geq 200$$
$$x_1, x_2, x_3, x_4, x_5 \geq 0 \text{ and integer}$$

EXAMPLE 5:

A machine shop's supervisor plans to expand by buying some new machines – M1 and M2. The owner estimated that every M1 purchased would increase profits by Rs. 1,200 per day and that every M2 would increase profits by Rs. 1,800 per day. The number of machines that the owner can purchase is limited by the cost (in rupees) of the machines and the available floor space (in square feet) in the shop. The purchase prices and space requirements for machines are as follows (Table 9.19):

TABLE 9.19

Data for Example 5

Machine	Floor Space Required	Purchase Prize
M1	10	400,000
M2	20	800,000

A budget of Rs.2,400,000 for machine purchases and 180 square feet of floor space is available. For this problem, develop an integer programming model. *(BT level 6)*

Solution: Decision variable: x_1 = Number of M1 machine
x_2 = Number of M2 machine
Objective function: maximize $z = 1{,}200x_1 + 1{,}800x_2$
Subject to constraints

$$400{,}000x_1 + 800{,}000x_2 \leq 2{,}400{,}000$$

$$10x_1 + 20x_2 \leq 180$$

$$x_1, x_2 \geq 0 \text{ and integer}$$

EXAMPLE 6:

An engineering student is planning to study 42 hours a week. He is considering seven courses. The number of hours (per week) required to successfully complete each course is shown in Table 9.20.

TABLE 9.20

Data for Example 6

Course Subject	Course1	Course2	Course3	Course4	Course5	Course6	Course7
Hours per week	10	8	6	9	6	4	8

Completing each of these courses increases the chances of clearing the aptitude test for engineering by the student. But the courses' contributions to this dream are different, as indicated in Table 9.21:

TABLE 9.21

Data for Example 6

Course Subject	Course1	Course2	Course3	Course4	Course5	Course6	Course7
Contributions	0.15	0.09	0.11	0.17	0.13	0.08	0.09

Formulate an integer programming model for this problem. *(BT level 6)*
Solution:

$$\text{Decision variable } xj = \begin{cases} 1, \text{if project } j \text{ is accepted} \\ 0, \text{if project } j \text{ is rejected} \end{cases} j = 1,2,3,4,5,6,7$$

Objective function:

$$\text{maximize } z = 0.15x_1 + 0.09x_2 + 0.11x_3 + 0.17x_4 + 0.13x_5 + 0.08x_6 + 0.09x_7$$

Subject to constraints

$$10x_1 + 8x_2 + 6x_3 + 9x_4 + 6x_5 + 4x_6 + 8x_7 \leq 42$$

$$x_j \geq 0 \text{ and integer}$$

9.6 Solution of Integer Programming Using the Branch-and-Bound Method

LO 6. Understand and apply the branch-and-bound method for solving the integer programming problem

The branch-and-bound algorithm is the most widely used method for solving both pure and mixed-integer problems in practice. Divide and conquer is the basic concept that underlies the branch-and-bound technique. As it is too complicated to solve the original "large" problem directly, the problem is broken down into smaller and smaller problems until the problem can be conquered. The division (branching) is done by dividing the whole set of feasible solutions into smaller and smaller subsets. Conquering (fathoming) is done partly by bounding how well the best solution in the subset can be and then by discarding the subset if its bound indicates that it cannot possibly contain an optimal solution to the original problem.

In its simplest form, branch and bound is just an organized way to deal with a hard problem and divide it into two or more small (and therefore easier) sub-problems. If these sub-problems are still too difficult, we will branch out again and further subdivide the problems. The process is repeated until each of the sub-problems can be solved easily. Branching is done in such a way that the solution of each of the sub-problems (and the selection of the best answer found) is equivalent to the solution of the original problem.

The iterative procedure of the branch-and-bound method is as follows:

1. Obtain an optimal solution for the given linear programming problem while ignoring the integer's restriction.

2. Test the integrity of the optimum solution obtained in step 1. If the solution is integers, the current solution is optimal for the whole programming problem. If the solution isn't an integer, go to the next step.

3. Considering the value of the objective function as upper bound, obtain the lower bound by rounding off to integral values of the decision variables.

4. Let the optimum value x^*_j of the variable x_j is not an integer. Then subdivide the given problem into two problems: (branch step).
   Sub-problem 1: Given the problem with an additional constraint $x^*_j \leq [x^*_j]$
   Sub-problem 2: Given problem with an additional constraint $x^*_j \geq [x^*_j] + 1$
   Where $[x^*_j]$ is the integer part of x^*_j

5. Solve the two sub-problems identified in step 4. Three cases may arise: (bound step)
   i. If the two sub-problems' optimum solutions are integral, then the solution required is one that gives a greater value of Z.

   ii. If the optimum solution of one sub-program is integral and the other sub-program does not have a feasible optimum solution, the solution required is the same as that of a sub-program with an integer-valued solution.

   iii. If the optimum solution of one sub-program is integral while that of the other sub-program is not integral, then record the integral valued solution and repeat step 3 and step 4 for the non-integer-valued sub-problem.

6. Repeat steps 3 to 5 until all integer-valued solutions are recorded.

7. Choose the solution among the valued integer solutions recorded that yield an optimum value of Z.

Examples based on the branch-and-bound method

EXAMPLE 1:

Use a branch-and-bound method to solve the following integer programming problem. *(BT level 3)*

$$\max Z = 14x_1 + 18x_2$$

Subject to

$$-2x_1 + 6x_2 \leq 12$$

$$14x_1 + 2x_2 \leq 70$$
$$2x_2 \leq 14$$
$$\text{and } x_1, x_2 \geq 0 \text{ and integer}$$

Solution: First, we ignore the integer restriction and solve the problem. The optimal solution to the problem can easily be found (as discussed in Chapter 2) as follows:

Solution is Max $Z = 126$ ($x_1 = 9/2$, $x_2 = 7/2$)

Since the solution obtained is not an integer solution, let us choose $x_1 = 9/2$ being the largest fractional value.

Consider the value of Z as an initial upper bound (i.e., $Z = 126$); the lower bound is obtained by rounding off the values of decision variables x_1 and x_2 to the nearest integers. Here we consider $x_1 = 4$ and $x_2 = 3$. Then the lower bound $Z_1 = 110$.

Now, divide the problem into the following two sub-problems considering two new constraints as, $x_1 \leq 4$ and $x_1 \geq 5$.

Sub-problem 1: Solution is found by adding $x_1 \leq 4$.

$$\text{Max } Z = 14x_1 + 18x_2$$

Subject to

$$-2x_1 + 6x_2 \leq 12$$

$$14x_1 + 2x_2 \leq 70$$
$$2x_2 \leq 14$$
$$x_1 \leq 4$$
$$\text{and } x_1, x_2 \geq 0$$

Solution is Max $Z_{\text{Sub Problem 1}} = 116$ ($x_1 = 4$, $x_2 = 10/3$)
and $Z_2 = 110$ ($x_1 = 4$, $x_2 = 3$) obtained by the rounded off solution values.

Sub-problem 2: Solution is found by adding $x_1 \geq 5$.

$$\text{max } Z = 14x_1 + 18x_2$$

Subject to

$$-2x_1 + 6x_2 \leq 12$$
$$14x_1 + 2x_2 \leq 70$$
$$2x_2 \leq 14$$
$$x_1 \geq 5$$
$$\text{and } x_1, x_2 \geq 0$$

Solution is Max $Z_{\text{Sub Problem 2}} = 70$ ($x_1 = 5$, $x_2 = 0$)
and $Z_3 = 70$ ($x_1 = 5$, $x_2 = 0$) obtained by the rounded off solution values.
This problem has an integer solution, so no further branching is required.

In sub-problem 1, ($x_2 = 10/3$) is not in integers, we subdivide it into two more sub-problems considering two new constraints as, $x_2 \leq 3$ and $x_2 \geq 4$.

Sub-problem 3: Solution is found by adding $x_2 \leq 3$.

$$\text{Max } Z = 14x_1 + 18x_2$$

Subject to

$$-2x_1 + 6x_2 \leq 12$$

$$14x_1 + 2x_2 \leq 70$$

$$2x_2 \leq 14$$

$$x_1 \leq 4$$

$$x_2 \leq 3$$

$$\text{and } x_1, x_2 \geq 0$$

Solution is Max $Z_{\text{Sub Problem 3}} = 110$ ($x_1 = 4$, $x_2 = 3$)
This problem has an integer solution, so no further branching is required.

Sub-problem 4: Solution is found by adding $x_2 \geq 4$.

$$\max Z = 14x_1 + 18x_2$$

Subject to

$$-2x_1 + 6x_2 \leq 12$$

$$14x_1 + 2x_2 \leq 70$$

$$2x_2 \leq 14$$

$$x_1 \leq 4$$

$$x_2 \geq 4$$

$$\text{and } x_1, x_2 \geq 0$$

This problem has an infeasible solution, so this branch is terminated.

The branch-and-bound algorithm thus terminated, and the optimal integer solution is

$$Z = 110 (x_1 = 4, x_2 = 3).$$

The entire branch-and-bound procedure for the given problem is shown in Figure 9.1.

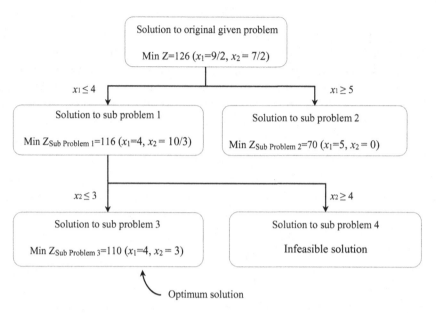

FIGURE 9.1
Branch and bound for Example 1.

EXAMPLE 2:

Use branch-and-bound method to solve the following integer programming problem. *(BT level 3)*

$$\min Z = 8x_1 + 18x_2$$

Subject to

$$2x_1 + 4x_2 \geq 14$$

$$4x_1 + 6x_2 \geq 34$$

and $x_1, x_2 \geq 0$ and integer

Solution: First, we ignore the integer restriction and solve the problem. The optimum solution to the given problem can easily be obtained (by the method as discussed in Chapter 2) as:
Solution is Min $Z = 68$ ($x_1 = 17/2$, $x_2 = 0$)

Since the solution obtained is not an integer solution, let us choose $x_1 = 17/2$ being the largest fractional value.

Consider the value of Z as an initial lower bound (i.e. $Z = 68$; the lower bound is obtained by rounding off the values of decision variables x_1 to the nearest integers). Here we consider $x_1 = 8$. Then the lower bound $Z_1 = 64$.

Now, divide the problem into the following two sub-problems considering two new constraints as, $x_1 \leq 8$ and $x_1 \geq 9$.

Sub-problem 1: Solution is found by adding $x_1 \leq 8$.

$$\min Z = 8x_1 + 18x_2$$

Subject to

$$2x_1 + 4x_2 \geq 14$$

$$4x_1 + 6x_2 \geq 34$$

$$x_1 \leq 8$$

$$\text{and } x_1, x_2 \geq 0$$

Solution is Min $Z_{\text{Sub Problem 1}} = 70 \ (x_1 = 8, x_2 = 1/3)$
and $Z_2 = 64 \ (x_1 = 8, x_2 = 0)$ obtained by the rounded off solution values.

Sub-problem 2: Solution is found by adding $x_1 \geq 9$.

$$\min Z = 8x_1 + 18x_2$$

Subject to

$$2x_1 + 4x_2 \geq 14$$

$$4x_1 + 6x_2 \geq 34$$

$$x_1 \geq 9$$

$$\text{and } x_1, x_2 \geq 0$$

Solution is Min $Z_{\text{Sub Problem 2}} = 72 \ (x_1 = 9, x_2 = 0)$
This problem has an integer solution, so no further branching is required.
In sub-problem 1, $(x_2 = 1/3)$ is not in integers, we subdivide it into two more sub-problems considering two new constraints as, $x_2 \leq 0$ and $x_2 \geq 1$.

Sub-problem 3: Solution is found by adding $x_2 \leq 0$.

$$\min Z = 8x_1 + 18x_2$$

Subject to

$$2x_1 + 4x_2 \geq 14$$

$$4x_1 + 6x_2 \geq 34$$

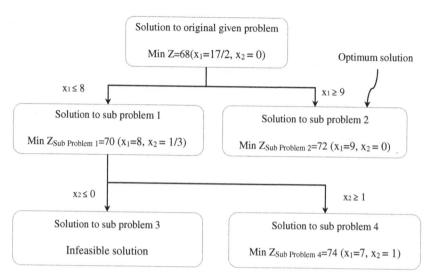

FIGURE 9.2
Branch and bound for Example 2.

$$x_1 \leq 8$$

$$x_2 \leq 0$$

$$\text{and } x_1, x_2 \geq 0$$

This problem has an infeasible solution, so this branch is terminated.
Sub-problem 4: Solution is found by adding $x_2 \geq 1$.

$$\min Z = 8x_1 + 18x_2$$

Subject to

$$2x_1 + 4x_2 \geq 14$$

$$4x_1 + 6x_2 \geq 34$$

$$x_1 \leq 8$$

$$x_2 \geq 1$$

$$\text{and } x_1, x_2 \geq 0$$

Solution is Min $Z_{\text{Sub Problem 4}} = 74$ ($x_1 = 7$, $x_2 = 1$)
This problem has an integer solution, so no further branching is required.
The branch-and-bound algorithm thus terminated, and the optimal integer solution is:
$Z = 72$ ($x_1 = 9$, $x_2 = 0$).
The entire branch-and-bound procedure for the given problem is shown in Figure 9.2.

9.7 Solution of Integer Programming Using the Cutting Plane Method

LO 7. *Understand and apply the cutting plane method for solving the integer programming problem*

Historically the cutting plane method was the first method to solve integer programming problems. R. E. Gomory first proposed a systematic method for finding an optimal integer solution for all integer programming problems. The optimal solution to an integer programming problem is first obtained by a simplex method that ignores the restriction of integral values. If all variables have integer values, the current solution will be the desired optimum integer solution. Otherwise, we are adding a new constraint to the problem so that the new set of feasible solutions includes all the original feasible integer solutions but does not include the initially found optimum non-integer solution. Then, using the simplex method, we solve the revised problem and see if we can get an integer solution. If not, we will add another fractional cut, and repeat the process until we find an integer solution. Since we never remove any feasible integer solution from consideration when adding functional cuts, the integer solution ultimately found must be optimal.

The basic steps involved in the method are as follows

Step 1. Solve the given problem using any linear programming solution method, ignore the integer condition.

Step 2. Check that the optimal solution is an integer. If yes, stop. The solution that results is the required one. If no, then proceed to step 3.

Step 3. Generate a new constraint (cutting plane) from the optimal fractional solution obtained in step 1. The constraint is created to rule out the obtained factional values, without excluding any integer solutions to the problem. In step 1, add the constraint to the problem, and then go back to step 1.

Examples based on the cutting plane method

EXAMPLE 1:

Use a cutting plane method to solve the following integer programming problem. *(BT level 3)*

$$\max Z = 12x_1 + 4x_2$$

Subject to

$$6x_1 + 4x_2 \leq 10$$

$$\text{and } x_1, x_2 \geq 0 \text{ and integer}$$

Solution: First the problem is converted to canonical form by adding the slack variable.

As the constraint-1 is of type '$\leq$' we should add slack variable S_1
After introducing slack variables, the problem is rewritten as follows

$$\max Z = 12x_1 + 4x_2 + 0S_1$$

Subject to

$$6x_1 + 4x_2 + S_1 = 10$$

and $x_1, x_2, S_1 \geq 0$ and integer

Now solve the previous problem by applying the simplex method (as discussed in Chapter 2). The final simplex table is presented in Table 9.22).

Since all $Zj - Cj \geq 0$ the current solution is optimum. The optimum solution is

$$\max Z = 20 \text{ and } \left(x_1 = 5/3, x_2 = 0\right)$$

But the solution obtained is not an integer solution.
To obtain the integer-valued solution, we proceed to construct Gomory's fractional cut, with the help of x_1-row as follows:

$$5/3 = 1x_1 + 23x_2 + 1/6S_1$$

$$\left(1 + 2/3\right) = \left(1 + 0\right)x_1 + \left(0 + 23\right)x_2 + \left(0 + 1/6\right)S_1$$

The fractional cut will become

$$-2/3 = Sg1 - 2/3x_2 - 1/6S_1 \rightarrow \left(\text{Cut} - 1\right)$$

TABLE 9.22

Final Simplex Table

		C_j	12	4	0	
B	C_B	X_B	x_1	x_2	S_1	Min Ratio
x_1	12	5/3	1	2/3	1/6	
Z = 20		Z_j	12	8	2	
		$Z_j - C_j$	0	4	2	

Adding this additional constraint at the bottom of the optimal simplex table. The new table so obtained is as follows (Table 9.23)

The minimum negative X_B is $-2/3$, and its row index is 2. So, the leaving basis variable is Sg1.

The maximum negative ratio is -6, and its column index is 2. So, the entering variable is x_2.

∴ The pivot element is $-2/3$, entering variable $= x_2$, leaving variable $=$ Sg1.

Perform the following row operations:

1. New row 2 = (Old row 2) x (– 3/2)
2. New row 1 = (Old row 1) – (2/3 New row 2)

The new table so obtained is shown below (Table 9.24)

Since all $Zj - Cj \geq 0$ hence, integer optimal solution arrives with the value of variables as

Max Z = 16 and ($x_1 = 1$, $x_2 = 1$).

The solution is also an integer solution.

TABLE 9.23

Simplex Table

		C_j	12	4	0	0
B	C_B	X_B	x_1	x_2	S_1	Sg1
x_1	12	5/3	1	2/3	1/6	0
Sg1	0	–2/3	0	–2/3	–1/6	1
Z = 20		Z_j	12	8	2	0
		$Z_j - C_j$	0	4	2	0
		$\dfrac{Z_j - C_j}{Sg1}$	--	–6	–12	

TABLE 9.24

Simplex Table

		C_j	12	4	0	0
B	C_B	X_B	x_1	x_2	S_1	Sg1
x_1	12	1	1	0	0	1
x_2	4	1	0	1	1/4	– 3/2
Z =16		Z_j	12	4	1	6
		$Z_j - C_j$	0	0	1	6

EXAMPLE 2:

Use a cutting plane method to solve the following integer programming problem. *(BT level 3)*

$$Max\, Z = 10x_1 + 16x_2$$

Subject to

$$2x_1 + 4x_2 \leq 16$$

$$8x_1 + 2x_2 \leq 20$$

$$and\, x_1, x_2 \geq 0\, and\, integer$$

Solution: First, the problem is converted to canonical form by adding the slack variable.

As the constraint-1 is of type '$\leq$' we should add slack variable S_1 and constraint-2 is of type '$\leq$' we should add slack variable S_2

After introducing slack variables, the problem is rewritten as follows

$$Max\, Z = 10x_1 + 16x_2 + 0S_1 + 0S_2$$

Subject to

$$2x_1 + 4x_2 + S_1 = 16$$

$$8x_1 + 2x_2 + S_2 = 20$$

$$and\, x_1, x_2, S_1, S_2 \geq 0\, and\, integer$$

Now solve the above problem by applying the simplex method (as discussed in Chapter 2). The final simplex table is presented in Table 9.25).

TABLE 9.25

Final Simplex Table

		C_j	10	16	0	0	
B	C_B	X_B	x_1	x_2	S_1	S_2	**Min Ratio**
x_2	16	22/7	0	1	2/7	−1/14	
x_1	10	12/7	1	0	−1/14	1/7	
$Z = 472/7$		Z_j	10	16	27/7	2/7	
		$Z_j − C_j$	0	0	27/7	2/7	

Since all $Zj - Cj \geq 0$ the current solution is optimum. The optimum solution is

$$\max Z = 472/7 \text{ and} \left(x_1 = 12/7, x_2 = 22/7\right).$$

But the solution obtained is not an integer solution.

To obtain the integer-valued solution, we proceed to construct Gomory's fractional cut, with the help of x_1-row as follows:

$$12/7 = 1x_1 - 1/14S_1 + 1/7S_2$$

$$\left(1 + 5/7\right) = \left(1 + 0\right)x_1 + \left(-1 + 13/14\right)S_1 + \left(0 + 1/7\right)S_2$$

The fractional cut will become

$$-5/7 = Sg1 - 13/14S_1 - 1/7S_2 \rightarrow \left(Cut - 1\right)$$

Adding this additional constraint at the bottom of the optimal simplex table. The new table so obtained is as follows (Table 9.26).

The minimum negative X_B is $-5/7$, and its row index is 3. So, the leaving basis variable is Sg1. The maximum negative ratio is -2, and its column index is 4. So, the entering variable is S_2.

∴ The pivot element is $1/7$, entering variable = S_2, leaving variable = Sg1

Perform the following row operations:

1. New row 3 = (Old row 3) × (−7)
2. New row 1 = (Old row 1) + (1/14 New row 3)
3. New row 2 = (Old row 2) − (1/7 New row 3)

The new table so obtained is as follows Table 9.27.

Since all $Zj - Cj \geq 0$ the current solution is optimum. The optimum solution is

Max $Z = 66$ and ($x_1 = 1, x_2 = 7/2$).

But the solution obtained is not an integer solution.

To obtain the integer-valued solution, we proceed to construct Gomory's fractional cut, with the help of x_2-row as follows:

TABLE 9.26

Simplex Table

		C_j	10	16	0	0	0
B	C_B	X_B	x_1	x_2	S_1	S_2	Sg1
x_2	16	22/7	0	1	2/7	−1/14	0
x_1	10	12/7	1	0	−1/14	1/7	0
Sg1	0	−5/7	0	0	−13/14	−1/7	1
$Z = 472/7$		Z_j	10	16	27/7	2/7	0
		$Z_j - C_j$	0	0	27/7	2/7	0
		$\dfrac{Z_j - C_j}{Sg1}$	−	−	−4.1538	−2	

$$7/2 = 1x_2 + 3/4S1 - 1/2Sg1$$

$$(3 + 1/2) = (1+0)x_2 + (0 + 3/4)S_1 + (-1 + 1/2)Sg1$$

The fractional cut will become

$$-1/2 = Sg2 - 3/4S_1 - 1/2Sg1 \rightarrow (Cut - 2).$$

Adding this additional constraint at the bottom of the optimal simplex table. The new table so obtained is as follows (Table 9.28)

The minimum negative X_B is $-1/2$, and its row index is 4. So, the leaving basis variable is Sg2.

The maximum negative ratio is -2.6667, and its column index is 3. So, the entering variable is S_1.

∴ The pivot element is $-3/4$, entering variable = S_1, leaving variable = Sg2

Perform the following row operations:

TABLE 9.27

Simplex Table

			C_j	10	16	0	0	0
B	C_B	X_B	x_1	x_2	S_1	S_2	Sg1
x_2	16	7/2	0	1	3/4	0	1/2
x_1	10	1	1	0	−1	0	1
S_2	0	5	0	0	13/2	1	−7
Z = 66		Z_j	10	16	2	0	2
		$Z_j - C_j$	0	0	2	2	2

TABLE 9.28

Simplex Table

			C_j	10	16	0	0	0	0
B	C_B	X_B	x_1	x_2	S_1	S_2	Sg1	Sg2
x_2	16	7/2	0	1	3/4	0	−1/2	0
x_1	10	1	1	0	−1	0	1	0
S_2	0	5	0	0	13/2	1	−7	0
Sg2	0	−1/2	0	0	−3/4	0	−1/2	1
Z = 66		Z_j	10	16	2	0	2	0
		$Z_j - C_j$	0	0	2	2	2	0
		$\dfrac{Z_j - C_j}{Sg2}$	−	−	−2.666	−	−4	−

1. New row 4 = (Old row 4) × (−4/3)
2. New row 1 = (Old row 1) − (3/4 New row 4)
3. New row 2 = (Old row 2) + (New row 4)
4. New row 3 = (Old row 3) + (13/2 New row 4)

The new table so obtained is as follows (Table 9.29)
Since all $Zj - Cj \geq 0$ the current solution is optimum. The optimum solution is

$$\text{Max } Z = 194/3 \text{ and } \left(x_1 = 5/3, x_2 = 3 \right).$$

But the solution obtained is not an integer solution.

To obtain the integer-valued solution, we proceed to construct Gomory's fractional cut, with the help of x_1-row as follows:

$$5/3 = 1x_1 + 5/3Sg1 - 4/3Sg2,$$

$$\left(1 + 2/3\right) = \left(1 + 0\right)x_1 + \left(1 + 2/3\right)Sg1 + \left(-2 + 2/3\right)Sg2.$$

The fractional cut will become
−2/3=Sg3−2/3Sg1−2/3Sg2→(Cut-3).Adding this additional constraint at the bottom of the optimal simplex table. The new table so obtained is as follows (Table 9.30).

The minimum negative X_B is −2/3, and its row index is 5. So, the leaving basis variable is Sg3.

The maximum negative ratio is −1, and its column index is 5. So, the entering variable is Sg1.

∴ The pivot element is −2/3, entering variable = Sg1, leaving variable = Sg3.
Perform the following row operations:

1. New row 5 = (Old row 5) × (−3/2)
2. New row 1 = (Old row 1) + (New row 5)
3. New row 2 = (Old row 2) − (5/3 New row 5)
4. New row 3 = (Old row 3) + (34/3 New row 5)
5. New row 4 = (Old row 4) − (2/3 New row 5)

TABLE 9.29

Simplex Table

		C_j	10	16	0	0	0	0
B	C_B	X_B	x_1	x_2	S_1	S_2	Sg1	Sg2
x_2	16	3	0	1	0	0	−1	1
x_1	10	5/3	1	0	0	0	5/3	−4/3
S_2	0	2/3	0	0	0	1	−34/3	26/3
S_1	0	2/3	0	0	1	0	2/3	−4/3
$Z = 194/3$		Z_j	10	16	0	0	2/3	8/3
		$Z_j - C_j$	0	0	0	0	2/3	8/3

The new table so obtained is as follows (Table 9.31).

Since all $Zj - Cj \geq 0$ the current solution is optimum. The optimum solution is Max $Z = 64$ and ($x_1 = 0$, $x_2 = 4$).

The solution is also an integer solution.

TABLE 9.30

Simplex Table

		C_j	10	16	0	0	0	0	
B	C_B	X_B	x_1	x_2	S_1	S_2	Sg1	Sg2	Sg3
x_2	16	3	0	1	0	0	−1	1	0
x_1	10	5/3	1	0	0	0	5/3	−4/3	0
S_2	0	2/3	0	0	0	1	−34/3	26/3	0
S_1	0	2/3	0	0	1	0	2/3	−4/3	0
Sg3	0	−2/3	0	0	0	0	−2/3	−2/3	1
Z = 194/3		Z_j	10	16	0	0	2/3	8/3	0
		$Z_j - C_j$	0	0	0	0	2/3	8/3	0
		$\dfrac{Z_j - C_j}{Sg3}$	−	−	−	−	−1	−4	−

TABLE 9.31

Simplex Table

		C_j	10	16	0	0	0	0	0
B	C_B	X_B	x_1	x_2	S_1	S_2	Sg1	Sg2	Sg3
x_2	16	4	0	1	0	0	0	2	−3/2
x_1	10	0	1	0	0	0	0	−3	5/2
S_2	0	12	0	0	0	1	0	20	−17
S_1	0	0	0	0	1	0	0	−2	1
Sg1	0	1	0	0	0	0	1	1	−3/2
Z = 64		Z_j	10	16	0	0	0	2	1
		$Z_j - C_j$	0	0	0	0	0	2	1

CHAPTER SUMMARY

- Dynamic programming is a very useful technique for making a sequence of interrelated decisions.
- Dynamic programming solves a relatively complex problem by breaking the problem into a series of simple problems.
- Dynamic programming requires the formulation of a suitable recursive relationship for each problem.
- The dynamic programming concept is largely based on the principle of optimality.
- The principle of optimality states, "An optimal policy (a sequence of decisions) has the property that whatever the initial state and decision are, the remaining decisions must constitute an optimal policy with regard to the state resulting from the first decision."
- Integer programming problems often occur when some or all of the decision variables have to be limited to integer values.
- The only difference between the integer linear programming problems and the linear programming problems is that one or more of the variables must be an integer.
- If some variables must be an integer, but not necessarily all, then we have a mixed-integer linear program. Most integer programming applications have 0-1 or binary variables.

The branch-and-bound method and the cutting plane method are popular methods for solving an integer programming problem.

Questions

1. What is dynamic programming? *(BT level 2)*
2. State the principle of optimality. *(BT level 1)*
3. What are the basic characteristics of a dynamic programming problem? *(BT level 2)*
4. How does dynamic programming differ conceptually from linear programming? *(BT level 5)*
5. Explain state and stage in the context of dynamic programming. *(BT level 2)*
6. What is integer programming? *(BT level 2)*
7. Differentiate between linear and integer programming. *(BT level 4)*
8. Explain the type of integer programming. *(BT level 2)*
9. Differentiate between pure and mixed-integer programming. *(BT level 4)*
10. Explain the zero-one programming problem. *(BT level 2)*
11. Name the methods used to solve the integer programming problem. *(BT level 1)*

Practice Problem

1. The sales manager for the plastic chair manufacturing industry has four travelling salespeople to be assigned to three different regions of the country. He decided that at least one salesperson should be assigned to each region and that each individual salesperson should be limited to one of the regions, but now he wants to determine how many salespeople should be assigned to the respective regions to maximize sales. Table 9.32 shows the estimated increase in sales (in appropriate units) in each region where different numbers of salesmen have been allocated:

TABLE 9.32

Data Practice Problem 1

No. of Salesmen	Zone 1	Zone 2	Zone 3
1	400	200	300
2	500	500	500
3	800	600	700
4	1,000	800	800

Use dynamic programming to solve this problem. *(BT level 3)*

2. A dealer must dispose of certain goods within five weeks. Market prices fluctuate from week to week. It is estimated that the chances of getting Rs. 4,000 for the whole stock are 40%, the chances of getting Rs. 5,000 are 35% and the chances of getting Rs. 6,000 are 25%. If the goods are not sold within the first four weeks, they will have to be disposed of at the prevailing market price in the fifth week. When are the stocks to be sold? *(BT level 3)*

3. The ship is to be loaded with a stock of three items. The maximum cargo weight that the ship can take is 6 and the details of the three items are as follows (Table 9.33):

TABLE 9.33

Data Practice Problem 3

Item	Weight	Value
1	3	10
2	4	15
3	2	5

Find the most valuable cargo load without exceeding the maximum cargo weight by using dynamic programming. *(BT level 3)*

4. Consider an electronic system consisting of four components, each of which must operate in order for the system to function. The system reliability can be improved by installing several parallel units in one or more of the components. The probability the system will work is the product of the probabilities that the respective components will work.

The probability and cost (in thousands of rupees) of the respective components are shown in Table 9.34:

TABLE 9.34

Data Practice Problem 4

Parallel Unit	Comp. 1		Comp. 2		Comp.3	
	Probability	Cost	Probability	Cost	Probability	Cost
1	0.4	1	0.5	2	0.6	1
2	0.5	3	0.6	4	0.7	1
3	0.7	4	0.7	5	0.8	2

Because of budget limitations, a maximum of Rs.10,000 can be spent. Use dynamic programming to determine how many parallel units should be installed on each of the three components to maximize the system's probability of functioning. *(BT level 3)*

5. Mr. 'X' produces pottery by hand. He has 24 hours a week to make bowls and vases. The bowl requires four hours of work, and the vase requires three hours of work. It takes 4 kg of special clay to make a bowl and 10 kg to make a vase; it takes 70 kg of clay per week. Mr. 'X' is selling his bowls for Rs. 150 and vases for Rs. 140. He wants to know how many of each item he makes every week to maximize his income. Formulate an integer programming model for this problem. *(BT level 6)*

6. Use the branch-and-bound method to solve the following integer programming problem. *(BT level 3)*

$$\max Z = 7x_1 + 5x_2$$

Subject to

$$7x_1 + 3x_2 \leq 84$$

$$2x_1 + 5x_2 \leq 94$$

$$\text{and } x_1, x_2 \geq 0 \text{ and integer}$$

7. Use a cutting plane method to solve the following integer programming problem. *(BT level 3)*

$$\max Z = 8x_1 + 14x_2$$

Subject to

$$x_1 + 2x_2 \leq 14$$

$$6x_1 + x_2 \leq 18$$

$$\text{and } x_1, x_2 \geq 0 \text{ and integer}$$

10

Goal Programming and Decision Making

Learning Outcomes: After studying this chapter, the reader should be able to

LO 1. Explain the concept of goal programming. (BT level 2)

LO 2. Formulate business problems involving multiple goals as goal programming problems. (BT level 6)

LO 3. Describe the graphical approach for solving the problem of goal programming. (BT level 2)

LO 4. Understand and apply the analytical hierarchy process to multi-criteria decision making. (BT level 2 and 3)

LO 5. Understand the decision-making process. (BT level 2)

LO 6. List the main components of a decision. (BT level 1)

LO 7. Describe the types of decision-making environments. (BT level 2)

LO 8. Make decisions under uncertainty and use decision criteria to suggest decisions. (BT level 3 and 5)

LO 9. Describe risk situations and use expected values to suggest decisions. (BT level 2 and 3)

LO 10. Construct decision trees and use them to decide alternatives. (BT level 6)

10.1 Goal Programming Introduction

LO 1. Explain the concept of goal programming

A single goal was either maximized or minimized in all the linear programming models outlined in Chapter 2. However, in today's business environment, maximizing profits or minimizing costs is not always the only goal set by a company. Maximizing total profits is often just one of several objectives, including conflicting objectives such as maximizing market share, maintaining full employment, providing quality environmental management, minimizing noise levels in the neighbourhood, a company that is at risk of a labour strike might want to avoid employee layoffs or a company that is likely to be fined for pollution offences, which might want to minimize pollution. So we see that management in the current business scenario has several contradictory objectives to achieve. That implies multi-dimensional decision criteria and multiple goals are involved in the decision. Goal programming takes on greater importance in this context as a powerful quantitative technique able to handle multiple decision criteria.

Goal programming is a variation of linear programming in that it considers more than one objective (called goals) in the objective function. Goal programming models are set-up as linear programming models in the same general format, with objective function and linear constraints. Model solutions are very similar to linear programming models. Goal programming is capable of dealing with multi-objective decision problems. A four-decade-old concept, it began in 1961 with the work of Charnes and Cooper and was refined and extended in the 1970s and 1980s by Lee and Ignizio.

Goal programming can be applied to almost unlimited management decision-making areas. It may be applied to media planning and product mix decisions in the field of marketing. It may be applied in finance for portfolio selection, capital budgeting, and financial planning. It can be applied in production to aggregate the planning and scheduling of the production. It can be used in the academic for assigning teaching schedules for the faculty and for the planning of university admission. It can be utilized for manpower planning in the HRD area. It can be applied to transportation systems in the area of public systems and medical planning is underway.

10.2 Goal Programming Formulation

The goals set by management can only be reached in traditional decision-making circumstances, at the cost of other goals. A hierarchy of importance among these goals must be defined such that goals of lower priority are addressed only after goals of higher priority are fulfilled. As it is not always possible to accomplish every goal to the degree that the decision maker wants, goal programming aims to achieve a reasonable amount of multiple objectives.

LO 2. Formulate business problems involving multiple goals as goal programming problems

The key difference between goal programming and linear programming is the objective function. Through goal programming, we attempt to eliminate deviations between set goals and what we can actually accomplish within the specified constraints, instead of attempting to explicitly maximize or minimize the objective function. These deviations are called slack and surplus variables in the linear programming simplex method. Since the coefficient is zero for all of these in the objective function, there is no effect on the optimal solution for slack and surplus variables. The deviational variables are usually the only variables in objective function in goal programming, and the aim is to minimize the sum of these deviation variables.

To illustrate the formulation of the goal programming problem, let us consider the following example.

$$Maximize\ Z = 20x_1 + 25x_2$$

Subject to

$$x_1 + 2x_2 \leq 40\ (\text{hours of labour})$$

$$2x_1 + 2x_2 \leq 100 \left(\text{raw material in kg}\right)$$

$$x_1, x_2 \geq 0,$$

where x_1 and x_2 are number of products A and B produced, respectively.

It's a standard linear programming model. The objective function, Z, is the total profit to be made by a firm from products A and B, given that Rs. 20 is the profit per product A and Rs. 25 is the profit per product B. The first constraint is on the labour available. It shows that product A requires one hour of labour and that product B requires two hours of labour and that there are 40 hours of labour available daily. The second constraint is for raw material and shows that both product A and product B require 2 kg of raw material and that the daily raw material limit is 100 kg.

But let's say that the company thinks that maximizing profit is not a reasonable goal, and management would like to achieve a satisfactory Rs. 50,000/- per day profit amount. We now have a goal programming problem where we want to find the production mix that achieves this goal as closely as possible. This basic case will provide a strong starting point for approaching more complex goal programs.

We first define two deviational variables:

d_1^- = underachievement of the profit target,

d_1^+ = overachievement of the profit target.

Now we can state the firms' problem as a *single-goal* programming model:
Minimize under or overachievement of profit target = $d_1^- + d_1^+$
Subject to

$$20x_1 + 25x_2 + d_1^- - d_1^+ = 50,000 \left(\text{profit goal constraint}\right)$$

$$x_1 + 2x_2 \leq 40 \left(\text{hours of labour}\right)$$

$$2x_1 + 2x_2 \leq 100 \left(\text{raw material in kg}\right)$$

$$x_1, x_2, d_1^-, d_1^+ \geq 0$$

Note that the first constraint states that the profit made, $20x_1 + 25x_2$, plus any underachievement of profit minus any overachievement of profit has to equal the target of Rs. 50,000. If the target profit is exactly achieved, we see that both d_1^- and d_1^+ are equal to zero. The objective function will also be minimized to zero. If the firm is concerned only with the underachievement of the target objective, how would the objective function change? It would be as follows: minimize underachievement = d_1^-. This is also a reasonable goal, as the firm would probably not be upset by the overachievement of its target.

In general, once all goals and constraints have been identified in a problem, management should evaluate each goal to see whether the underachievement or overachievement of that goal is an acceptable situation. If overachievement is appropriate, the related d_1^+ variable may be removed from the objective function. If the under-realization is perfect, the d_1^- variable should be removed. If management specifically seeks to accomplish a goal, both d_1^- and d_1^+ must appear in the objective function.

Let's look now at the situation in which the management of companies needs to accomplish several goals, each one of which is equal in priority.

Goal 1: To achieve a *satisfactory* profit level of Rs. 50,000/- per day

Goal 2: To avoid overtime in the firm

Goal 3: Avoid keeping more than 100 kg of raw material on hand each day.

Here the deviational variables can be defined as follows:

d_1^- = underachievement of the profit target,

d_1^+ = overachievement of the profit target,

d_2^- = idle time in the firm,

d_2^+ = over time in the firm,

d_3^- = amount of raw material less than 100 kg, and

d_3^+ = amount of raw material in excess of 100 kg.

Management is unconcerned as to whether there is overachievement of the profit goal, overtime at the company, or raw material in excess of 100 kg.; hence, d_1^+, d_2^+, and d_3^+ may be omitted from the objective function. The new objective function and constraints are

$$Minimize \text{ total deviation} = d_1^- + d_2^- + d_3^-$$

Subject to

$$20x_1 + 25x_2 + d_1^- - d_1^+ = 50,000 \text{ (profit goal constraint)}$$

$$x_1 + 2x_2 + d_2^- - d_2^+ = 40 \text{ (hours of labour constraint)}$$

$$2x_1 + 2x_2 + d_3^- - d_3^+ = 100 \text{ (raw material constraint)}$$

$$\text{All } x_i, d_i \geq 0$$

In summary from the previous illustration, the following are the major steps in the formulation of goal programming problem.

1. Identify the decision variables.
2. Formulate all the goals of the problem.
3. Reduce the number of goals by removing a few goals which are insignificant or redundant.
4. Specify priority levels for the goals.
5. Express each goal in the form of a constraints equation by adding a variable for negative and positive deviations (d_i^- and d_i^+).
6. Establish the objective function and constrained.

Examples based on goal programming formulation

EXAMPLE 1:

The manufacturer of ABC makes Deluxe and Supreme two types of air coolers. Deluxe takes one hour to assemble, while the Supreme takes two hours to assemble. Normal assembly is limited to 48 hours per week. The net profit from Deluxe and Supreme is Rs. 1,000 and Rs. 2,000, respectively. In order of priority, the management has set out the following goals.

Goal 1: Maximize total profit (Rs. 30,000/- per week)

Goal 2: Minimize overtime operation of the assembly line

Formulate this problem as a goal programming problem. *(BT level 6)*

Solution: Let the decision variables be

x_1 = number of deluxe air coolers produced each week,

x_2 = number of supreme air coolers produced each week.

Goals in priority as desired by the management of ABC.

Goal 1: Maximize total profit

Goal 2: Minimize the overtime operation of the assembly line

Here, the deviational variables can be defined as follows:

d_1^- = underachievement of the profit target,

d_1^+ = overachievement of the profit target,

d_2^- = slack time in the firm, and

d_2^+ = over time in the firm.

Management is unconcerned about whether there is overachievement of the profit goal or slack time in the firm; hence, d_1^+ and d_2^- may be omitted from the objective function. The objective function and constraints are

$$\text{Minimize total deviation } z = d_1^- + d_2^+$$

Subject to

$$1,000\,x_1 + 2,000\,x_2 + d_1^- - d_1^+ = 30,000 \,\left(\text{profit goal constraint}\right)$$

$$x_1 + 2x_2 + d_2^- - d_2^+ = 48 \,\left(\text{overtime assembly constraint}\right)$$

$$\text{All } x_i, d_i \geq 0$$

EXAMPLE 2:

ABC survey consultant has received a contract to conduct a survey. The firm must appoint interviewers to carry out the survey. Interviews are conducted by phone and in person. One person can conduct 40 telephone interviews or 20 personal interviews a day. It costs Rs. 1,200 a day for a telephone interviewer and Rs. 3,200 a day for a personal interviewer. The following three goals, which are set out in order of priority, have been established by the firm to ensure a representative survey:

a. At least 2,000 total interviews should be conducted.
b. An interviewer should conduct only one type of interview each day. The firm wants to maintain its daily budget of Rs. 20,000.
c. At least 1,000 interviews should be by telephone.

Formulate a goal programming model of this problem. *(BT level 6).*
Solution: Let the decision variables be

x_1 = number of telephone interviewers,
x_2 = number of personal interviewers.

Goals in priority as desired by the management of ABC.

Goal 1: At least 2,000 total interviews should be conducted.
Goal 2: An interviewer should conduct only one type of interview each day. The firm wants to maintain its daily budget of Rs. 20,000.
Goal 3: At least 1,000 interviews should be by telephone.

Here, the deviational variables can be defined as follows:

d_1^- = underachievement of the total interviews target
d_1^+ = overachievement of the total interviews target
d_2^- = under use of daily budget
d_2^+ = over use of daily budget
d_3^- = underachievement of the telephone interview target
d_3^+ = overachievement of the telephone interview target

The objective function and constraints are

$$\text{Minimize total deviation } z = P_1 d_1^- + P_2 d_2^+ + P_3 d_3^-$$

Subject to
$$40 x_1 + 20 x_2 + d_1^- - d_1^+ = 2,000 \ (\text{goal 1 constraint})$$

$$1,200 x_1 + 3,200 x_2 + d_2^- - d_2^+ = 20,000 \ (\text{goal 2 constraint})$$

$$1,200\,x_1 + d_3^- - d_3^+ = 1,000 \ \left(\text{goal 3 constraint}\right)$$

$$\text{All } x_i,\, d_i \geq 0$$

In the objective function the symbol P_1 designates the minimization of d_1^- as the goal of first priority. When this model is solved, the first step is to minimize the d_1^- value before addressing any other goal. Likewise, symbols P_2 and P_3 are added to the objective function.

EXAMPLE 3:

ABC firm's marketing team is trying to decide how to better promote the product during the two weeks prior to its launch. Four types of advertisements were defined by the marketing department: TV ads, radio ads, the internet, and newspaper ads. The costs are Rs. 50,000/- for each TV ad, Rs. 30,000/- for each radio ad, Rs. 36,000/- for the internet for one month, and Rs. 60,000/- for each newspaper ad. It has been estimated that the audience reached for each form of advertising is 40,000 for each TV ad, 30,000 for each radio ad, 30,000 for each internet ad, and 20,000 for each newspaper ad. The total monthly budget for ads is Rs. 900,000/-. The following goals have been established and ranked:

1. The number of people reached is expected to be at least 1,000,000.
2. There should be no excess of the overall monthly advertisement budget.
3. Together, the number of advertisements will be at least five on either TV or radio.
4. There should be no more than ten advertisements for any advertising type.

Formulate this as a goal programming problem. *(BT level 6).*

Solution: Let the decision variables be

x_1 = number of TV ad

x_2 = number of Radio ad

x_3 = number of internet ad

x_4 = number of a newspaper ad

Goals in priority as desire by the marketing department of ABC.

Goal 1. The number of people reached is expected to be at least 1,000,000.

Goal 2. There should be no excess of the overall monthly advertisement budget.

Goal 3. Together, the number of advertisements will be at least five on either TV or radio.

Goal 4. There should be no more than ten advertisements for any advertising type.

Here, the deviational variables can be defined as follows:

d_1^- = underachievement of the number of people reached

d_1^+ = overachievement of the number of people reached

d_2^- = under use of monthly advertising budget

d_2^+ = over use of monthly advertising budget

d_3^- = underachievement of the number of ads on either TV or radio

d_3^+ = overachievement of the number of ads on either TV or radio

$d_4^-, d_5^-, d_6^-, d_7^-$ = underachievement of any one type of advertising

$d_4^+, d_5^+, d_6^+, d_7^+$ = overachievement of any one type of advertising

The objective function and constraints are

$$\text{Minimize total deviation } z = P_1 d_1^- + P_2 d_2^+ + P_3 d_3^- + P_4 d_4^+ + P_4 d_5^+ + P_4 d_6^+ + P_4 d_7^+$$

Subject to

$$40,000\, x_1 + 30,000\, x_2 + 30,000\, x_3 + 20,000\, x_4 + d_1^- - d_1^+ = 1,000,000 \; \text{(goal 1 constraint)}$$

$$50,000\, x_1 + 30,000\, x_2 + 36,000\, x_3 + 60,000\, x_4 + d_2^- - d_2^+ = 900,000 \; \text{(goal 2 constraint)}$$

$$x_1 + x_2 + d_3^- - d_3^+ = 5 \; \text{(goal 3 constraint)}$$

$$x_1 + d_4^- - d_4^+ = 10 \; \text{(goal 4 constraint)}$$

$$x_2 + d_5^- - d_5^+ = 10 \; \text{(goal 4 constraint)}$$

$$x_3 + d_6^- - d_6^+ = 10 \; \text{(goal 4 constraint)}$$

$$x_4 + d_7^- - d_7^+ = 10 \; \text{(goal 4 constraint)}$$

$$\text{All } x_i, d_i \geq 0$$

In the objective function the symbol P_1 designates the minimization of d_1^- as the goal of first priority. When this model is solved, the first step is to minimize the d_1^- value before addressing any other goal. Likewise, symbols P_2, P_3 and P_4 are added to the objective function.

10.3 Graphical Solution Method of Goal Programming

LO 3. Describe the graphical approach for solving the problem of goal programming

The graphical goal programming solution procedure is similar to that for linear programming discussed in Chapter 2. The only difference is that the goal programming method requires a different solution for each priority level. In linear programming, the approach is used to maximize or minimize an objective function with one goal, while in goal programming; the total deviation from a set of multiple goals is minimized. The deviation from the goal with the highest priority is minimized to the maximum extent possible prior to minimizing deviation from the next goal.

The following are the major steps in the graphical solution method of goal programming problems.

1. Formulate the goal programming problem
2. Create a graph of all the goals in terms of decision variables. Arbitrarily pick two sets of points for each goal equation. Plot and connect each set of points in a straight line. Indicate the positive and negative deviation variable by the arrow for each goal line.
3. Identify the goal line which corresponds to the highest priority goal. Locate the feasible area at the first-priority level for this goal.
4. Move to the next highest-priority goal(s) and work out the best solution(s) space for the goal(s) at this next high-priority level, ensuring that the best solution(s) are not degrading the solution(s) already achieved for the higher priority goal.
5. Repeat step 4 until all priority levels have been investigated.
6. Identify the optimal solution for the most acceptable best value, as set out in Step 5.

Examples based on the graphical solution method of goal programming

EXAMPLE 1:

ABC Company produces two products, A and B. Each A product requires 20 man-hours, while each B requires 30 man-hours. There are 60 hours of work available every week. A and B both earn a profit of Rs. 200/-. In order of priority, the management of ABC Company has set out the following goals.

1. Avoid overtime utilization of man-hours
2. Achieve a weekly profit of at least Rs. 2,000
3. Supply of at least six products of product B every week to one of its old customers
    a) Formulate this problem as a goal programming problem and *(BT level 6)*.
    b) Solve it using a graphical method. *(BT level 3)*.

Solution:

a) Let the decision variables be
x_1 = number of product A produce each week
x_2 = number of product B produce each week

Goals in priority as desired by the management of ABC.

Goal 1: avoid overtime utilization of man-hours

Goal 2: achieve a weekly profit of at least Rs. 2,000

Goal 3: supply of at least six product of product B every week to one of its old customer

Here, the deviational variables can be defined as follows:

d_1^- = work in regular time

d_1^+ = work in overtime

d_2^- = underachievement of the profit target

d_2^+ = overachievement of the profit target

d_3^- = underachievement of product B

d_3^+ = overachievement of product B

The first goal is to be reached if the overtime is minimized, i.e. the positive deviation d_1^+ is minimized. The second goal would be accomplished if profit is maximized and will occur if the negative deviation d_2^- is minimized. When the underperformance of the product B d_3^- is minimized, the third goal is achieved.

The objective function and constraints are

$$\text{Minimize total deviation } z = P_1 d_1^+ + P_2 d_2^- + P_3 d_3^-$$

Subject to

$$20 x_1 + 30 x_2 + d_1^- - d_1^+ = 60 \text{ (goal 1 constraint)}$$

$$200 x_1 + 200 x_2 + d_2^- - d_2^+ = 2{,}000 \text{ (goal 2 constraint)}$$

$$x_2 + d_3^- - d_3^+ = 6 \text{ (goal 3 constraint)}$$

$$\text{All } x_i, d_i \geq 0$$

b) As there are only two decisions variable x1 and x2, there would be two axes, x1 on the horizontal axis and x2 on the vertical axis, in the graph of different goals. To graph this model, the deviational variables are set at zero in each goal constraint, and we graph each subsequent equation on a set of coordinates, as discussed in Chapter 2 for the linear programming graphical method. Figure 10.1 is a graph of the three-goal constraints for this model.

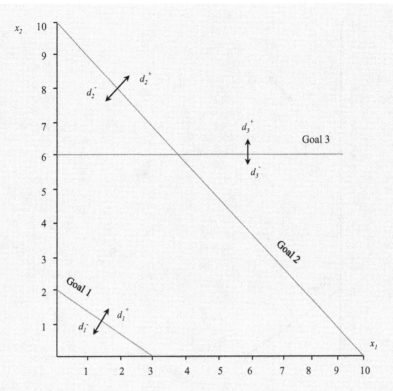

FIGURE 10.1
Graphical Solution For Example 1.

In this example, we first consider the first-priority goal of minimizing d_1^+. The relationship of d_1^- and d_1^+ to the goal constraint is shown in Figure 10.2. The area below the goal 1 constraint line represents possible values for d_1^-, and the area above the line represents values for d_1^+. To achieve the goal of minimizing d_1^+ the area above the constraint line corresponding to d_1^+ is removed, leaving the shaded area as a feasible solution area. We may set $d_1^+ = 0$ as its minimum value. The region satisfying $d_1^+ = 0$ and $x_1, x_2 \geq 0$ is shown shaded in Figure 10.2. Any pair of values x_1, x_2 in the region shall satisfy goal 1.

Next, we're looking at the second-priority goal, minimizing d_2^-. It can be easily noted that d_2^- cannot be set at zero, i.e. its minimum value, as the previous priority objective is degraded. The minimum d_2^- value without impacting the higher priority goals previously achieved is $x_1 = 3$ and $x_2 = 0$, and this point is shown in Figure 10.2.

Next, let's consider the last-priority goal, minimizing d_3^-. But we see that any d_3^- minimization would degrade at least one higher priority goal already achieved. Consequently, the optimal solution to the problem is

$$x_1 = 3, \ x_2 = 0, \ d_1^+ = 0, \ d_2^- = 1{,}400, \ d_3^- = 6.$$

Optimal values of the objective function simply indicate that the highest-priority objective (Goal 1) has been achieved while the second and third-priority objectives have not been achieved.

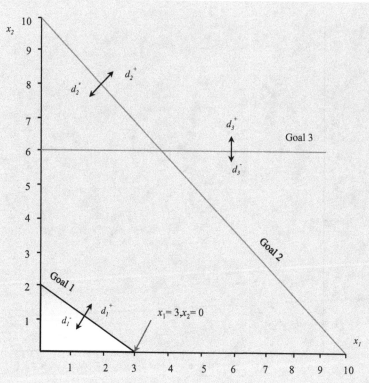

FIGURE 10.2
Graphical Solution for Example 1.

EXAMPLE 2:

Solve the following goal programming problem graphically. *(BT level 3)*

$$\text{Minimize total deviation } z = P_1 d_1^- + P_2 d_2^- + P_3 d_3^+ + P_4 d_4^-$$

Subject to

$$70 x_1 + 60 x_2 + d_1^- - d_1^+ = 420 \ (\text{goal 1 constraint})$$

$$15 x_1 + 30 x_2 + d_2^- - d_2^+ = 120 \ (\text{goal 2 constraint})$$

$$60 x_1 + 50 x_2 + d_3^- - d_3^+ = 300 \ (\text{goal 3 constraint})$$

$$x_2 + d_4^- - d_4^+ = 8 \ (\text{goal 4 constraint})$$

$$\text{All } x_i, \, d_i \geq 0$$

The goals have been listed in order of priority.

Solution: Let the decision variables be x_1 and x_2

As there are only two decisions variable $x1$ and $x2$, there would be two axes, $x1$ on the horizontal axis and $x2$ on the vertical axis, in the graph of different goals. To graph this model, the deviational variables are set at zero in each goal constraint, and we graph each subsequent equation on a set of coordinates, as discussed in Chapter 2 for the linear programming graphical method. Figure 10.3 is a graph of the four goal constraints for this model.

In this example, we first consider the first-priority goal, minimizing d_1^-. The relationship of d_1^- and d_1^+ to the goal constraint is shown in Figure 10.3. The area below the goal 1 constraint line represents possible values for d_1^- and the area above the line represents values for d_1^+. In order to achieve the goal of minimizing d_1^- the area below the constraint line corresponding to d_1^- is eliminated. The feasible area is the region above the goal 1 constraint line.

Next, we consider the second-priority goal, minimizing d_2^-. The area below the goal 2 constraint line represents possible values for d_2^- and the area above the line represents values for d_2^+. To achieve the goal of minimizing d_2^- the area below the constraint line corresponding to d_2^- is eliminated. The feasible area is the region above the goal 2 constraint line. Notice that by eliminating the area for d_2^-, we do not affect the first-priority goal of minimizing d_1^-.

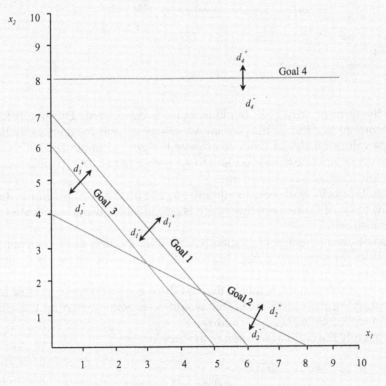

FIGURE 10.3
Graphical Solution for Example 2.

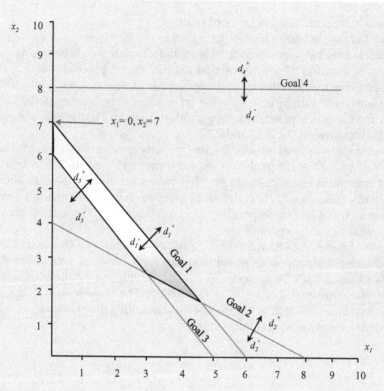

FIGURE 10.4
Graphical Solution for Example 2.

Next, the third-priority goal, minimizing d_3^+ is considered. The area below the goal 3 constraint line represents possible values for d_3^- and the area above the line represents values for d_3^+. In order to achieve the goal of minimizing d_3^+ the area above the constraint line corresponding to d_3^+ is eliminated. The feasible area is the region below the goal 3 constraint line.

The fourth priority goal seeks to minimize d_4^-. To do this requires eliminating the area below the goal 4 constraint line which is not possible given the previous higher priority constraints.

Any point inside the feasible region bounded by the first three constraints will meet the three most priority goals. This feasible region is shown by the shaded area in Figure 10.4.

The optimal solution must satisfy the first three goals and come as close as possible to satisfying the fourth goal. This would be a point shown on the graph in Figure 10.4 with coordinates $x_1 = 0$ and $x_2 = 7$.

Consequently the optimal solution to the problem is

$$x_1 = 0, x_2 = 7, d_1^- = 0, d_2^- = 0, d_2^+ = 90, d_3^+ = 50, d_4^- = 1.$$

10.4 The Analytic Hierarchy Process (AHP)

LO 4. Understand and apply the analytical hierarchy process to multi- criteria decision making

The AHP, developed by Thomas Saaty (1980), is an effective method for dealing with difficult decision making that can help decision makers to set priorities that make the right decision. It is a way of choosing the best decision-making choices when a decision maker has multiple objectives or criteria for a decision maker. AHP requires the decision maker to make decisions on the significance of each criterion and to specify a preference, using each criterion for each alternative decision. The output of AHP is the priority of decision-making alternatives based on the overall preferences of decision makers.

The AHP considers a set of assessment criteria, as well as a set of alternative options for making the best decisions. It is important to note that since some of the criteria might be contrasting, it is not generally true that the best option is the one that optimizes each criterion but the one that achieves the most appropriate trade-off between the different criteria. The AHP generates a weight for each evaluation criterion according to the pairwise comparison of the criteria by the decision maker. The higher the weight, the more important the corresponding criterion is.

The calculations made by the AHP are always guided by the experience of the decision maker, and the AHP can, therefore, be considered as a tool capable of converting the assessments (both qualitative and quantitative) made by the decision maker into a multi-criteria ranking. Moreover, the AHP is simple because there is no need to build a complex system of experts with the knowledge of the decision maker embedded in it. AHP has the advantage of handling situations in which the individual decision makers unique subjective judgments play an important part in the decision-making process.

10.4.1 Steps in AHP

The basic procedure to carry out the AHP consists of the following steps:

Step 1. Structuring a decision problem and selection of criteria

The first step is to decompose the problem of decision into its constituent parts. This structure, in its simplest form, includes a goal or objective at the highest level, criteria (and subcriteria) at the intermediate level, while the lowest level contains the alternatives. The arrangement of all elements in a hierarchy offers a general image of complicated relationships and allows the decision maker to decide if the elements in each level are of the same magnitude to be measured precisely.

Step 2. Making pairwise comparisons

Pairwise comparisons are the basic building blocks of AHP. In AHP, the decision maker decides how well every option "scores" using pairwise comparisons on a criterion. The decision maker compares two alternatives (i.e., a pair) according to one criterion and shows a choice in a pairwise comparison. Such comparisons are made using a preference scale that assigns numerical values to the different preference levels. The standard preference scale used for AHP is shown in Table 10.1. This scale has been established by experienced AHP researchers as a rational basis for comparing two objects or alternatives. Every scale rating is based on a comparison of two items.

TABLE 10.1

Scale for Pairwise Comparisons

Verbal Judgment	Numeric Value
Equally preferred	1
Equally to moderately preferred	2
Moderately preferred	3
Moderately to strongly preferred	4
Strongly preferred	5
Strongly to very strongly preferred	6
Very strongly preferred	7
Very strongly to extremely preferred	8
Extremely preferred	9

A score is given for each pairing within each criterion, again on a scale between 1 and 9, while the other alternative in the pairing is assigned a score equal to the inverse of that valuation. Each score tracks how well the "X" alternative meets the "Y" criteria. The scores are then weighted and summed.

Step 3. Developing preferences within criteria

The next step is to prioritize the alternatives for decision making under each criterion. It means we want to assess which criteria or alternatives are the most preferred, which is the second most preferred, and so on. That aspect of AHP is called synthesization. The exact mathematical procedure needed to perform the synthesis is beyond this text's scope. But a good approximation of the synthesis results is given by the following three-stage method:

1. Sum the values in each column of the pairwise comparison matrix.
2. Divide each number in the column by the corresponding sum of the column. It results in a standardized matrix. Note that the sum of the values at each column is 1.
3. Calculate the average of the elements in each row of the normalized pairwise comparison matrix; these averages set the criteria priorities.

The AHP synthesization method gives priority to each criterion as regards its contribution to the overall goal.

Step 4. Calculating an overall score for each alternative decision based on the magnitude of their scores by multiplying the criteria preference vector by the criteria matrix and rank the alternative decisions

Consistency:

One important factor in the AHP process is the consistency of the decision makers' pairwise decisions. With multiple pairwise comparisons, it is difficult to achieve the perfect consistency. In addition, it can be assumed that there will be some degree of inconsistency in almost every set of pairwise comparisons. AHP offers a tool for calculating the degree of consistency between pairwise comparisons made by the decision maker to deal with the consistency problem. If the degree of consistency is not appropriate, the decision maker will evaluate and update the pairwise comparisons before continuing with the AHP analysis.

TABLE 10.2

RI Values

n	2	3	4	5	6	7	8	9	10
RI	0	0.58	0.90	1.12	1.24	1.32	1.41	1.45	1.51

By computing a consistency ratio, AHP provides a measure of consistency for the pairwise comparisons. This ratio is designed in such a way that a value greater than 0.10 indicates inconsistencies in pairwise judgments. Thus if the consistency ratio is 0.10 or less, the consistency of the pairwise comparisons is considered fair and the AHP method will proceed with the measurement of the synthesis.

Use the following formulae to calculate the consistency ratio.

$$CI = \frac{\lambda_{max} - n}{n - 1},\tag{10.1}$$

$$CR = \frac{CI}{RI},\tag{10.2}$$

where λ_{max} is the largest eigenvalue, n is the rank of the matrix, CI is the consistency index and RI is the random index. The *RI* has the values shown in Table 10.2, depending on the number of items, *n*, being compared.

Example based on AHP

EXAMPLE 1:

Mr. X wishes to buy a new car and is considering three models – Model A, Model B, and Model C. Mr. X has defined three preference criteria on which he will base his decision: price, safety, and durability. Mr. X has developed the following pairwise comparison matrices (Table 10.3) for the three criteria:

Mr. X has prioritized his decision criteria according to the following pairwise comparisons (Table 10.4).

Using AHP, develop an overall ranking of the three cars Mr. X is considering. *(BT level 3)*

Solution: The first step in AHP is to create a graphic representation of the problem in terms of the ultimate goal, the criteria to be used, and the options for decision making. Such a graph represents the problem hierarchy. Figure 10.5 shows the hierarchy for the car selection problem.

Now we can measure the priority of each criterion in terms of its contribution to the overall goal of selecting the best car using the pairwise comparison matrix given in the problem.

We carry out this three-step procedure for the criteria pairwise comparison matrix.

TABLE 10.3

Pairwise Comparison Matrixes

Car	Price		
	A	B	C
A	1	4	7
B	1/4	1	3
C	1/7	1/3	1

Car	Safety		
	A	B	C
A	1	1/4	1/8
B	4	1	1/5
C	8	5	1

Car	Durability		
	A	B	C
A	1	4	1
B	1/4	1	1/3
C	1	3	1

TABLE 10.4

Pairwise Comparisons for Decision Criteria

Criteria	Price	Safety	Durability
Price	1	4	6
Safety	1/4	1	3
Durability	1/6	1/3	1

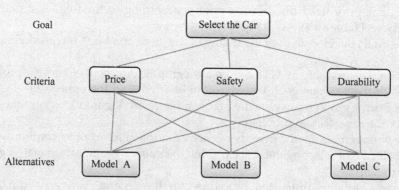

FIGURE 10.5

Hierarchy for the Car Selection Problem.

Step 1. Sum the values in each column (Table 10.5).
Step 2. Divide each element of the matrix by its column total (Table 10.6).
Step 3. Average the elements in each row to determine the priority of each criterion (Table 10.7).

Applying the aforementioned steps, we develop priority for all the pairwise comparison matrices for criteria safety and durability as follows (Table 10.8).

Next, we rank the criteria by applying synthesization steps to the decision criteria pairwise comparison matrix and obtained the following matrix (Table 10.9).

The overall score for each car is now determined by multiplying the values in the preference vector criteria by the preceding matrix criteria and summing the products as follows:

$$\text{Car model A score} = 0.701(0.685) + 0.070(0.221) + 0.457(0.093) = 0.538,$$

TABLE 10.5

Step 1

Car	Price		
	A	B	C
A	1	4	7
B	1/4	1	3
C	1/7	1/3	1
Sum	1.392	5.333	11

TABLE 10.6

Step 2

Car	Price		
	A	B	C
A	0.718	0.750	0.636
B	0.179	0.187	0.272
C	0.102	0.062	0.090

TABLE 10.7

Step 3

Car	Price			Priority
	A	B	C	
A	0.718	0.750	0.636	0.701
B	0.179	0.187	0.272	0.212
C	0.102	0.062	0.090	0.084

TABLE 10.8

Priority

Car	Safety			Priority
	A	B	C	
A	0.076	0.040	0.094	0.070
B	0.307	0.160	0.150	0.205
C	0.615	0.800	0.754	0.723
Car	Durability			Priority
	A	B	C	
A	0.444	0.500	0.429	0.457
B	0.111	0.125	0.143	0.126
C	0.444	0.375	0.429	0.416

TABLE 10.9

Synthesization

Criteria	Price	Safety	Durability	Priority
Price	0.706	0.750	0.600	0.685
Safety	0.176	0.187	0.300	0.221
Durability	0.117	0.062	0.100	0.093

TABLE 10.10

Car Priority

Car	Priority
Model A	0.538
Model C	0.256
Model B	0.202

$$\text{Car model B score} = 0.212(0.685) + 0.205(0.221) + 0.126(0.093) = 0.202,$$

$$\text{Car model C score} = 0.084(0.685) + 0.723(0.221) + 0.416(0.093) = 0.256.$$

Ranking these priorities, we have the AHP ranking of the decision alternatives (Table 10.10):

The AHP priorities show that the car model A is preferred.

Consistency:

We test the consistency of the comparisons for the three-car selection criteria in pairs to show how the CI is computed. This matrix, shown as follows, is multiplied by the chosen vector for the criteria:

$$\begin{bmatrix} 1 & 4 & 6 \\ \dfrac{1}{4} & 1 & 3 \\ \dfrac{1}{6} & \dfrac{1}{3} & 1 \end{bmatrix} \times \begin{bmatrix} 0.685 \\ 0.221 \\ 0.093 \end{bmatrix}$$

The product of the multiplication is computed as follows:

$$1(0.685) + 4(0.221) + 6(0.093) = 2.127$$

$$1/4(0.685) + 1(0.221) + 3(0.093) = 0.671,$$

$$1/6(0.685) + 1/3(0.221) + 1(0.093) = 0.280.$$

Next, we divide each of these values by the corresponding weights from the criteria preference vector as follows:

$$2.127 / 0.685 = 3.105,$$

$$0.671 / 0.221 = 3.036,$$

$$0.280 / 0.093 = 3.010.$$

Now, compute the average of the previous values; this average is denoted λmax.

$$\lambda \max = \frac{3.105 + 3.036 + 3.010}{3} = 3.050$$

Next, compute the CI as follows:

$$CI = \frac{\lambda_{\max} - n}{n - 1}$$

$$= \frac{3.050 - 3}{3 - 1} = 0.025.$$

Now compute the consistency ratio, as follows:

$$CR = \frac{CI}{RI},$$

with $n = 3$ criteria, we have RI = 0.58 (from Table 10.2):

$$CR = \frac{0.025}{0.58} = 0.0431.$$

A consistency ratio of 0.10 or less is deemed appropriate, as stated earlier. Since the pairwise comparisons for the car selection criteria show CR = 0.0431, we can assume that the degree of consistency in the pairwise comparisons is appropriate.

Remember that we evaluated the degree of consistency in this instance only for the pairwise comparisons in the preferential matrix of the decision criteria. This does not mean we checked the consistency for the whole AHP. We will also need to test the pairwise comparisons for each of the three individual criterion matrices before we can be sure that the entire AHP was consistent with this problem.

10.5 Introduction to Decision Making

To a large extent, the successes or failures that a person experiences in his or her life depend on the decisions he or she makes. Resources are limited, and human needs and desires are infinite and diversified, and each person wants to fulfil their

LO 5. Understand the decision-making process

needs. Here, the theory of decisions helps to make a certain decision in order to meet their needs most adequately. Decisions are made to achieve those goals and goals. Hundreds of operational decisions are taken to achieve local outcomes that contribute to the company's overall strategic goal. These local outcomes are generally calculated not directly by benefit but by consistency, cost-effectiveness, performance, productivity, and so on. It is an essential goal for individual operational units and individual operational managers to produce good results for local outcomes. However, all these decisions are interrelated and must be coordinated with a view to achieving the overall goals of the company.

A decision is the result of a process intended to assess the relative benefits or merits of a set of alternatives available to determine the most desired course of action for implementation. A decision can, in general, be characterized as the selection of an act by the decision maker, deemed to be better in compliance with any of the pre-designed criteria from among the many options available.

10.5.1 Decision-Making Components

Regardless of the form of difficulty with decision making, there are certain important elements that are similar to all of these problems. They are as follows:

LO 6. List the main components of a decision

1) Courses of action: There are a number of possible courses of action established by a decision. These are often referred to as acts, actions, or strategies which are under control and are known to the decision maker.

2) State of nature: Actual events that may occur in the future, known as states of nature. At the time of the decision, the decision maker does not know and has no power over the state of nature in the future.

3) Payoff: The combination of a course of action and an event is correlated with a payoff, which determines the net profit to the decision maker resulting from a combination of decisions and events.

4) Payoff table: The payoff table is a way of organizing and explaining the payoffs of the various decisions, despite the specific states of nature of the decision-making problem.

10.6 Decision-Making Environments

LO 7. Describe the types of decision-making environments

A quantitative technique called decision analysis is illustrated in this section. Decision analysis is a common method that can be applied to a variety of different forms of organizational decision making. The types of decisions that people make depend on the amount of knowledge or facts they have about the situation. We are addressing the following decision-making environments.

LO 8. Make decisions under uncertainty and use decision criteria to suggest decisions

a) Decision making under uncertainty

b) Decision making under risk

10.6.1 Decision Making under Uncertainty

When there are several states of nature and the manager cannot reliably assess the probabilities or if there is basically no data on probabilities, the condition is called uncertain decision making. Under conditions of uncertainty, only payoffs are known, and the probability of each state of nature is not known.

When the decision scenario has been arranged into a payoff table, a number of criteria are available for making the final decision. Such decision criteria, which will be set out in this section, shall include the following:

1. Maximax (optimistic)

2. Maximin (pessimistic)

3. Hurwicz (criterion of realism)

4. Laplace (equally likely)

5. Minimax regret

Let's take a look at each of the above models

1. Maximax (optimistic):

   The decision maker selects the decision with the maximax criterion which will result in the maximum (best) of the maximum (best) payoffs. (This is indeed how the name derives of this criterion, a maximum of a maximum.) The maximax criterion is very optimistic. The decision maker assumes that for each alternative decision the most favourable state of nature will occur. By using the optimistic criterion for minimizing problems where lower payoffs (e.g., cost) are better, you'd look at the best (minimum) payoff for each alternative and choose the best (minimum) alternative. This is the term as a minimin criterion.

   The maximax criterion includes the following steps:

   Step 1. For each alternative, determine the maximum possible payoff.

   Step 2. Choose the alternative that meets the maximum of maximum payoffs mentioned above.

2. Maximin (pessimistic):

   The decision maker selects the decision with the maximin criterion that will reflect the maximum of the minimum payoffs. The decision maker assumes the minimum payoff will occur for every alternative decision. The maximum of those minimum payoffs is selected. This criterion ensures the payoff is at least the maximin value (the best of the worst). Choosing any other alternative could result in a worse (lower) payoff. In the case of problems involving minimization (cost), this approach identifies an alternative that minimizes the maximum payoff.

   The maximin criterion includes the following steps:

   Step 1. Determine the minimum assured payoff for each alternative.

   Step 2. Select that alternative which corresponds to the maximum of the above minimum payoffs.

   For each alternative, the maximax and maximin criteria consider only one extreme payoff, while all other payoffs are ignored. The next criterion concerns both of these extremes.

3. Hurwicz (criterion of realism):

   The Hurwicz criterion strikes a compromise between the maximax and maximin criteria. The principle behind this criterion is that the decision maker is either not totally optimistic (as assumed by the maximax criterion) or completely pessimistic (as assumed by the maximin criterion). It is often referred to as the weighted average. First of all, the realistic coefficient, α is chosen; this measures the decision makers' degree of optimism. This coefficient is between 0 and 1. If α is 1, the decision maker is 100% optimistic for the future. At α, the decision maker is 100% pessimistic about the future. The advantage is that the decision maker can build on personal sentiments regarding relative optimism and pessimism. The weighted average is computed as follows:

$$\text{Weighted average} = \alpha\left(\text{best in row}\right) + \left(1 - \alpha\right)\left(\text{worst in row}\right). \qquad (10.3)$$

The Hurwicz criterion includes the following steps:

Step 1. Select the decision maker's appropriate degree of optimism (or pessimism). When α is the optimism coefficient, $1 - \alpha$ is the pessimism coefficient.

Step 2. For each alternative, determine maximum and minimum pay. Multiply the maximum payoff by α, and multiply the minimum payoff by $1 - \alpha$.

4. Laplace (equally likely)

   One criterion that uses all payoffs for each alternative is the Laplace, also referred to as the equally likely decision criterion. This includes finding the average payoff per alternative and selecting the best or highest average alternative. The equally likely approach assumes that all probabilities of occurrence are equal for the states of nature, and thus every state of nature is equally likely.

   The Laplace criterion includes the following steps:

   Step 1. Assign equal chances ($1/n$) for each strategy payoff (with n possible payoffs)

   Step 2. Determine for each alternative the expected payoff value and select the alternative that is equal to the maximum of these payoffs.

5. Minimax regret

   It is based on the concept of regret (opportunity loss) and calls for the action to be taken to minimize the greatest regret. Opportunity loss refers to the difference between a given state of nature's optimal profit or payoff and the actual payoff received for a particular decision. That is to say, it is the amount lost by not choosing the best alternative in a given state of nature.

   The minimax criterion includes the following steps:

   Step 1. Determine the amounts of regret for a payoff of each alternative for a particular event.

   Step 2. Determine for each alternative the maximum amount of regret.

   Step 3. Select the alternative that matches the minimum regrets.

Examples based on decision making under uncertainty

EXAMPLE 1:

Mr. X is the president of the profit-making company XYZ. The problem Mr. X identifies is whether to expand its product line by producing a new product and marketing it. Mr. X decides that his alternatives are to construct (1) a large new plant for the production of a new product, (2) a small plant, or (3) not to develop a new product line.

Mr. X has already assessed the potential profits associated with the various outcomes. With good competitive market conditions, he believes that a big new plant would result in a net profit of Rs. 2,000,000 for his company. The conditional value, if the market has poor competitive conditions, would be Rs. 1,000,000 net losses. A small plant would have resulted in a net profit of Rs. 1,000,000 with good competitive market conditions, but a net loss of Rs. 500,000 would have occurred under poor competitive conditions. At the end of the day, doing nothing would result in Rs. 0 profit on either market.

(a) Make the payoff table describes this decision situation *(BT level 6)*, (b) determine the best decision using each of the following decision criteria *(BT level 3)*, and (c) compare the decision from the following decision criteria and support the best decision. *(BT level 4)*

1. Maximax (Optimistic) 2. Maximin (Pessimistic)
3. Hurwicz (Criterion of realism) assume $\alpha = 0.7$ 4. Laplace (Equally likely)
5. Minimax regret

Solution:

a) The following payoff in Table 10.11 describes the given decision situation. A negative sign is included for losses.

b) The best decision is determined using different decision criteria as follows:

1. Maximax (optimistic): The decision is chosen which will result in a maximum of the maximum payoffs. The company would thus be optimistic for this example that good competitive conditions will prevail in the future, resulting in the following maximum payoffs and decisions:

$$\text{Big plant } 2,000,000 \leftarrow (\text{Maximum})$$

$$\text{Small plant } 1,000,000$$

Not developing the new product line 0

Decision: Construct a big plant to manufacture a new product

2. Maximin (pessimistic): The decision maker selects the decision with the maximin criterion which will reflect the maximum of minimum payoffs. The

TABLE 10.11

Payoff Table

Alternative	State of Nature	
	Good Market Competitive Conditions (Rs.)	Poor Market Competitive Conditions (Rs.)
Big plant	2,000,000	−1,000,000
Small plant	1,000,000	−500,000
Not developing the new product line	0	0

decision maker assumes there will be a minimum payoff for each alternative decision, of which the maximum is chosen as follows:

$$\text{Big plant} - 1,000,000$$

$$\text{Small plant} - 500,000$$

Not developing the new product line 0 (maximum)
Decision: Not developing the new product line

3. Hurwicz (criterion of realism): A compromise is made between the maximax and maximin criteria. Given that $\alpha = 0.7$ so $1 - \alpha = 0.3$.

$$\text{Big plant}: 0.7(2,000,000) - 0.3(1,000,000) = 1,100,000\,(\text{Maximum})$$

$$\text{Small plant}: 0.7(1,000,000) - 0.3(500,000) = 550,000$$

Not developing the new product line: 0
Decision: Construct a big plant to manufacture a new product

4. Laplace (equally likely): It weighs equally on each state of nature, assuming that the states of nature are equally likely to occur. Since our example contains two states of nature, we assign each one a weight of 0.50. Next, we multiply these weights for each decision by payoff and choose the alternative with the maximum of these weighted values.

$$\text{Big plant}: 0.5(2,000,000) - 0.5(1,000,000) = 500,000\,(\text{Maximum})$$

$$\text{Small plant}: 0.5(1,000,000) - 0.5(500,000) = 250,000$$

Not developing the new product line: 0
Decision: Construct a big plant to manufacture a new product

5. Minimax regret: By choosing the alternative decision that reduces the maximum regret, the decision maker attempts to avoid regret. First, select the maximum payoff under each state of nature; then, subtract from these amounts all other payoffs under the respective states of nature, as follows (Table 10.12):

The maximum regret has to be determined for each decision, and the decision corresponding to the minimum of these regret values is chosen as follows:

$$\text{Big plant}: 1,000,000\,(\text{Minimum})$$

$$\text{Small plant}: 1,000,000\,(\text{Minimum})$$

TABLE 10.12

Payoff

State of Nature	
Good Market Competitive Conditions (Rs.)	**Poor Market Competitive Conditions (Rs.)**
2,000,000 – 2,000,000 = 0	0 – (–1,000,000) = 1,000,000
2,000,000 – 1,000,000 = 1,000,000	0 – (–500,000) = 500,000
2,000,000 – 0 = 2,000,000	0 – 0 = 0

Not developing the new product line: 2,000,000

Decision: Construct a big plant or small plant

c) The decisions indicated by the decision criteria can be summarized as follows:

1. Maximax (optimistic): Construct a big plant to manufacture a new product

2. Maximin (pessimistic): Not developing the new product line

3. Hurwicz (criterion of realism): Construct a big plant to manufacture a new product

4. Laplace (equally likely): Construct a big plant to manufacture a new product

5. Minimax regret: Construct a big plant to manufacture new product/ Construct small plant

Four of the five decision criteria designated the decision to construct the large plant most often.

EXAMPLE 2:

A decision maker faced with four decision alternatives and four states of nature develops the following profit payoff in Table 10.13:

TABLE 10.13

Profit Payoff

Alternative	State of Nature			
	SN 1	SN 2	SN 3	SN 4
A1	10	5	6	1
A2	7	6	4	3
A3	5	6	6	7
A4	4	6	7	9

a) What is the recommended decision using the maximax (optimistic), maximin (pessimistic), and minimax regret approaches? *(BT level 3).*

b) Assume the payoff table provides cost and not profit payoffs. What is the recommended decision using the maximax (optimistic), maximin (pessimistic), and minimax regret approaches? *(BT level 5)*

Solution: a)

1. Maximax (optimistic): The decision is chosen which will result in a maximum of the maximum payoffs. Thus, for this example, the company would be optimistic that good competitive conditions will prevail in the future, resulting in the following maximum payoffs and decisions:

   A1   10 (Maximum)

   A2   7

   A3   7

   A4   9

   Decision: A1

2. Maximin (pessimistic): The decision maker selects the decision with the maximin criterion which will reflect the maximum of minimum payoffs. For each alternative decision, the decision maker assumes that there will be a minimum payoff, of which the maximum is selected as follows:

   A1   1

   A2   3

   A3   5 (Maximum)

   A4   4

   Decision: A3

3. Minimax regret: By choosing the alternative decision that reduces the maximum regret, the decision maker attempts to avoid regret. First, select the maximum payoff under each state of nature; then subtract from these amounts all other payoffs under the respective states of nature, as follows (Table 10.14):

TABLE 10.14

Payoff

Alternative	State of Nature			
	SN 1	SN 2	SN 3	SN 4
A1	$10 - 10 = 0$	$6 - 5 = 1$	$7 - 6 = 1$	$9 - 1 = 8$
A2	$10 - 7 = 3$	$6 - 6 = 0$	$7 - 4 = 3$	$9 - 3 = 6$
A3	$10 - 5 = 5$	$6 - 6 = 0$	$7 - 6 = 1$	$9 - 7 = 2$
A4	$10 - 4 = 6$	$6 - 6 = 0$	$7 - 7 = 0$	$9 - 9 = 0$

It is necessary to determine the maximum regret for each decision, and the decision corresponding to the minimum of these regret values is chosen as:

A1   8

A2   6

A3   5 (Minimum)

A4   6

Decision: A3

b)

1. Maximax (optimistic): By using the optimistic criterion for minimizing problems (cost) where lower payoffs are better, determine the best (minimum) payoff for each alternative and choose the best (minimum) alternative.

A1   1 ←——————— (Minimum)

A2   3

A3   5

A4   4

Decision: A1

2. Maximin (pessimistic): In the case of problems involving minimization (cost), this approach identifies an alternative that minimizes the maximum payoff.

A1   10

A2   7 (Minimum)

A3   7 (Minimum)

A4   9

Decision: A2, A3

3. Minimax regret: First selects the minimum payoff under each state of nature; then subtract these amounts from all other payoffs under the respective states of nature, as follows (Table 10.15):

It is necessary to determine the maximum regret for each decision, and the decision corresponding to the minimum of these regret values is chosen as:

TABLE 10.15

Payoff

Alternative	State of Nature			
	SN 1	SN 2	SN 3	SN 4
A1	$10 - 4 = 6$	$5 - 5 = 0$	$6 - 4 = 2$	$1 - 1 = 0$
A2	$7 - 4 = 3$	$6 - 5 = 1$	$4 - 4 = 0$	$3 - 1 = 2$
A3	$5 - 4 = 1$	$6 - 5 = 1$	$6 - 4 = 2$	$7 - 1 = 6$
A4	$4 - 4 = 0$	$6 - 5 = 1$	$7 - 4 = 3$	$9 - 1 = 8$

A1	6
A2	3 (Minimum)
A3	6
A4	8

Decision: A2

10.6.2 Decision Making under Risk

LO 9. Describe risk situations and use expected values to suggest decisions

If the decision maker chooses a number of choices, the probability of which can be estimated, it is referred to as decision under risk. The probability of different outcomes can be objectively estimated from historical data. In some cases, decision makers may be able to allocate probabilities to different outcomes on the basis of their knowledge and judgment.

After the arrangement of the decision situation into a payoff table, multiple criteria are required to take the final decision. Such criteria for the decision which will be discussed in this section include,

1. Expected Monetary Value (EMV)
2. Expected Opportunity Loss (EOL)
3. Expected Value of Perfect Information (EVPI)

Let's take a look at each of the above models

1. Expected Monetary Value (EMV):
   The decision maker must first estimate the probability of occurrence of each state of nature in this criterion. After these estimates have been made, it is possible to determine the estimated value for each alternative decision. The expected value, or the mean value, is the long-run average value of that decision. The EMV for an alternative is only the sum of the possible payoffs of the alternative, each weighted by the probability that the payoff will occur.

   The expected value of alternative x, written symbolically as EV (x), may be expressed simply as the following:

$$EV(x) = \sum_{i=0}^{n} x_i P(x_i), \qquad (10.4)$$

where
   x_i = payoff for the alternative in state of nature i,
   $P(x_i)$ = probability of state of nature i,
   n = the number of states of nature.
The EVM criterion consists of the following steps:

Step 1. Determine the expected conditional profit for each act.

Step 2. Determine the EVM for each act.

Step 3. Choose the act that corresponds to the optimal EVM.

2. Expected opportunity Loss (EOL):
   An alternative approach to maximizing EMV is to minimize EOL. In order to use this criterion, we multiply the probabilities by regret (i.e. opportunity loss) for each decision result instead of multiplying the decision results by the probability of its occurrence, as we did for the EMV.

   The EOL criterion consists of the following steps:

   Step 1. Determine the opportunity loss values for each event by first finding the maximum payoff for that event, and then taking the difference between that conditional profit value and each conditional profit for that event.

   Step 2. The EOL is then computed for each alternative by multiplying the opportunity loss by the probability and adding it together.

   Step 3. Select the act that is equal to the minimum opportunity loss.

3. Expected Value of Perfect Information (EVPI):

   Additional information on future events is sometimes available or may be purchased, allowing the decision maker to make a better decision. It would, however, be insane to pay more for this information than the extra profit from it. The information has a certain maximum value, which is the limit of what the decision maker would be willing to spend. This information value can be expressed as an expected value – hence its name, which is the EVPI.

   The EVPI is computed as follows:

$$EVPI = EVwPI - maximum\,EMV \qquad (10.5)$$

   where,
   EVwPI = $\sum$ (best payoff in state of nature i) (probability of the state of nature i)

Examples based on decision making under risk

EXAMPLE 1:

Let's consider *Example 1* from the previous section again.

Mr. X is the president of the profit-making company XYZ. The problem Mr. X identifies is whether to expand its product line by producing a new product and marketing it. Mr. X decides that his alternatives are to construct (1) a large new plant for the production of a new product, (2) a small plant, or (3) not to develop a new product line.

Mr. X has already assessed the potential profits associated with the various outcomes. With good competitive market conditions, he believes that a big new plant would result in a net profit of Rs. 2,000,000 for his company. The conditional value, if the market has poor competitive conditions, would be Rs. 1,000,000 net losses. A small plant would have resulted in a net profit of Rs. 1,000,000 with good competitive market conditions, but a net loss of Rs. 500,000 would have occurred under poor competitive conditions. At the end of the day, doing nothing would result in Rs. 0 profit on either market.

a) Make the payoff table describes this decision situation *(BT level 6)*

b) Suppose that Mr. X now assumes that a good market competitiveness probability is exactly the same as a poor market competition; in other words, every state of nature has a probability of 0.50. Which alternative would give the highest EMV? *(BT level 5)*

c) Calculate the EOL for each alternative with a probability of 0.50 for each state of nature. *(BT level 3)*

d) Mr. X approached a marketing firm that would help him to make a decision. For the details, this firm charges Rs. 500,000. What would you advise Mr. X? Will he hire a firm to do the marketing study? *(BT level 5)*

Solution:

a) The following payoff in Table 10.16 describes the given decision situation. A negative sign is included for losses.

$$EV(\text{big plant}): (2,000,000)(0.5)+(-1,000,000)(0.5) = 500,000(\text{Maximum})$$

$$EV(\text{small plant}): (1,000,000)(0.5)+(-500,000)(0.5) = 250,000$$

$$EV(\text{not developing the new product line}): (0)(0.5)+(0)(0.5) = 0$$

Decision: Construct a big plant to manufacture new product, given the greatest expected monetary value

b) First, calculate the table of opportunity loss by calculating the opportunity loss for not selecting the best alternative for each state of nature. Opportunity loss for any state of nature, or column, is determined by subtracting each payoff in the column from the best payoff in the same column. (Table 10.17).

TABLE 10.16

Payoff

Alternative	State of Nature	
	Good Market Competitive Conditions (Rs.)	Poor Market Competitive Conditions (Rs.)
Big plant	2,000,000	−1,000,000
Small plant	1,000,000	−500,000
Not developing the new product line	0	0

TABLE 10.17

Payoff

State of Nature	
Good Market Competitive Conditions (Rs.)	**Poor Market Competitive Conditions (Rs.)**
2,000,000 – 2,000,000 = 0	0 – (–1,000,000) = 1,000,000
2,000,000 – 1,000,000 = 1,000,000	0 – (–500,000) = 500,000
2,000,000 – 0 = 2,000,000	0 – 0 = 0

Using these opportunity losses, we compute the EOL for each alternative by multiplying the probability of each state of nature times the appropriate opportunity loss value and adding these together:

$$\text{EOL}(\text{big plant}): (0)(0.5)+(1,000,000)(0.5)=500,000(\text{Minimum})$$

$$\text{EOL}(\text{small plant}): (1,000,000)(0.5)+(500,000)(0.5)=750,000$$

$$\text{EOL}(\text{not developing the new product line}): (2,000,000)(0.5)+(0)(0.5)=1,000,000$$

Decision: Construct a big plant to manufacture new product, given the greatest expected monetary value

c) The expected value with perfect information is

$$\text{EVwPI}=(2,000,000)(0.5)+(0)(0.5)=1,000,000.$$

Thus if we had perfect information, the payoff would average Rs. 1,000,000

The maximum EMV without additional information is Rs. 500,000 (from part b of this example):

$$\text{EVPI}=\text{EVwPI}-\text{maximum EMV}$$

$$=1,000,000-500,000$$

$$=\text{Rs.}\,500,000.$$

The EVPI, Rs. 500,000, is the maximum amount that Mr. X would pay to purchase perfect information from a marketing firm. As the firm is also charging the same amount for information, Mr. X should hire the firm to make the marketing study.

EXAMPLE 2:

A decision maker faced with three decision alternatives and three states of nature develops the following profit (x 1,000) payoff in Table 10.18:

a) Compute the expected value for each decision and select the best one. *(BT level 3)*

b) Determine how much the company will be willing to pay to a market analysis firm to gain more insight into potential business conditions. *(BT level 3)*

Solution: a)

$$EV(A1):(80)(0.2)+(45)(0.5)+(-25)(0.3)=31(Maximum)$$

$$EV(A2):(35)(0.2)+(20)(0.5)+(10)(0.3)=20$$

$$EV(A3):(22)(0.2)+(18)(0.5)+(15)(0.3)=17.9$$

$$Decision:Alternative\ A1$$

b) The expected value with perfect information is

$$EVwPI=(80)(0.2)+(45)(0.5)+(15)(0.3)=43.$$

The maximum EMV without additional information is 31 (from part a of this example)

$$EVPI=EVwPI-maximum\,EMV$$

$$=43-31=12$$

Since perfect information is rare, it would probably pay less than 12 (x 1,000).

TABLE 10.18

Payoff

Alternative	State of Nature		
	SN 1 Probability (0.2)	SN 2 Probability (0.5)	SN 3 Probability (0.3)
A1	80	45	−25
A2	35	20	10
A3	22	18	15

10.7 Decision Trees

Decision trees provide a useful means of viewing the problem visually and then organizing the computational work already mentioned in the two previous sections. These trees are especially useful when a series of decisions need to be made. A decision tree is a diagram composed of nodes and

LO 10. Construct decision trees and use them to decide alternatives

branches. In the decision tree, the user determines the expected value of each outcome and makes a decision based on the expected values. If we have probabilities for the different states of nature, we can use those probabilities to calculate the EMVs for each strategy. This strategy is at the core of the decision tree. A decision tree has the primary advantage of offering an illustration (or a picture) of the decision-making process. It helps to accurately evaluate the expected values required and to understand the decision-making process.

All decision trees are similar in that they contain nodes for the decision and nodes for the state of nature:

- Decision node from which one of several alternatives can be selected
- A state-of-nature node which will result in one state of nature

As their name suggests, decision trees represent the different sequences of tree-style outcomes and decisions, spanning from left to right. The tree thus shows sequential decisions and outcomes. Lines or branches from the squares (decision nodes) are alternatives, and branches from the circles are the states of nature.

Decision tree analysis consists of the following steps:

Step 1. Draw the decision tree.

Step 2. Assign probabilities to the states of nature.

Step 3. Analyse the decision tree by computing EMVs for each state of nature node. This is done by working backward from the right side of the tree to the decision-making nodes on the left. Also, an alternative with the best EMV is selected for each decision node.

Step 4. The final step is to choose a decision or a series of decisions where there is more than one stage of decision making that produces the highest EMV.

Following this process from right to left via a tree, we, ultimately, return to the originating node, and the value of this is the cumulative predicted value of making the best decisions.

Examples based on decision tree

EXAMPLE 1:

Let's consider *Example 1* from the previous section again. The payoff table is given in Table 10.19. Using this information construct a decision tree to represent the situation and use it to advise Mr. X. *(BT level 3 and 6)*

TABLE 10.19

Payoff

Alternative	State of Nature	
	Good Market Competitive Conditions (Rs.) Probability (0.5)	Poor Market Competitive Conditions (Rs.) Probability (0.5)
Big plant	2,000,000	−1,000,000
Small plant	1,000,000	−500,000
Not developing the new product line	0	0

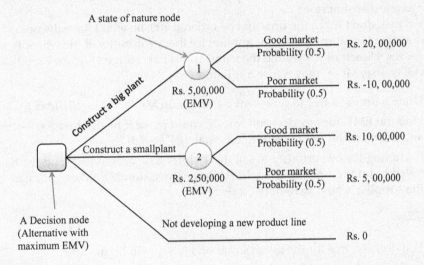

FIGURE 10.6
Decision Tree for Example 1.

Solution: The decision tree for the given situation is shown in Figure 10.6.

$$\text{EMV}\left(\text{big plant}\right):\left(2,000,000\right)\left(0.5\right)+\left(-1,000,000\right)\left(0.5\right)=500,000\left(\text{Maximum}\right)$$

$$\text{EMV}\left(\text{small plant}\right):\left(1,000,000\right)\left(0.5\right)+\left(-500,000\right)\left(0.5\right)=250,000$$

$$\text{EMV}\left(\text{not developing the new product line}\right):\left(0\right)\left(0.5\right)+\left(0\right)\left(0.5\right)=0$$

Decision: Mr. X should choose to construct a big plant

An Introduction to Optimization Techniques

EXAMPLE 2:

Mr. 'X' has a market stall of ready-made clothes in a small town. The local hockey team has unexpectedly reached the semi-finals of a big tournament. A day before the semi-final is to be played, a distributor offers him a good price consignment of the team's T-shirts but says he can either have 300 or 500 and has to commit to the contract immediately.

When the team reaches the final, a probability that the coach has put at 0.6, and Mr. 'X' has ordered 500 T-shirts, he will be able to sell them all at a profit of Rs. 150 each. If the team fails to reach the final, and he has ordered 500, he will not sell any of them this season but will be able to store them and sell them each next season to a profit of Rs. 80, unless the team changes their strip in which case it will only make a profit of Rs. 50 per T-shirt. The likelihood that the team will change their strip for the next season is 0.70. Rather than keeping the T-shirts, he could sell it for Rs. 60 per T-shirt to one discount store.

When they don't make the final and he ordered 300, he won't have the option to sell to the discount store because the amount for them will be small. He will only sell them for Rs. 80 each next season if the team strip is not changed and for Rs. 50 each if it is. Of course, Mr 'X' may refuse the T-shirt bid.

a) Draw a decision tree to represent the situation Mr. 'X' faces. *(BT level 6)*
b) Find the EMV for any decision Mr. 'X' could make if he ordered 500 T-shirts and the team did not make it to the final. *(BT level 3)*
c) Excluding the opportunity to sell the T-shirts if he ordered 500 and the team failed to hit the final, find the EMV for any decision Mr. 'X' might make on the supplier's offer to him. *(BT level 5)*

Solution:

a) The decision tree for the given situation is shown in Figure 10.7.
b) EMV (Store) = (25,000) (0.7) + (40,000) (0.3) = Rs. 29,500
   Since this figure is lower than the value of selling the T-shirts to the discount chain, Rs. 60, Mr. 'X' should sell the T-shirts at this stage.
c) EMV (Order 500) = (75,000) (0.6) + (29,500) (0.4) = Rs. 56,800

In determining the EMV of the decision to order 300 T-shirts we will take into account the possibility that the team strip will be changed, as well as the possibility that the team will reach the final. This includes applying the probability multiplication rule:

$$EMV(Order\,300) = (45,000)(0.6) + (15,000)(0.7)(0.4) + (24,000)(0.3)(0.4) = Rs.34,080.$$

We would advise Mr. 'X' to order 500 T-shirts because the EMV is higher for that strategy, Rs. 56,800 then the EMV for ordering 300 T-shirts, Rs. 34,080, and the EMV for not ordering, Rs. 0.

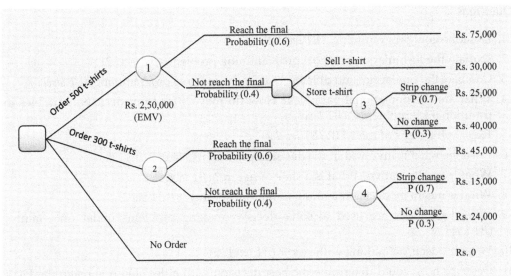

FIGURE 10.7
Decision Tree for Example 2.

CHAPTER SUMMARY

- Goal programming is a powerful tool to tackle multiple and incompatible goals some of which may be non-economic in nature.
- The goal programming model contains one or more goal equations and an objective function designed to minimize deviations from the goals.
- A variation of the graphical solution procedure in linear programming can be used with two decision variables to solve goal programming problems.
- The analytic hierarchy process (AHP) is an approach to decision making with multi-criteria.
- AHP is a decision-making approach based on pairwise comparisons of elements in a hierarchy for multi-criteria.
- Decision theory is an analytical and systematic approach to the study of decision making.
- With decision making under uncertainty, there are several possible events, but we do not know which will occur and cannot even give them probabilities.
- In decision making under uncertainty, payoff tables are constructed to compute such criteria as maximax, maximin, the criterion of realism, equally likely, and minimax regret.
- With decision making under risk, there are several possible events, and we can give each a probability.
- Such methods as determining EMV, EOL, and the EVPI are used in decision making under risk.
- Decision trees are another option, particularly for larger decision problems when one decision must be made before other decisions can be made.

Questions

1. Describe goal programming. *(BT level 2)*
2. Explain the formulation of a goal programming problem. *(BT level 2)*
3. Compare the similarities and differences in linear and goal programming. *(BT level 5)*
4. What are the deviational variables? How do they differ from decision variables in traditional LP problems? *(BT level 2)*
5. Explain the steps of the AHP. *(BT level 2)*
6. Describe what is involved in the decision process. *(BT level 2)*
7. What is an alternative? What is a state of nature? *(BT level 2)*
8. What is meant by decision making under uncertainty? *(BT level 2)*
9. What techniques are used to solve decision-making problems under uncertainty? *(BT level 2)*
10. What is "decision making with risk"? *(BT level 2)*
11. Describe how you'd determine the best decision with a decision tree using the EMV criterion. *(BT level 2)*

Practice Problem

1. ABC Corporation blends three raw materials to produce two products: X and Y. Each ton of product X is a mixture of 2/5 tons of material 1 and 3/5 tons of material 3. A ton of product Y is a blend of 1/2 tons of material 1, 1/5 tons of material 2, and 3/10 tons of material 3. The production of ABC is limited by the limited availability of the three raw materials. For the current production period, ABC has the following quantities of each raw material: 1, 22 tons; 2, 7 tons; 3, 23 tons. Management intends to achieve the following priority level goals:
   Goal 1: Produce at least 35 tons of products X.
   Goal 2: Produce at least 20 tons of products Y.

2. Treating the amounts of each material available as constraints, formulate a goal programming model to determine the optimal product mix. *(BT level 6)*

   The municipal cooperation of the city has received a central grant of Rs. 3.5 crore to expand its public recreation facilities. Representatives of municipal cooperation demanded four different types of facilities – gymnasiums, gardens, badminton courts, and swimming pools. In fact, eight gyms, four gardens, six badminton courts, and eight swimming pools have been requested by the various communities in the city. Each facility costs a certain amount requires a certain number of acres, and a certain amount is expected to be used as follows (Table 10.20):

   Cooperation has established 45 acres of land for construction (although more land could be located, if necessary). Cooperation has set out the following objectives, which are set out in order of priority:

   Goal 1: The cooperation wishes to spend the entire grant as the government must reimburse any amount not spent.

   Goal 2: The cooperation seeks to make use of facilities for a total of at least 20,000 people each week.

   Goal 3: The department wants to avoid having to secure more than the already-located 45 acres of land.

TABLE 10.20

Data Practice Problem 2

Facilities	Cost (Rs.)	Acres	People Usage/Week
Gymnasiums	4,800,000	3	2,000
Garden	1,800,000	7	3,500
Badminton courts	1,200,000	2	1,000
Swimming pools	2,400,000	4	1,500

Formulate a goal programming model to determine how many of each type of facility should be constructed to best achieve the city's goals. *(BT level 6)*

3. Solve the following goal programming model graphically. *(BT level 3)*

$$Minimize\ P_1 d_1^+, P_2 d_2^-, P_3 d_3^-$$

Subject to

$$2x_1 + x_2 + d_1^- - d_1^+ = 40$$

$$x_1 + d_2^- - d_2^+ = 15$$

$$x_2 + d_3^- - d_3^+ = 25$$

$$All\ x_i, d_i \geq 0$$

4. Vikrant was considering entering a master's degree in one of two post-graduate schools. When asked how he compared the two schools in terms of reputation, he replied that he very strongly preferred school A over school B.

a) Set up the pairwise comparison matrix for this problem. *(BT level 6)*

b) Determine the priorities for the two schools relative to this criterion. *(BT level 3)*

5. A university HR personnel department has narrowed the search for a new Assistant Professor hire to three candidates: AP (1), AP (2), and AP (3). The final selection is based on three criteria: personal interview (*PI*), experience (*E*), and research paper (*RP*). The department uses matrix **A** (given below) to establish the preferences among the three criteria. After interviewing the three candidates and compiling the data regarding their experiences and research papers, the matrices A_{PI}, A_E, and A_{RP} are constructed. Which of the three candidates should be hired? Assess the consistency of the data. *(BT level 5)* (Table 10.21)

6. The owners, Mr. Sharma and Mrs. Sharma, are considering expanding a coffee bar business. They could do this by investing in new sites or by franchising operation for fast-food entrepreneurs who would pay them a fee. The estimated profits for each strategy depend on future coffee demand, which could increase, stabilize, or decline. Another possibility for Mr. Sharma and Mrs. Sharma is to accept the Rs. 50 lakhs offer made for their business by a major international fast-food company. The expected profits are shown in Table 10.22.

TABLE 10.21

Data Practice Problem 5

A_{PI}	AP (1)	AP (2)	AP (3)
AP (1)	1	4	5
AP (2)	1/4	1	1/6
AP (3)	1/5	6	1
A_{RP}	AP (1)	AP (2)	AP (3)
AP (1)	1	1/3	2
AP (2)	3	1	1/3
AP (3)	1/2	3	1

A_E	AP (1)	AP (2)	AP (3)
AP (1)	1	1/4	3
AP (2)	4	1	1/3
AP (3)	1/3	3	1
A	PI	E	RP
PI	1	3	1/5
E	1/3	1	1/6
RP	5	6	1

TABLE 10.22

Data Practice Problem 6

Alternative	State of Nature (Expected Profit in Rs.)		
	Increase	Remain Stable	Decline
Invest in new site	15,000,000	10,000,000	−5,500,000
Franchising	8,000,000	6,000,000	0
Sell	5,000,000	5,000,000	5,000,000

a) Determine the best decision using each of the following decision criteria and

b) Compare the decision from the following decision criteria and support the best decision. *(BT level 3 and 6)*

1. Maximax (Optimistic) 2. Maximin (Pessimistic)

3. Hurwicz (Criterion of realism) assume $\alpha = 0.7$ 4. Laplace (Equally likely)

5. Minimax regret

7. The following payoff table shows a profit for a decision analysis problem with two decision alternatives and three states of nature (Table 10.23):

Suppose that the decision maker obtained the probability assessments $P(S1) = 0.65$, $P(S2) = 0.15$, and $P(S3) = 0.20$. Use the expected value approach to determine the optimal decision. *(BT level 3)*

TABLE 10.23

Data Practice Problem 7

Alternative	State of Nature (Expected Profit in Rs.)		
	S1	S2	S3
A1	2,500,000	1,000,000	250,000
A2	1,000,000	1,000,000	750,000

8. Private Bank management was concerned about the potential loss that could occur in the event of a physical disaster, such as power failure or fire. The bank estimated the loss from one of these incidents could amount to as much as Rs. 50 million, including losses due to interrupted service and customer relationships. The installation of an emergency power generator at its headquarters is one project the bank considers. The cost of the emergency generator is Rs. 4,500,000 and no losses from this type of incident will be incurred when it is installed. However, if the generator is not installed, there is a 15% chance that a power outage will occur in the next year. If an outage occurs, there is a probability of 0.05 that the resulting losses will be very large, or around Rs. 45 million in lost earnings. Alternatively, a probability of only slight losses of around Rs. 1 million is estimated at 0.95. Decide whether the bank should install the new power generator using decision tree analysis. *(BT level 4)*

11

Optimization Modelling with an Open Source Tool (Excel)

Learning Outcomes: After studying this chapter, the reader should be able to

LO 1. *Understand the use of open source tool Excel for solving optimization problems. (BT level 2)*

LO 2. *Make a model of different optimization problems in Excel. (BT level 6)*

LO 3. *Solve different optimization problems using Excel. (BT level 3)*

11.1 Introduction

The problems of optimization are real-world problems that we encounter in many areas, such as mathematics, engineering, science, business, and economics. In these problems, we find an optimal or most efficient way to use limited resources to achieve the objective of the situation. This could maximize the profit, minimize the cost, minimize the total distance travelled, or minimize the total time needed to complete a project. For the given problem, a mathematical description called a mathematical model is formulated to represent the situation. While optimization problems involve numbers, the amount of time you actually spend performing calculations during your subject study can be reduced by using readily available technology, specifically an appropriate calculator and appropriate computer software.

A spreadsheet is a two-dimensional rectangular array. Each of the rectangles is called a cell. You can type four types of information into a cell: number, fraction, function, and text. It is identified on the spreadsheet by its column and row location, which are designated by letter and number (i.e. cell, A1), respectively. Numbers are usually stored in the rows and columns of the spreadsheet in Excel. The mathematical model used in the spreadsheet is called the spreadsheet model. Major spreadsheet packages come with an integrated optimization tool called the Solver. In this chapter, we show how Excel spreadsheet modelling and the Solver can be used to find the optimal solution to optimization problems.

LO 1. Understand the use of open source tool Excel for solving optimization problems

In addition to solving equations, the Excel solver enables us to find solutions to all types of optimization problems (single or multiple variables, with or without constraints). The basic features of Excel are known to virtually everyone in the business world, but relatively few know some of its more powerful features. Excel comes with built-in tools to optimize the spreadsheet, called solvers. Modelling of spreadsheets is the process of entering inputs and decision variables into a spreadsheet and then linking them appropriately, using formulae, in order to obtain outputs.

The following are just a few of the reasons why a spreadsheet is growing in importance.

- The data is often submitted to the modeller in a spreadsheet.
- Data can easily be turned into information on the spreadsheet using formulae and embedded functions.
- Using the spreadsheet charting and graphing functions, data, and information can easily be converted into informative visual displays.

Using a specialized mathematical programming package, you can also solve optimization problems without using a spreadsheet. The package includes a partial list of those packages: LINDO, MPSX, CPLEX, and MathPro. These packages are typically used by researchers and companies interested in solving extremely large problems that do not fit conveniently into a spreadsheet.

Two important add-ins to Excel are Solver and ToolPak for Analysis. Both of these are a part of Excel but need to be activated or loaded before you can first use them. Follow these steps to load those add-ins:

1. For Excel 2010, click the File tab, click Options, and then click Add-Ins.

   For Excel 2007, click the Microsoft Office button, click Excel Options, and then click Add-Ins.

2. In the Manage box, select Excel Add-Ins, and click Go.

3. Check the boxes next to Analysis ToolPak and Solver Add-In, and then click OK.

The Data tab now displays Solver and Data Analysis every time Excel is started.

Examples based on the linear programming graphical method using Excel

EXAMPLE 1:

Make the model of the following linear programming problem and solve using the graphical method in a spreadsheet. *(BT level 6 and 3)*

$$\max Z = 500\,X1 + 150\,X2$$

Subject to

$$10\,X1 + 2\,X2 \le 200$$

$$2\,X1 + 2\,X2 \le 120$$

$$X1, X2 \ge 0$$

Solution: First write the function in Excel, as shown in Figure 11.1.

1. After writing the objective function and constraints in the spreadsheet, let's calculate the values for the C1 and C2 constraints. You can calculate values by setting the value of another variable to zero. Like for C1, at X1 = 0, X2's value would be 100 and at X2 = 0, X1's value would be 20. In the same way, calculate the C2 constraint. Write these values in the spreadsheet (Figure 11.2; from cell A7 to B15).

2. Now, let's plot a graph. Select the values of the C1 constraint; go to insert and plot the scatter chart with the smoothed lines. Maybe you can see your chart, as shown in Figure 11.3.

   Right-click on the chart, go to Select Data, see two series, remove Series 2, and Edit Series 1 (Figure 11.4)

   Name the series as C1 and change its X value to C1's A column values (from cell A9 to cell A10) and Y to C1's B column values (from cell B9 to cell B10). And click on OK. You'll see your chart like the one shown in Figure 11.5:

   Again, go to Select Data by right-clicking on the chart and add another series. Name it C2 and in X values select constraints C2 X column values (from cell A14 to cell A15) and in Y values select constraints C2 Y column values (from cell B14 to cell B15). You'll see your chart like the one shown in Figure 11.6:

▲	A	B	C	D	E	F
1						
2		X1	X2			
3	Z	500	150			
4	C1	10	2	<=	200	
5	C2	2	2	<=	120	
6						

FIGURE 11.1
Function in Excel.

7	C1	
8	X1	X2
9	0	100
10	20	0
11		
12	C2	
13	X1	X2
14	0	60
15	60	0
16		

FIGURE 11.2
First Step in Excel.

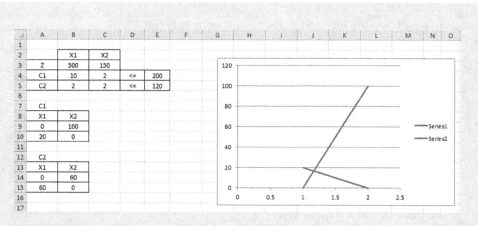

FIGURE 11.3
Second Step in Excel.

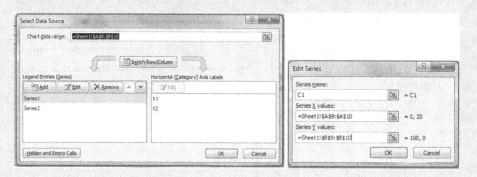

FIGURE 11.4
Data Setting in Excel.

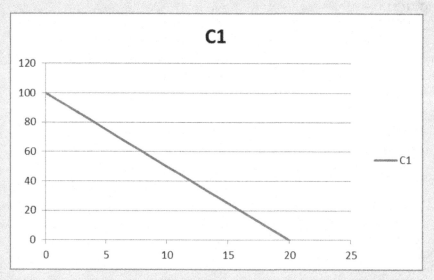

FIGURE 11.5
Constraint C1 Plot in Excel.

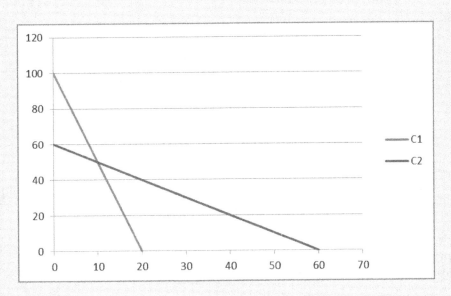

FIGURE 11.6
Constraint C2 Plot in Excel.

3. Now identify the feasible region from constraints and mark point O, A, B, and C. The feasible region is shaded as Click on Insert tab > Shapes > Freeform tool. Click on one of the corners of the feasible region, say corner A; click on the second one, say corner O; then third, say corner C; then corner B; and move the cursor to the beginning point (Figure 11.7).

4. Now calculate the objective function values for each corner point of the feasible region.

    • The coordinate values of (X1, X2) at point O are (0, 0). Therefore, the objective function value would be 0.

    • The coordinate values of (X1, X2) at point A are (0, 60). Therefore, the objective function value would be (500 * 0 + 150 * 60) = 9,000.

    • The coordinate values of (X1, X2) at point B are (20, 0). Therefore, the objective function value would be (500 * 20 + 150 * 0) = 10,000

    • At point C, to find the coordinate values at point Z, we need to find the intersection point of the constraint lines C1 and C2. To calculate the intersection of two equations, use the function MMULT() and MINVERSE().

    • Use the following formula by selecting cell U 22 and U 23 = MMULT(MINVERSE(O24:P25),R24:R25). Before using this formula, select the highlighted yellow cells X1 and X2 (from cell U22 to U23) and use this function to calculate the intersection point of the two equations and press CTRL+SHIFT+ENTER.

Applying the previous formula, we get (X1, X2) as (10, 50). Hence, the objective function value would be (500 × 10 + 150 × 50) = 12,500 (Figure 11.8).

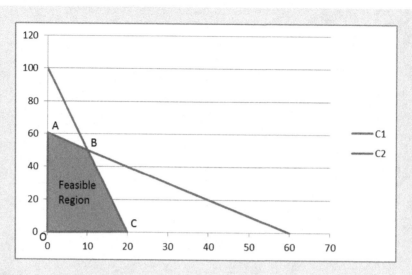

FIGURE 11.7
Marking the Feasible Region in Excel.

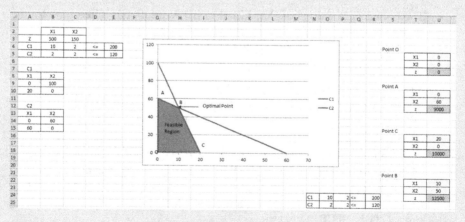

FIGURE 11.8
Solution in Excel.

EXAMPLE 2:

Make the model of the following linear programming problem and solve using the graphical method in a spreadsheet. (*BT level 6 and 3*)

$$\min Z = 22X1 + 6X2$$

Subject to,

$$5X1 + 1X2 \geq 15$$

$$3X1 + 3X2 \leq 18$$

$$3X1 - 4X2 \leq 0$$

$$X1, X2 \geq 0.$$

Solution: First write the function in Excel like the one shown in Figure 11.9.

1. Now by following the steps as discussed in previous Example 1, plot a graph and obtain a feasible region, as shown in Figure 11.10.

2. Now calculate the objective function values for each corner point of the feasible region.
   - Point A is the intersection of C1 and C2. The coordinate values of (X1, X2) are determined as follows.

     Use the following formula by selecting cell X3 and X4 = MMULT(MINVERSE(R3:S4),U3:U4). Before using this formula, select the highlighted yellow cells X1 and X2 (from cell X3 to X4), and use this

	A	B	C	D	E
1					
2		X1	X2		
3	Z	22	6		
4	C1	5	1	>=	15
5	C2	3	3	<=	18
6	C3	3	-4	<=	0
7					

FIGURE 11.9
Function in Excel.

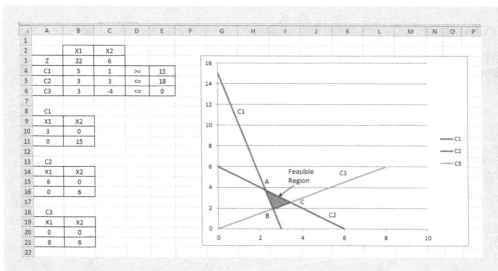

FIGURE 11.10
Feasible Region Plot in Excel.

function to calculate the intersection point of two equations and press CTRL+SHIFT+ENTER.

Applying the above formula, we get (X1, X2) as (2.25, 3.75). Hence, the objective function value would be (22 × 2.25 + 6 × 3.75) = 72.

- Point B is the intersection of C1 and C3. The coordinate values of (X1, X2) are determined as follows.

Use the following formula by selecting cell X9 and X10 = MMULT(MINVERSE(R9:S10),U9:U10). Before using this formula, select the highlighted yellow cells X1 and X2 (from cell X9 to X10) and use this function to calculate the intersection point of two equations and press CTRL+SHIFT+ENTER.

The coordinate values by applying the above formula we get (X1, X2) as (2.609, 1.957). Hence, the objective function value would be (22 × 2.309 + 6 ×1.957) = 69.13

- Point C is the intersection of C2 and C3. The coordinate values of (X1, X2) are determined as follows.

Use the following formula by selecting cell X15 and X16 = MMULT(MINVER SE(R15:S16),U15:U16). Before using this formula, select the highlighted yellow cells X1 and X2 (from cell X15 to X16) and use this function to calculate the intersection point of two equations and press CTRL+SHIFT+ENTER.

Applying the above formula, we get (X1, X2) as (3.429, 2.571). Hence, the objective function value would be (22 × 3.429+ 6 × 2.571) = 90.86

Hence point B is the optimum point with (X1, X2) as (2.609, 1.957). (Figure 11.11)

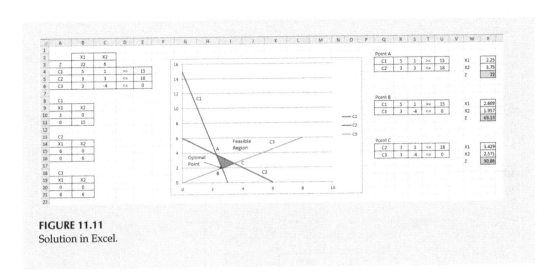

FIGURE 11.11
Solution in Excel.

Examples based on the linear programming simplex method using Excel

EXAMPLE 1:

Make the model of the following linear programming problem and solve using the simplex method in a spreadsheet. *(BT level 6 and 3)*

$$\max Z = 500\,X1 + 150\,X2$$

Subject to

$$10\,X1 + 2\,X2 \le 200$$

$$2\,X1 + 2\,X2 \le 120$$

$$X1, X2 \ge 0$$

Solution: First write the function in Excel, as shown in Figure 11.12.

1. For the objective function value, enter the formula (in cell B11) for computing Z = = SUMPRODUCT(B9:C9,B10:C10). This formula uses the coefficient values and the solution values for the X1 and X2 variables to be solved.

2. Similarly, enter the formula for LHS of the Constraints C1 & C2.

Enter = SUMPRODUCT(E9:F9,$E$6:$F$6) into cell B14 and = SUMPRODUCT(E10:F10,$E$6:$F$6) into cell B15. (Figure 11.13).

Now in the Data tab click Solver. The solver box appears as follows (Figure 11.14).

Set the target cell for the objective function Z value – i.e., $B$11.

Check the Equal to Max.

▲	A	B	C	D	E	F
1						
2		X1	X2			
3	Z	500	150			
4	C1	10	2	<=	200	
5	C2	2	2	<=	120	
6						

FIGURE 11.12
Function in Excel.

▲	A	B	C	D	E	F
1						
2		X1	X2			
3	Z	500	150			
4	C1	10	2	<=	200	
5	C2	2	2	<=	120	
6						
7						
8		X1	X2			
9		500	150			
10	Solution					
11	Z	0				
12						
13		LHS	RHS			
14	C1	0	200			
15	C2	0	120			
16						

FIGURE 11.13
Function in Excel.

For the changing variable cell, select the solution values of the variables – i.e., $B$10:$C$10.

For the subject to the constraints, LHS <= RHS – i.e., click on the Add option and select $B$14:$B$15 <= $C$14:$C$15.

Click on Make unconstrained variable non-negative.

Select a solution method as Simplex LP.

3. Now click the Solve button.

4. After selecting the solve button, the solver results appear in a window; the default option has a keep solver solution. Click on the Answer in the Reports section on the right-hand side. Finally, click the OK button to get the results (Figure 11.15).

5. Finally, the Excel Solver gives the solution values for variables X & Y and for the objective function Z (Figure 11.16).

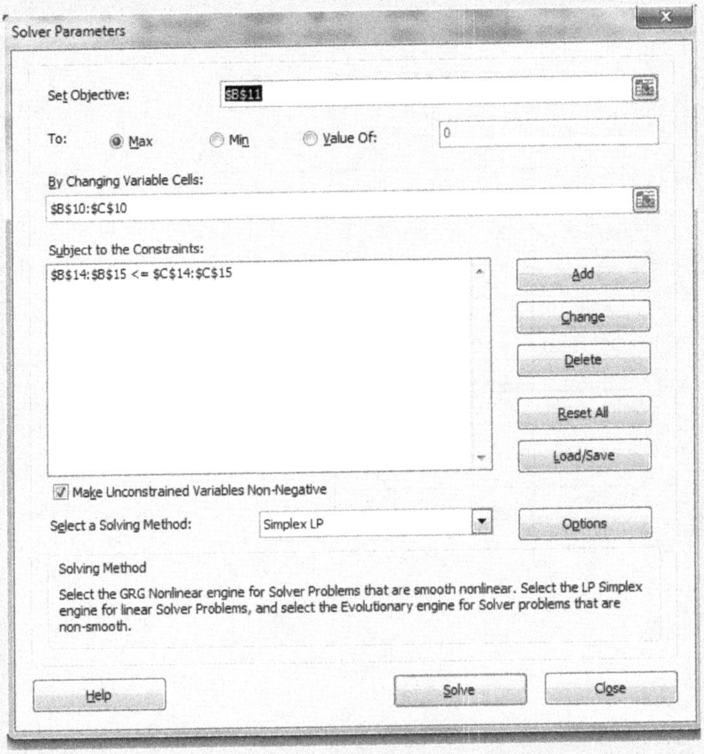

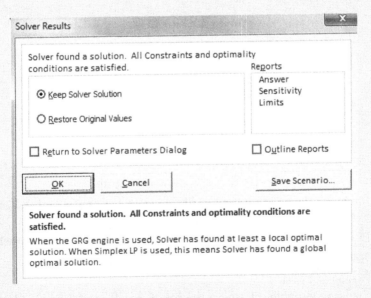

FIGURE 11.14
Solver Parameters in Excel.

FIGURE 11.15
Solver Results Window in Excel.

⊿	A	B	C	D	E	F
1						
2		X1	X2			
3	Z	500	150			
4	C1	10	2	<=	200	
5	C2	2	2	<=	120	
6						
7						
8		X1	X2			
9		500	150			
10	Solution	10	50			
11	Z	12500				
12						
13		LHS	RHS			
14	C1	200	200			
15	C2	120	120			
16						

FIGURE 11.16
Solution in Excel.

EXAMPLE 2:

Make the model of the following linear programming problem and solve using the simplex method in a spreadsheet. *(BT level 6 and 3)*

$$\min Z = 4X1 + 6X2$$

Subject to

$$2X1 + 2X2 \geq 12$$

$$14X1 + 2X2 \geq 28$$

$$X1, X2 \geq 0$$

Solution: First write the function in Excel, as shown in Figure 11.17.

1. Now by following the first two steps as discussed in previous Example 1, we get the following results (Figure 11.18).

2. Now in the Data tab click Solver. The solver box appears as follows (Figure 11.19).

   Set the target cell for the objective function Z value – i.e., $B$11.

   Check the Equal to Min.

   For the changing variable cell, select the solution values of the variables – i.e., $B$10:$C$10.

⊿	A	B	C	D	E
1					
2		X1	X2		
3	Z	4	6		
4	C1	2	2	>=	12
5	C2	14	2	>=	28
6					

FIGURE 11.17
Function in Excel.

⊿	A	B	C	D	E
1					
2		X1	X2		
3	Z	4	6		
4	C1	2	2	>=	12
5	C2	14	2	>=	28
6					
7					
8		X1	X2		
9		4	6		
10	Solution				
11	Z	0			
12					
13		LHS	RHS		
14	C1	0	12		
15	C2	0	28		
16					

FIGURE 11.18
Function in Excel.

For the subject to the constraints, LHS <= RHS – i.e., click on the Add option and select $B$14:$B$15 <= $C$14:$C$15.

Click on Make unconstrained variable non-negative.

Select a solution method as Simplex LP.

Now click the Solve button.

3. After selecting the solve button, the solver results appear in a window; the default option has a keep solver solution. Click on the Answer in the Reports section on the right-hand side. Finally, click the OK button to get the results.

4. Finally, the Excel Solver gives the solution values for variables X & Y and for the objective function Z (Figure 11.20).

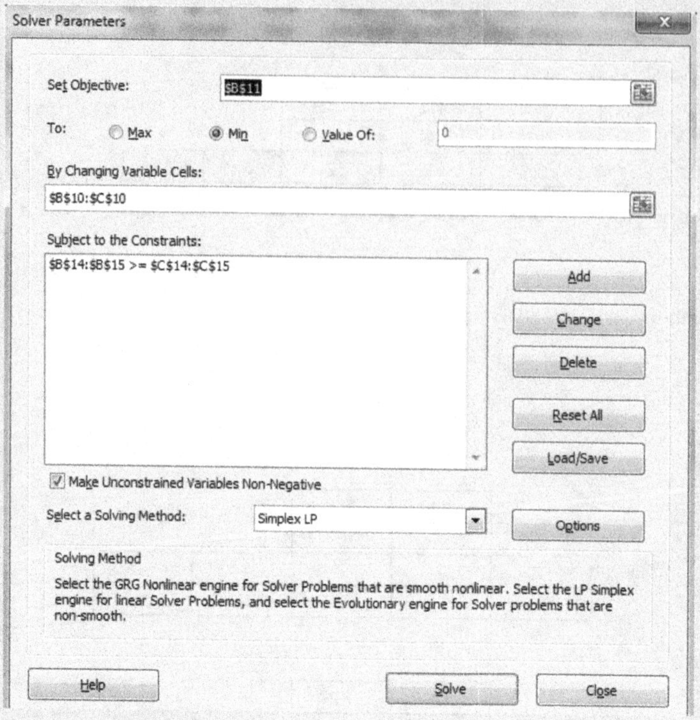

FIGURE 11.19
Solver Parameters in Excel.

▲	A	B	C	D	E	F
1						
2		X1	X2			
3	Z	4	6			
4	C1	2	2	>=	12	
5	C2	14	2	>=	28	
6						
7						
8		X1	X2			
9		4	6			
10	Solution	6	0			
11	Z	24				
12						
13		LHS	RHS			
14	C1	12	12			
15	C2	84	28			
16						

FIGURE 11.20
Solution in Excel.

Examples based on transportation and assignment problems using Excel

EXAMPLE 1:

Make the model of the following transportation problem (Table 11.1) and solve using Excel Solver so that the total transportation cost will be minimum. *(BT level 6 and 3)*

Solution:

1. First, write the problem in Excel, as shown in Figure 11.21.
2. Now insert the following functions and make the model as shown next.

   Function insert in cell E12 = SUM(B12:D12)

   Function insert in cell E13 = SUM(B13:D13)

   Function insert in cell E14 = SUM(B14:D14)

   Function insert in cell B15 = SUM(B12:B14)

   Function insert in cell C15 = SUM(C12:B14)

   Function insert in cell D15 = SUM(D12:B14)

   For the objective function value, enter the formula (in cell B18) for computing total transportation cost = SUMPRODUCT(B4:D6,B12:D14) (Figure 11.22).
3. Now in the Data tab click Solver. The solver box appears as follows (Figure 11.23).

   Set the target cell for the objective function value – i.e., $B$18.

   Check the Equal to Min.

   For the changing variable cell, select the solution values of the variables – i.e., $B$12:$D$14.

   For the subject to the constraints, click on the Add option and add the following constraints:

   $B$15 = $B$16,

   $C$15 = $C$16,

   $D$15 = $D$16,

   $E$12 <= $F$12,

   $E$13 <= $F$13,

   $E$14 <= $F$14.

TABLE 11.1

Data Example 1

Unit cost	W1	W2	W3	Factory Capacity
F1	4	3	2	200
F2	7	3	2	400
F3	8	6	4	400
Warehouse requirement	400	300	300	

⊿	A	B	C	D	E	
1						
2						
3	Unit cost	W1	W2	W3	Factory Capacity	
4	F1	4	3	2	200	
5	F2	7	3	2	400	
6	F3	8	6	4	400	
7	Warehouse requirement	400	300	300		
8						

FIGURE 11.21
Function in Excel.

⊿	A	B	C	D	E	F
1						
2						
3	Unit cost	W1	W2	W3	Factory Capacity	
4	F1	4	3	2	200	
5	F2	7	3	2	400	
6	F3	8	6	4	400	
7	Warehouse requirement	400	300	300		
8						
9						
10						
11	Unit cost	W1	W2	W3	Factory supply	Factory Capacity
12	F1				0	200
13	F2				0	400
14	F3				0	400
15	Warehouse demand	0	0	0		
16	Warehouse requirement	400	300	300		
17						
18	Total transportation cost	0				
19						

FIGURE 11.22
Function in Excel.

Click on Make unconstrained variable non-negative.

Select a solution method as Simplex LP.

Now click the Solve button.

4. After selecting the solve button, the solver results appear in a window; the default option has a keep solver solution. Click on the Answer in the Reports section on the right-hand side. Finally, click the OK button to get the results.

5. Finally, the Excel Solver gives the solution values for variables and for the objective function (Figure 11.24).

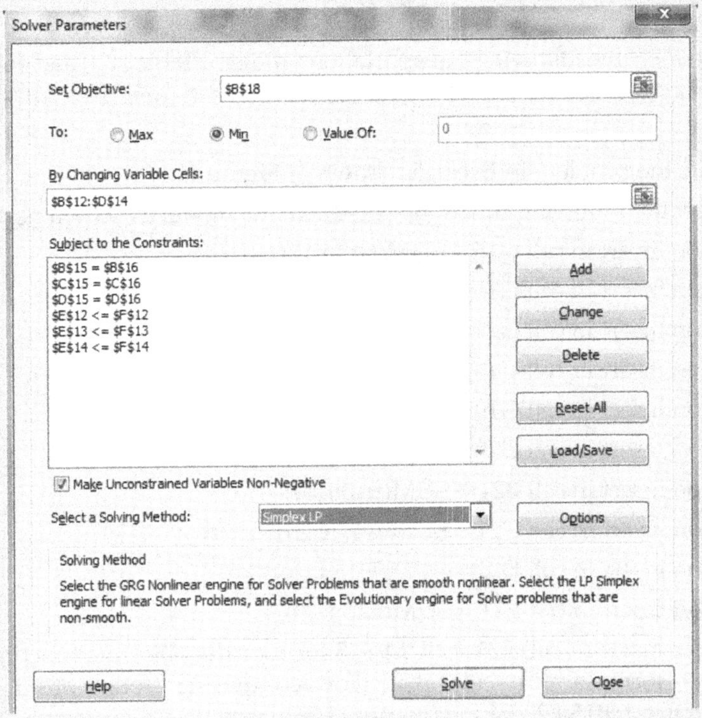

FIGURE 11.23
Solver Parameters in Excel.

	A	B	C	D	E	F
1						
2						
3	Unit cost	W1	W2	W3	Factory Capacity	
4	F1	4	3	2	200	
5	F2	7	3	2	400	
6	F3	8	6	4	400	
7	Warehouse requirement	400	300	300		
8						
9						
10						
11	Unit cost	W1	W2	W3	Factory supply	Factory Capacity
12	F1	200	0	0	200	200
13	F2	0	300	100	400	400
14	F3	200	0	200	400	400
15	Warehouse demand	400	300	300		
16	Warehouse requirement	400	300	300		
17						
18	Total transportation cost	4300				
19						

FIGURE 11.24
Solution in Excel.

EXAMPLE 2:

Make the model of the following transportation problem (Table 11.2) and solve using Excel Solver so that total profit will be maximum. *(BT level 6 and 3)*

Solution:

1. First, write the problem in Excel, as shown in Figure 11.25.
2. Now insert the following functions and make the model as shown next.

   Function insert in cell G15 = SUM(B15:F15)

   Function insert in cell G16 = SUM(B16:F16)

   Function insert in cell G17 = SUM(B17:F17)

   Function insert in cell G18 = SUM(B18:F18)

   Function insert in cell G19 = SUM(B19:F19)

   Function insert in cell G20 = SUM(B20:F20)

   Function insert in cell B21 = SUM(B15:B20)

   Function insert in cell C21 = SUM(C15:C20)

   Function insert in cell D21 = SUM(D15:D20)

   Function insert in cell E21 = SUM(E15:E20)

   Function insert in cell F21 = SUM(F15:F20)For the objective function value, enter the formula (in cell B24) for computing total profit cost = SUMPRODUCT(B15:F20,B4:F9) (Figure 11.26).
3. Now in the Data tab click Solver. The solver box appears as follows (Figure 11.27).
4. Set the target cell for the objective function value – i.e., $B$24.

   Check the Equal to Max.

TABLE 11.2

Data Example 2

Profit	D1	D2	D3	D4	D5	Factory Capacity
S1	12	11	5	10	4	850
S2	1	6	6	3	17	650
S3	15	4	6	14	12	630
S4	19	9	9	16	1	650
S5	11	6	14	12	7	459
S6	13	11	16	4	15	290
Warehouse requirement	790	680	600	390	840	

	A	B	C	D	E	F	G
1							
2							
3	Profit	D1	D2	D3	D4	D5	Factory Capacity
4	S1	12	11	5	10	4	850
5	S2	1	6	6	3	17	650
6	S3	15	4	6	14	12	630
7	S4	19	9	9	16	1	650
8	S5	11	6	14	12	7	459
9	S6	13	11	16	4	15	290
10	Warehouse requirement	790	680	600	390	840	
11							

FIGURE 11.25
Function in Excel.

G15 f_x =SUM(B15:F15)

	A	B	C	D	E	F	G	H	I
1									
2									
3	Profit	D1	D2	D3	D4	D5	Factory Capacity		
4	S1	12	11	5	10	4	850		
5	S2	1	6	6	3	17	650		
6	S3	15	4	6	14	12	630		
7	S4	19	9	9	16	1	650		
8	S5	11	6	14	12	7	459		
9	S6	13	11	16	4	15	290		
10	Warehouse requirement	790	680	600	390	840			
11									
12									
13									
14	Profit	D1	D2	D3	D4	D5	Supply	Supply Capacity	
15	S1						0	850	
16	S2						0	650	
17	S3						0	630	
18	S4						0	650	
19	S5						0	459	
20	S6						0	290	
21	Demand	0	0	0	0	0			
22	Demand requirement	790	680	600	390	840			
23									
24	Total profit	0							

FIGURE 11.26
Function in Excel.

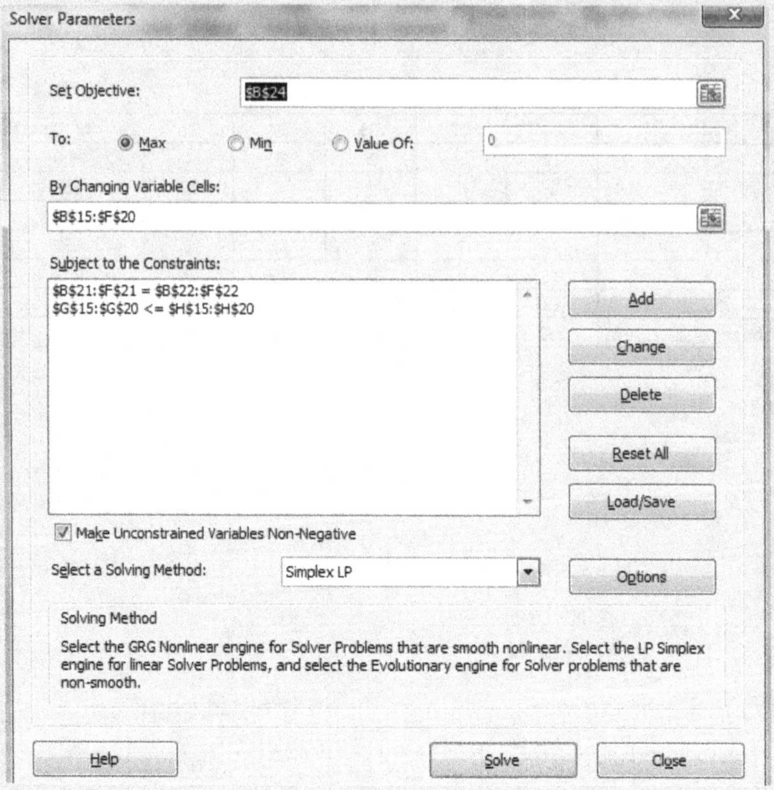

FIGURE 11.27
Solver Parameters in Excel.

For the changing variable cell, select the solution values of the variables – i.e., $B$15:$F$20.

For a subject to the constraints, click on the Add option and add the following constraints:

$B$21:$F$21 = $B$22:$F$22,

$G$25:$G$20 <= $H$15:$H$20.

Click on Make unconstrained variable non-negative.

Select a solution method as Simplex LP.

Now click the Solve button.

5. After selecting the solve button, the solver results appear in a window; the default option has a keep solver solution. Click on the Answer in the Reports section on the right-hand side. Finally, click the OK button to get the results.

6. Finally, the Excel Solver gives the solution values for variables and for the objective function (Figure 11.28).

	A	B	C	D	E	F	G	H	I
3	Profit	D1	D2	D3	D4	D5	Factory Capacity		
4	S1	12	11	5	10	4	850		
5	S2	1	6	6	3	17	650		
6	S3	15	4	6	14	12	630		
7	S4	19	9	9	16	1	650		
8	S5	11	6	14	12	7	459		
9	S6	13	11	16	4	15	290		
10	Warehouse requirement	790	680	600	390	840			
11									
12									
13									
14	Profit	D1	D2	D3	D4	D5	Supply	Supply Capacity	
15	S1	0	680	0	0	0	680	850	
16	S2	0	0	0	0	650	650	650	
17	S3	140	0	0	390	41	571	630	
18	S4	650	0	0	0	0	650	650	
19	S5	0	0	459	0	0	459	459	
20	S6	0	0	141	0	149	290	290	
21	Demand	790	680	600	390	840			
22	Demand requirement	790	680	600	390	840			
23									
24	Total profit	49849							
25									

FIGURE 11.28
Solution in Excel.

EXAMPLE 3:

Make the model of the following assignment problem (Table 11.3) and solve using Excel Solver so that the total cost will be minimum. *(BT level 6 and 3)*

Solution: The given problem is a balanced assignment problem.

1. First, write the problem in Excel, as shown in Figure 11.29.

2. Now insert the following functions and make the model as shown next.

Function insert in cell F12 = SUM(B12:E12)

Function insert in cell F13 = SUM(B13:E13)

Function insert in cell F14 = SUM(B14:E14)

Function insert in cell F15 = SUM(B15:E15)

Function insert in cell B16 = SUM(B12:B15)

Function insert in cell C16 = SUM(C12:B15)

Function insert in cell D16 = SUM(D12:B15)

Function insert in cell E16 = SUM(E12:B15)For the objective function value, enter the formula (in cell B19) for computing total cost = SUMPRODUCT(B4:E7,B12:E15) (Figure 11.30).

TABLE 11.3

Data Example 3

Unit cost	Job 1	Job 2	Job 3	Job 4
Worker 1	9	11	18	10
Worker 2	4	9	6	7
Worker 3	11	13	12	10
Worker 4	7	14	10	8

▲	A	B	C	D	E
1					
2					
3	Assign cost	Job 1	Job 2	Job 3	Job 4
4	Worker 1	9	11	18	10
5	Worker 2	4	9	6	7
6	Worker 3	11	13	12	10
7	Worker 4	7	14	10	8
8					

FIGURE 11.29
Function in Excel.

B19		f_x	=SUMPRODUCT(B4:E7,B12:E15)				
▲	A	B	C	D	E	F	G
1							
2							
3	Assign cost	Job 1	Job 2	Job 3	Job 4		
4	Worker 1	9	11	18	10		
5	Worker 2	4	9	6	7		
6	Worker 3	11	13	12	10		
7	Worker 4	7	14	10	8		
8							
9							
10							
11	Assign cost	Job 1	Job 2	Job 3	Job 4	Assignment	Available
12	Worker 1					0	1
13	Worker 2					0	1
14	Worker 3					0	1
15	Worker 4					0	1
16	Assignment	0	0	0	0		
17	Available	1	1	1	1		
18							
19	Total cost	0					
20							

FIGURE 11.30
Function in Excel.

3. Now in the Data tab click Solver. The solver box appears as follows (Figure 11.31).

   Set the target cell for the objective function value – i.e., $B$19.

   Check the Equal to Min.

   For the changing variable cell, select the solution values of the variables – i.e., $B$12:$E$15.

   For a subject to the constraints, click on the Add option and add the following constraints:

   $B$12:$E$15 = binary,

   $B$16:$E$16 = $B$17:$E$17,

   $F$12:$F$15 = $G$12:$G$15.

   Click on Make unconstrained variable non-negative.

   Select a solution method as Simplex LP.

   Now click the Solve button.

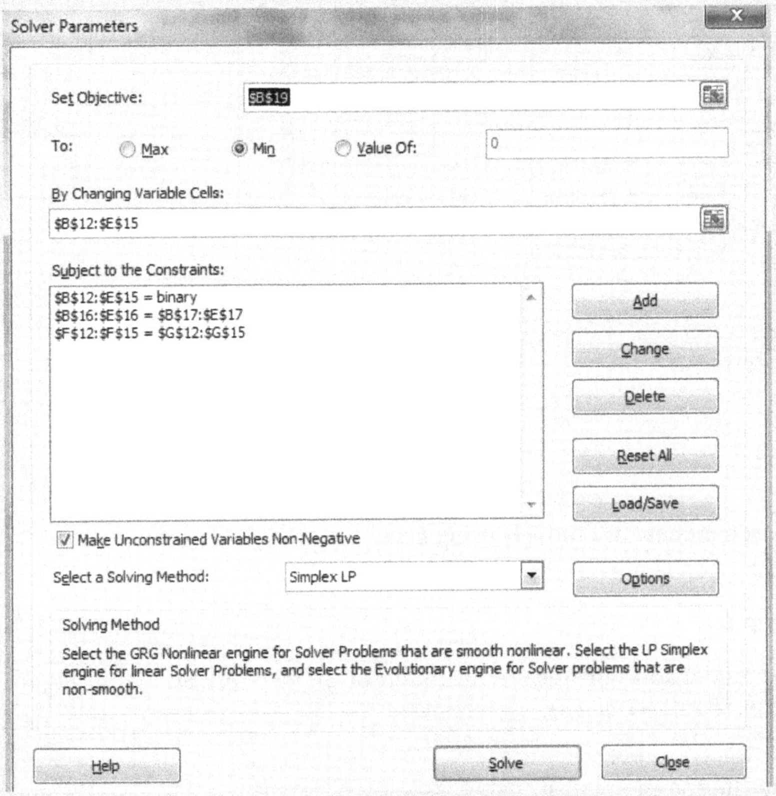

FIGURE 11.31
Solver Parameters in Excel.

4. After selecting the solve button, the solver results appear in a window; the default option has a keep solver solution. Click on the Answer in the Reports section on the right-hand side. Finally, click the OK button to get the results.

5. Finally, the Excel Solver gives the solution values for variables and for the objective function (Figure 11.32).

	A	B	C	D	E	F	G	I
1								
2								
3	Assign cost	Job 1	Job 2	Job 3	Job 4			
4	Worker 1	9	11	18	10			
5	Worker 2	4	9	6	7			
6	Worker 3	11	13	12	10			
7	Worker 4	7	14	10	8			
8								
9								
10								
11	Assign cost	Job 1	Job 2	Job 3	Job 4	Assignment	Available	
12	Worker 1	0	1	0	0	1	1	
13	Worker 2	0	0	1	0	1	1	
14	Worker 3	0	0	0	1	1	1	
15	Worker 4	1	0	0	0	1	1	
16	Assignment	1	1	1	1			
17	Available	1	1	1	1			
18								
19	Total cost	34						
20								

FIGURE 11.32
Solution in Excel.

Examples based on network models using Excel

EXAMPLE 1:

Make the model and find the shortest path of the network shown in Figure 11.33 using Excel. *(BT level 6 and 3)*

Solution:

1. First, write the problem in Excel, as shown next (Figure 11.34). We enter any maximum number, such as 100, into the cells which are not on the FROM TO path in the network.

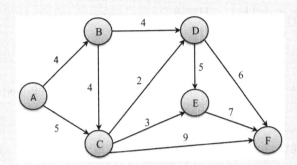

FIGURE 11.33
Network Diagram.

▲	A	B	C	D	E	F	G
1							
2				To			
3	From	A	B	C	D	E	F
4	A	100	4	5	100	100	100
5	B	100	100	4	4	100	100
6	C	100	100	100	2	3	9
7	D	100	100	100	100	5	6
8	E	100	100	100	100	100	7
9	F	100	100	100	100	100	100
10							

FIGURE 11.34
Function in Excel.

2. Now insert the following functions and make the model as shown next.

Function insert in cell H14 = SUM(B14:G14)

Drag above function up to cell H19

Function insert in cell B20 = SUM(B14:B19)

Drag above function up to cell G20

Function insert in cell I14 == H14-B20

Function insert in cell I15 == H15-C20

Function insert in cell I16 == H16-D20

Function insert in cell I17 == H17-E20

Function insert in cell I18 == H18-F20

Function insert in cell I19 == H19-G20

For the total distance value, enter the formula (in cell C23) = SUMPRODUCT(B4:G9,B14:G19) (Figure 11.35).

3. Now in the Data tab click Solver. The solver box appears as follows (Figure 11.36).

Set the target cell for the objective function value – i.e., $C$23.

Check the Equal to Min.

For the changing variable cell, select the solution values of the variables – i.e., $B$14:$G$19.

	C23			f_x	=SUMPRODUCT(B4:G9,B14:G19)						
	A	B	C	D	E	F	G	H	I	J	K
1											
2					To						
3	From	A	B	C	D	E	F				
4	A	100	4	5	100	100	100				
5	B	100	100	4	4	100	100				
6	C	100	100	100	2	3	9				
7	D	100	100	100	100	5	6				
8	E	100	100	100	100	100	7				
9	F	100	100	100	100	100	100				
10											
11											
12					To						
13	From	A	B	C	D	E	F	Total out	Out-In		
14	A							0	0	=	1
15	B							0	0	=	0
16	C							0	0	=	0
17	D							0	0	=	0
18	E							0	0	=	0
19	F							0	0	=	-1
20	Total in	0	0	0	0	0	0				
21											
22											
23	Total Distance		0								

FIGURE 11.35
Function in Excel.

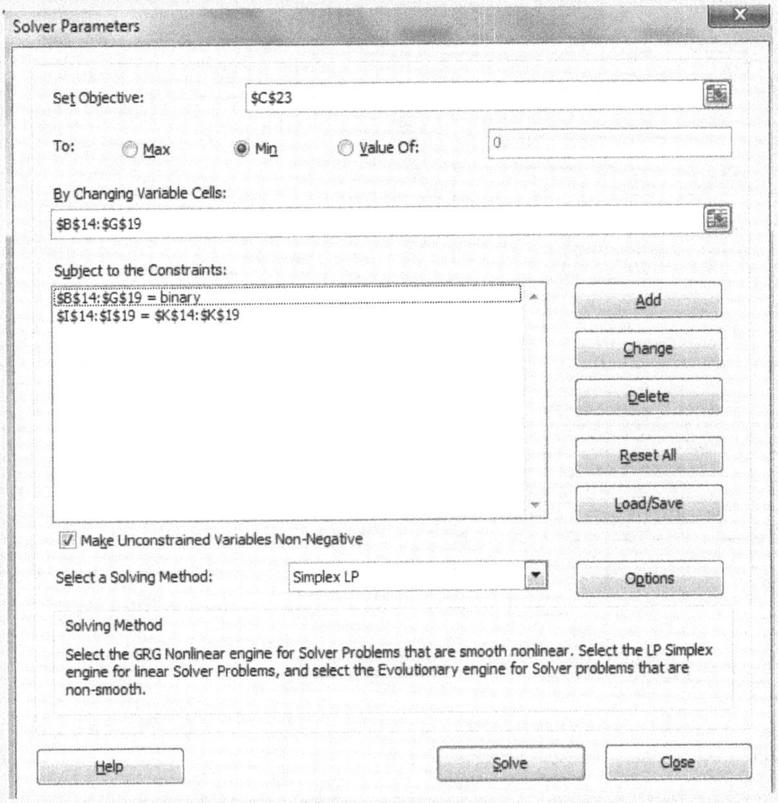

FIGURE 11.36
Solver Parameters in Excel.

For the subject to the constraints, click on the Add option and add the following constraints:

$B$14:$G$19 = binary,

$I$14:$I$19 = $K$14:$K$19.

Click on Make unconstrained variable non-negative.

Select a solution method as Simplex LP.

Now click the Solve button.

4. After selecting the solve button, the solver results appear in a window; the default option has a keep solver solution. Click on the Answer in the Reports section on the right-hand side. Finally, click the OK button to get the results.

5. Finally, the Excel Solver gives the solution values for variables and for the objective function (Figure 11.37).

⊿	A	B	C	D	E	F	G	H	I	J	K
1											
2					To						
3	From	A	B	C	D	E	F				
4	A	100	4	5	100	100	100				
5	B	100	100	4	4	100	100				
6	C	100	100	100	2	3	9				
7	D	100	100	100	100	5	6				
8	E	100	100	100	100	100	7				
9	F	100	100	100	100	100	100				
10											
11											
12					To						
13	From	A	B	C	D	E	F	Total out	Out-In		
14	A	0	0	1	0	0	0	1	1	=	1
15	B	0	0	0	0	0	0	0	0	=	0
16	C	0	0	0	1	0	0	1	0	=	0
17	D	0	0	0	0	0	1	1	0	=	0
18	E	0	0	0	0	0	0	0	0	=	0
19	F	0	0	0	0	0	0	0	-1	=	-1
20	Total in	0	0	1	1	0	1				
21											
22											
23	Total Distance		13			Path	A-C-D-F				
24											

FIGURE 11.37
Solution in Excel.

EXAMPLE 2:

Make the model and find the maximum flow of the network shown in Figure 11.38 using Excel. *(BT level 6 and 3)*

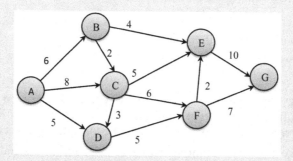

FIGURE 11.38
Network Diagram.

Solution:

1. First, write the problem in Excel and insert the following functions and make the model as shown next.

   Function insert in cell I4 = SUMIF(B4:B15,H4,D4:D15)

   Function insert in cell I5 = SUMIF(B4:B15,H5,D4:D15)-SUMIF(C4:C15,H5,D4:D15)

   Function insert in cell I6 = SUMIF(B4:B15,H6,D4:D15)-SUMIF(C4:C15,H6,D4:D15)

   Function insert in cell I7 = SUMIF(B4:B15,H7,D4:D15)-SUMIF(C4:C15,H7,D4:D15)

   Function insert in cell I8 = SUMIF(B4:B15,H8,D4:D15)-SUMIF(C4:C15,H8,D4:D15)

   Function insert in cell I9 = SUMIF(B4:B15,H9,D4:D15)-SUMIF(C4:C15,H9,D4:D15)

   The SUMIF functions calculate the Net Flow of each node.

   Function insert in cell D18 = I4 (Figure 11.39).

2. Now in the Data tab click Solver. The solver box appears as follows (Figure 11.40).

   Set the target cell for the objective function value – i.e., $D$18.

   Check the Equal to Max.

   For the changing variable cell, select the solution values of the variables – i.e., $D$4:$D$15.

I4			f_x	=SUMIF(B4:B15,H4,D4:D15)						
	A	B	C	D	E	F	G	H	I	J	K
1											
2											
3		From	To	Flow			Capacity		Node	Net flow		Requirement
4		A	B	0	<=	6		A	0		
5		A	C	0	<=	8		B	0	=	0
6		A	D	0	<=	5		C	0	=	0
7		B	C	0	<=	2		D	0	=	0
8		B	E	0	<=	4		E	0	=	0
9		C	D	0	<=	3		F	0	=	0
10		C	E	0	<=	5					
11		C	F	0	<=	6					
12		D	F	0	<=	5					
13		E	G	0	<=	10					
14		F	E	0	<=	2					
15		F	G	0	<=	7					
16											
17											
18		Maximum flow		0							

FIGURE 11.39
Network Diagram

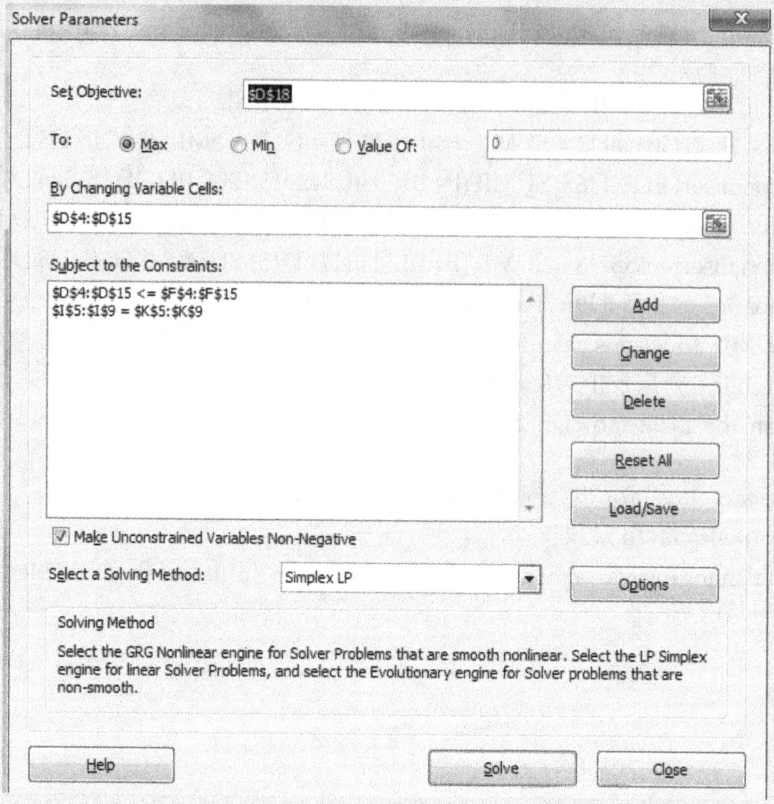

FIGURE 11.40
Solver Parameters in Excel.

For the subject to the constraints, click on the Add option and add the following constraints:

$D$4:$D$15 <= $F$4:$F$15,

$I$5:$I$9 = $K$5:$K$9.

Click on Make unconstrained variable non-negative.

Select a solution method as Simplex LP.

Now click the Solve button.

3. After selecting the solve button, the solver results appear in a window; the default option has a keep solver solution. Click on the Answer in the Reports section on the right-hand side. Click the OK button to get the results.

4. Finally, the Excel Solver gives the solution values for variables and for the objective function (Figure 11.41).

	A	B	C	D	E	F	G	H	I	J	K
1											
2											
3		From	To	Flow		Capacity		Node	Net flow		Requirement
4		A	B	4	<=	6		A	17		
5		A	C	8	<=	8		B	0	=	0
6		A	D	5	<=	5		C	0	=	0
7		B	C	0	<=	2		D	0	=	0
8		B	E	4	<=	4		E	0	=	0
9		C	D	0	<=	3		F	0	=	0
10		C	E	5	<=	5					
11		C	F	3	<=	6					
12		D	F	5	<=	5					
13		E	G	10	<=	10					
14		F	E	1	<=	2					
15		F	G	7	<=	7					
16											
17											
18		Maximum flow		17							
19											

FIGURE 11.41
Solution in Excel.

EXAMPLE 3:

A salesman wants to visit five cities, starting with city 1 where he is stationed. The distances between various cities are given in Table 11.4. Make the model and determine a path through the four other cities and return to his home city in such a way that he has to travel the minimum distance using Excel Solver. *(BT level 6 and 3)*

TABLE 11.4

Data Example 3

From City	To City				
	1	2	3	4	5
1	–	9	7	8	6
2	9	–	9	4	5
3	7	9	–	7	8
4	8	4	7	–	5
5	6	5	8	5	–

Solution:

1. First, write the problem in Excel and insert the following functions and make the model as shown next.

   Write sequence as 1-2-3-4-5-1 in cells B10 to G10

       Function insert in cell B11 = INDEX($B$3:$F$7,B10,C10)

       Function insert in cell C11 = INDEX($B$3:$F$7,C10,D10)

       Function insert in cell D11 = INDEX($B$3:$F$7,D10,E10)

       Function insert in cell E11 = INDEX($B$3:$F$7,E10,F10)

       Function insert in cell F11 = INDEX($B$3:$F$7,F10,G10)

   The INDEX functions calculate the distance between two cities in sequence.

       Function insert in cell D13 = SUM(B11:F11) (Figure 11.42).

2. Now in the Data tab click Solver. The solver box appears as follows (Figure 11.43).

   Set the target cell for the objective function value – i.e., $D$13.

   Check the Equal to Min.

   For the changing variable cell, select the solution values of the variables – i.e., $B$10:$F$10.

   For the subject to the constraints, click on the Add option and add the following constraints:

       $B$10:$F$10 = AllDifferent.

D13			f_x	=SUM(B11:F11)		
	A	B	C	D	E	F	G
1							
2		City 1	City 2	City 3	City 4	City 5	
3	City 1	100	9	7	8	6	
4	City 2	9	100	9	4	5	
5	City 3	7	9	100	7	8	
6	City 4	8	4	7	100	5	
7	City 5	6	5	8	5	100	
8							
9							
10	Sequence	1	2	3	4	5	1
11	Distance	9	9	7	5	6	
12							
13			Total Distance	36			
14							

FIGURE 11.42
Function in Excel.

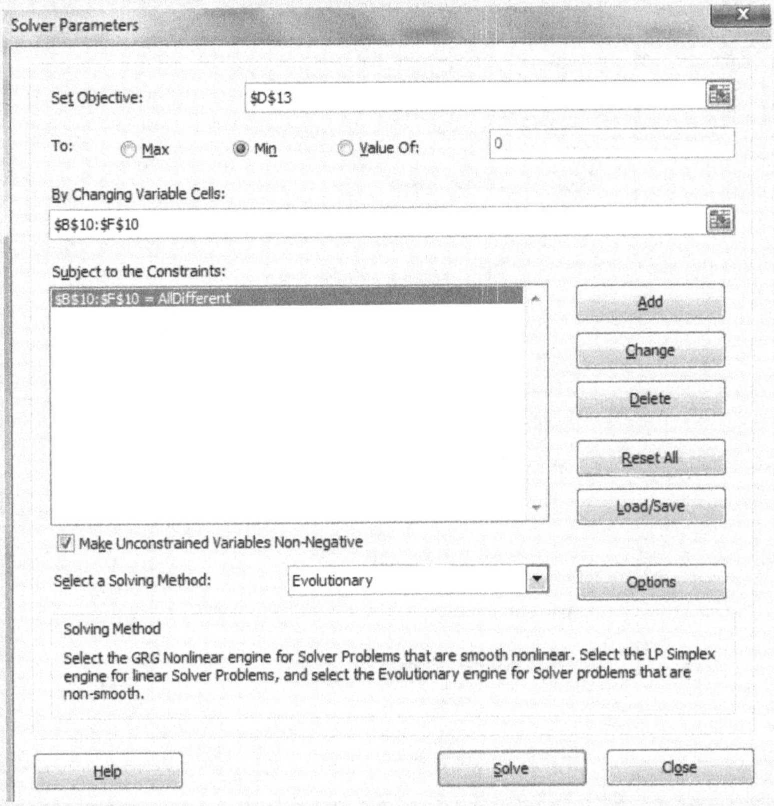

FIGURE 11.43
Solver Parameters in Excel.

Click on Make unconstrained variable non-negative.

Select a solution method as Evolutionary.

Select options and in All Methods mode write Max Time (Seconds): 20 and click OK (Figure 11.44).

Select Stop on Show Trial Solution.

Now click the Solve button.

3. After selecting the solve button, the solver results appear in a window; the default option has a keep solver solution. Click on the Answer in the Reports section on the right-hand side. Finally, click the OK button to get the results.

4. Finally, the Excel Solver gives the solution values for variables and for the objective function (Figure 11.45).

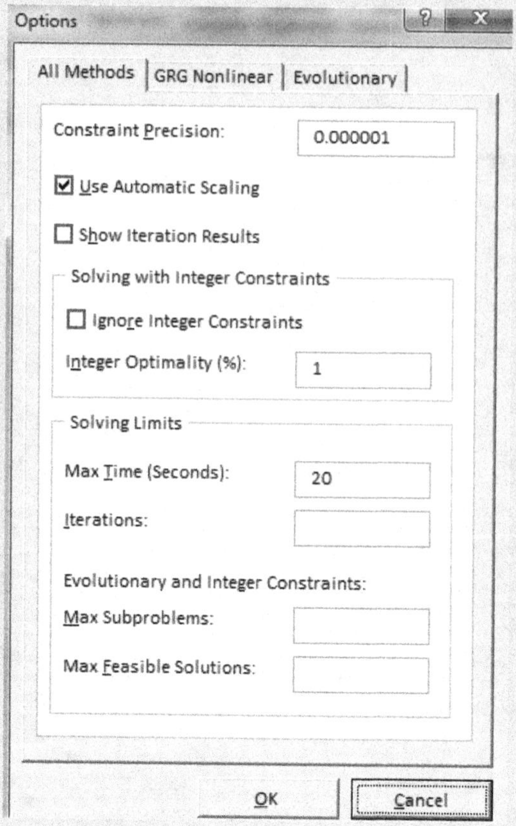

FIGURE 11.44
Solver Options in Excel.

	A	B	C	D	E	F	G
1							
2		City 1	City 2	City 3	City 4	City 5	
3	City 1	100	9	7	8	6	
4	City 2	9	100	9	4	5	
5	City 3	7	9	100	7	8	
6	City 4	8	4	7	100	5	
7	City 5	6	5	8	5	100	
8							
9							
10	Sequence	3	4	2	5	1	3
11	Distance	7	4	5	6	7	
12							
13			Total Distance	29			
14							

FIGURE 11.45
Solution in Excel.

Example based on sequencing using Excel

EXAMPLE 1:

Six jobs need to be processed on two work centres W_1 and W_2, in the order W_1 W_2. The processing times (in hours) for six jobs are given in Table 11.5.

Make the model and determine the optimal job sequence to minimize the total elapsed time for this job group. Also, determine the total elapsed time using Excel. (*BT level 6 and 3*)

Solution:

1. First, write the problem in Excel and insert the following functions and make the model as shown next (Figure 11.46).

   Write sequence as 1-2-3-4-5-1 in cells B11 to G11

TABLE 11.5

Data Example 1

Job	W_1	W_2
A	7	7
B	6	5
C	10	11
D	4	9
E	8	10
F	14	17

N9			f_x	=MAX(N7,M10)														
	A	B	C	D	E	F	G	H	I	J	K	L	M	N	O	P	Q	R
1																		
2	Job	W_1	W_2															
3	A	7	7									W_1	49	49	49	49	49	49
4	B	6	5									W_2	59	59	59	59	59	59
5	C	10	11															
6	D	4	9									Start on W1	0	49	98	147	196	245
7	E	8	10									Finish on W1	49	98	147	196	245	294
8	F	14	17															
9												Start on W2	49	108	167	226	285	344
10					Sequence							Finish on W2	108	167	226	285	344	403
11	Job	1	2	3	4	5	6	Total		Required								
12	A	1	1	1	1	1	1	6	=	1								
13	B	1	1	1	1	1	1	6	=	1								
14	C	1	1	1	1	1	1	6	=	1								
15	D	1	1	1	1	1	1	6	=	1								
16	E	1	1	1	1	1	1	6	=	1								
17	F	1	1	1	1	1	1	6	=	1								
18	Total	6	6	6	6	6	6											
19		=	=	=	=	=	=											
20	Required	1	1	1	1	1	1											
21																		

FIGURE 11.46
Function in Excel.

Function insert in cell H12 = SUM(B12:G12) and copy this function up to cell H17

Function insert in cell B18 = SUM(B12:B17) and copy this function up to cell G18

Function insert in cell M3 = SUMPRODUCT($B$3:$B$8, B$12:B$17) and copy this function up to cell R3

Function insert in cell M4 = SUMPRODUCT($C$3:$C$8,B$12:B$17) and copy this function up to cell R4

Enter 0 in M6 cell for job 1.

Function insert in cell N6 = M7 and copy this function up to cell R6

Function insert in cell M7 = M6 + M3 and copy this function up to cell R7

Function insert in cell M9 = M7

Function insert in cell M10 = M9 + M4 and copy this function up to cell R10

Function insert in cell N9 = MAX(N7, M10) and copy this function up to cell R9

2. Now in the Data tab click Solver. The solver box appears as follows (Figure 11.47).

FIGURE 11.47
Solver Parameters in Excel.

Set the target cell for the objective function value – i.e., $R$10.

Check the Equal to Min.

For the changing variable cell, select the solution values of the variables – i.e., $B$12:$G$17.

For the subject to the constraints, click on the Add option and add the following constraints:

$B$12:$G$17 = binary,

$B$18:$G$18 = $B$20:$G$20,

$H$12:$H$17 = $J$12:$J$17.

Click on Make unconstrained variable non-negative.

Select a solution method, such as GRG Nonlinear.

Now click the Solve button.

3. After selecting the solve button, the solver results appear in a window; the default option has a keep solver solution. Click on the Answer in the Reports section on the right-hand side. Finally, click the OK button to get the results.

4. Finally, the Excel Solver gives the solution values for variables and for the objective function (Figure 11.48).

Job	W_1	W_2
A	7	7
B	6	5
C	10	11
D	4	9
E	8	10
F	14	17

	4	8	7	10	14	6
W_1	4	8	7	10	14	6
W_2	9	10	7	11	17	5

Start on W1	0	4	12	19	29	43
Finish on W1	4	12	19	29	43	49

Start on W2	4	13	23	30	43	60
Finish on W2	13	23	30	41	60	65

Sequence

Job	1	2	3	4	5	6	Total		Required
A	0	0	1	0	0	0	1	=	1
B	0	0	0	0	0	1	1	=	1
C	0	0	0	1	0	0	1	=	1
D	1	0	0	0	0	0	1	=	1
E	0	1	0	0	0	0	1	=	1
F	0	0	0	0	1	0	1	=	1
Total	1	1	1	1	1	1			
	=	=	=	=	=	=			
Required	1	1	1	1	1	1			

Sequence	D	E	A	C	F	B

FIGURE 11.48
Solution in Excel.

Example based on integer programming using Excel

EXAMPLE 1:

Make the model and solve the following integer programming problem using Excel. (BT level 6 and 3)

$$\max Z = 10x_1 + 16x_2$$

Subject to,

$$2x_1 + 4x_2 \leq 16$$
$$8x_1 + 2x_2 \leq 20$$

and

$$x_1, x_2 \geq 0 \text{ and integer.}$$

Solution: First, write the function in Excel, as shown in Figure 11.49.

1. For the objective function value, enter the formula (in cell B11) for computing Z == SUMPRODUCT(B9:C9,B10:C10). This formula uses the coefficient values and the solution values for variables X1 and X2, which are supposed to be solved.

2. Similarly, enter the formula for LHS of the Constraints C1 & C2.
   Enter = SUMPRODUCT(E9:F9,$E$6:$F$6) into cell B14 and = SUMPRODUCT(E10:F10,$E$6:$F$6) into cell B15 (Figure 11.50).

3. Now in the Data tab click Solver. The solver box appears as follows (Figure 11.51).
   Set the target cell for the objective function Z value – i.e., $B$11.
   Check the Equal to Max

◢	A	B	C	D	E
1					
2		X1	X2		
3	Z	10	16		
4	C1	2	4	<=	16
5	C2	8	2	<=	20
6					

FIGURE 11.49
Function in Excel.

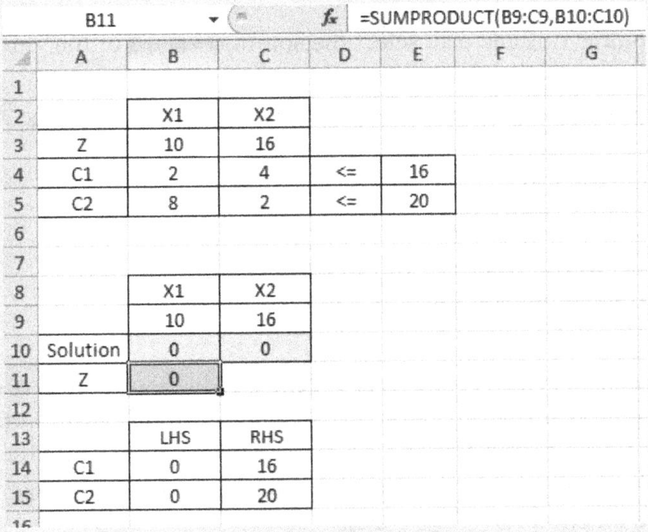

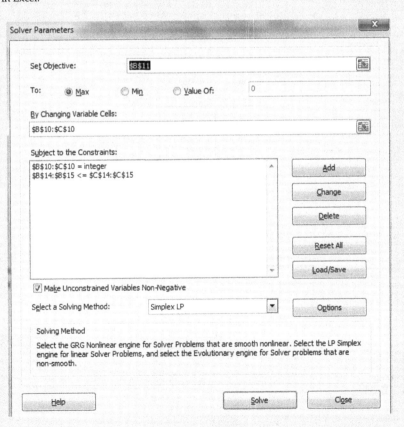

FIGURE 11.50
Function in Excel.

FIGURE 11.51
Solver Parameters in Excel.

For the changing variable cell, select the solution values of the variables – i.e., $B$10:$C$10.

For the subject to the constraints, LHS <= RHS – i.e., click on the Add option and select

$B$10:$C$10 = integer,

$B$14:$B$15 <= $C$14:$C$15.

Click on Make unconstrained variable non-negative.

Select a solution method as Simplex LP.

Now click the Solve button.

4. After selecting the solve button, the solver results appear in a window; the default option has a keep solver solution. Click on the Answer in the Reports section on the right-hand side. Finally, click the OK button to get the results.

5. Finally, the Excel Solver gives the solution values for variables X and Y and for the objective function Z (Figure 11.52).

	A	B	C	D	E
1					
2		X1	X2		
3	Z	10	16		
4	C1	2	4	<=	16
5	C2	8	2	<=	20
6					
7					
8		X1	X2		
9		10	16		
10	Solution	0	4		
11	Z	64			
12					
13		LHS	RHS		
14	C1	16	16		
15	C2	8	20		

FIGURE 11.52
Solution in Excel.

Example based on decision making using Excel

EXAMPLE 1:

Make the following model (Table 11.6) and determine the best decision for each of the following decision criteria using Excel. *(BT level 6 and 3)*

1. Maximax (optimistic) 2. Maximin (pessimistic)

3. Laplace (equally likely) 4. Minimax regret

Solution:

1. First, write the problem in Excel as shown next (Figure 11.53).

2. For maximax, add the following functions, and Excel gives the solution as shown next.

   Function insert in cell E12 = MAX(C12:D12) and copy this function up to cell E14

   Function insert in cell E15 = MAX(E12:E14). (Figure 11.54)

3. For Maximin, add the following functions, and Excel gives the solution as shown next.

TABLE 11.6

Data Example 1

Alternative	State of Nature	
	Good Market Competitive Conditions (Rs.)	Poor Market Competitive Conditions (Rs.)
Big plant	2,000,000	−1,000,000
Small plant	1,000,000	−500,000
Not developing the new product line	0	0

	A	B	C	D
1				
2			State of Nature	
3		Alternative	Good market competitive conditions (Rs.)	Poor market competitive conditions (Rs.)
4		Big Plant	20, 00,000	- 10, 00,000
5		Small Plant	10, 00,000	- 5, 00,000
6		Not developing the new product line	0	0
7				

FIGURE 11.53
Function in Excel.

E15	▾	*fx*	=MAX(E12:E14)		
	A	B	C	D	E
9	Maximax				
10			State of Nature		
11		Alternative	Good market competitive conditions (Rs.)	Poor market competitive conditions (Rs.)	Maximum
12		Big Plant	2000000	-1000000	2000000
13		Small Plant	1000000	-500000	1000000
14		Not developing the new product line	0	0	0
15				Maximax	2000000

FIGURE 11.54
Function for Maximax in Excel.

E22	▾	*fx*	=MAX(E19:E21)		
	A	B	C	D	E
16	Maximin				
17			State of Nature		
18		Alternative	Good market competitive conditions (Rs.)	Poor market competitive conditions (Rs.)	Minimum
19		Big Plant	2000000	-1000000	-1000000
20		Small Plant	1000000	-500000	-500000
21		Not developing the new product line	0	0	0
22				Maximin	0

FIGURE 11.55
Function for Maximin in Excel.

Function insert in cell E19 = MIN(C19:D19) and copy this function up to cell E21

Function insert in cell E22 = MAX(E19:E21). (Figure 11.55)

4. For Laplace, add the following functions, and Excel gives the solution as shown next.

Function insert in cell E27 = AVERAGE(C27:D27) and copy this function up to cell E29

Function insert in cell E30 = MAX(E27:E29). (Figure 11.56)

5. For minimax regret, add the following functions, and Excel gives the solution as shown next.

Function insert in cell C42 = MAX(C$35:C$37)-C35 and copy this function up to cell C44

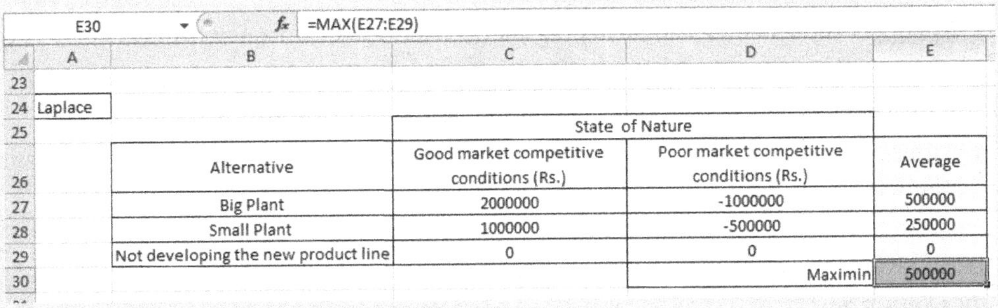

E30	▾	*fx*	=MAX(E27:E29)	

	A	B	C	D	E
23					
24	Laplace				
25			State of Nature		
26		Alternative	Good market competitive conditions (Rs.)	Poor market competitive conditions (Rs.)	Average
27		Big Plant	2000000	-1000000	500000
28		Small Plant	1000000	-500000	250000
29		Not developing the new product line	0	0	0
30				Maximin	500000

FIGURE 11.56
Function for Laplace in Excel.

Function insert in cell D42 = MAX(D\$35:D\$37)-D35 and copy this function up to cell D44

Previous functions determine regrets

Function insert in cell E42 = MAX(C42:D42) and copy this function up to cell E44

Function insert in cell E45 = MIN(E42:E44). (Figure 11.57)

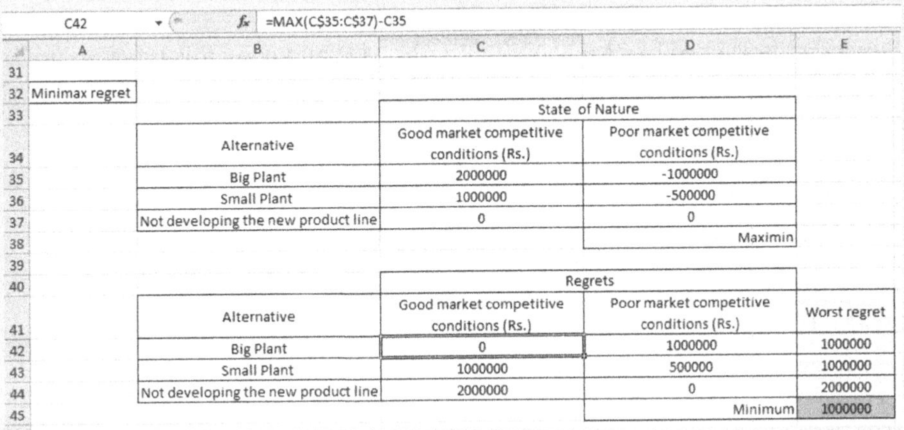

C42	▾	*fx*	=MAX(C\$35:C\$37)-C35	

	A	B	C	D	E
31					
32	Minimax regret				
33			State of Nature		
34		Alternative	Good market competitive conditions (Rs.)	Poor market competitive conditions (Rs.)	
35		Big Plant	2000000	-1000000	
36		Small Plant	1000000	-500000	
37		Not developing the new product line	0	0	
38				Maximin	
39					
40			Regrets		
41		Alternative	Good market competitive conditions (Rs.)	Poor market competitive conditions (Rs.)	Worst regret
42		Big Plant	0	1000000	1000000
43		Small Plant	1000000	500000	1000000
44		Not developing the new product line	2000000	0	2000000
45				Minimum	1000000

FIGURE 11.57
Function for Minimax Regret in Excel.

The decisions indicated by the decision criteria can be summarized as follows:

1. Maximax (optimistic): Construct big plant to manufacture a new product
2. Maximin (pessimistic): Not developing the new product line
3. Laplace (equally likely): Construct big plant to manufacture a new product
4. Minimax regret: Construct big plant to manufacture new product/Construct small plant

EXAMPLE 2:

Make the following model (Table 11.7) and calculate the EMV, EOL, and EVPI for each alternative with a probability of 0.50 for each state of nature using Excel. *(BT level 6 and 3)*
Solution:

1. First, write the problem in Excel, as shown next (Figure 11.58).
2. For EMV, add the following functions, and Excel gives the solution as shown next.

   Function insert in cell E12 = SUMPRODUCT(C12:D12,$C$8:$D$8)and copy this function up to cell E14

   Function insert in cell E15 = MAX(E12:E14). (Figure 11.59)
3. For EOL, add the following functions, and Excel gives the solution as shown next.

   Function insert in cell C21 = MAX(C$4:C$6) – C4 and copy this function up to cell C23

   Function insert in cell D21 = MAX(D$4:D$6) – D4 and copy this function up to cell D23

   Previous functions determine regrets

   Function insert in cell E21 = SUMPRODUCT(C21:D21,$C$8:$D$8) and copy this function up to cell E23

   Function insert in cell E24 = MIN(E21:E23). (Figure 11.60)

TABLE 11.7

Data Example 2

Alternative	State of Nature	
	Good Market Competitive Conditions (Rs.)	**Poor Market Competitive Conditions (Rs.)**
Big plant	2,000,000	–1,000,000
Small plant	1,000,000	–500,000
Not developing the new product line	0	0

⯅	A	B	C	D
1				
2			State of Nature	
3		Alternative	Good market competitive conditions (Rs.)	Poor market competitive conditions (Rs.)
4		Big Plant	2000000	-1000000
5		Small Plant	1000000	-500000
6		Not developing the new product line	0	0
7				
8		Probability	0.5	0.5

FIGURE 11.58
Function in Excel.

⯅	A	B	C	D	E
7					
8		Probability	0.5	0.5	
9					
10			State of Nature		
11		Alternative	Good market competitive conditions (Rs.)	Poor market competitive conditions (Rs.)	EMV
12		Big Plant	2000000	-1000000	500000
13		Small Plant	1000000	-500000	250000
14		Not developing the new product line	0	0	0
15				Maximum	500000

FIGURE 11.59
Function for EMV in Excel.

E21			*fx*	=SUMPRODUCT(C21:D21,$C$8:$D$8)	

⯅	A	B	C	D	E
7					
8		Probability	0.5	0.5	
9					
18					
19			Regrets		
20		Alternative	Good market competitive conditions (Rs.)	Poor market competitive conditions (Rs.)	EOL
21		Big Plant	0	1000000	500000
22		Small Plant	1000000	500000	750000
23		Not developing the new product line	2000000	0	1000000
24				Minimum	500000

FIGURE 11.60
Function for EOL in Excel.

4. For EVPI, add the following functions, and Excel gives the solution as shown next (Figure 11.61).

Function insert in cell C17 = MAX(C12:C14)

Function insert in cell D17 = MAX(D12:D14)

Function insert in cell I12 = SUMPRODUCT(C8:D8,C17:D17)

Function insert in cell I13 = E15

Function insert in cell I14 = I12-I13

		Probability	0.5	0.5					
			State of Nature						
		Alternative	Good market competitive conditions (Rs.)	Poor market competitive conditions (Rs.)	EMV				
		Big Plant	2000000	-1000000	500000				EVwPI	1000000
		Small Plant	1000000	-500000	250000			EVwoPI (Max EMV)	500000
		Not developing the new product line	0	0	0				EVPI	500000
				Maximum	500000				
		Best outcome	2000000	0					

FIGURE 11.61
Function for EVPI in Excel.

EXAMPLE 3:

Using AHP in Excel, developed the model of the following pairwise comparison matrixes (Table 11.8) and determine an overall ranking of the three cars. *(BT level 6 and 3)*

Car	Price		
	A	B	C
A	1	4	7
B	1/4	1	3
C	1/7	1/3	1

Car	Safety		
	A	B	C
A	1	1/4	1/8
B	4	1	1/5
C	8	5	1

Car	Durability		
	A	B	C
A	1	4	1
B	1/4	1	1/3
C	1	3	1

TABLE 11.8
Data Example 3

Criteria	Price	Safety	Durability
Price	1	4	6
Safety	1/4	1	3
Durability	1/6	1/3	1

Solution:

1. First, write the problem in Excel for the criteria matrix and insert the following functions and make the model as shown next.
   Function insert in cell B7 = SUM(B3:B5) and copy this function up to cell D7
   Function insert in cell G3 = B3/B$7 and copy this function up to cell I5
   Function insert in cell G7 = SUM(G3:G5) and copy this function up to cell J7
   Function insert in cell L3 = MMULT(B3:D3,J$3:J$5)/J3 and copy this function up to cell L5
   Function insert in cell L7 = SUM(L3:L5)
   Function insert in cell O3 = L7/O2
   Function insert in cell O4 = (O3 – O2)/(O2 – 1)
   Value of RI is taken from the table given in Q and R column
   Function insert in cell O7 = O4/O5 (Figure 11.62).
2. In a similar manner, the price, safety, and durability criteria models are obtained and shown next (Figure 11.63).

	A	B	C	D	E	F	G	H	I	J	K	L	M	N	O	P	Q	R
1																		
2	Criteria	Price	Safety	Durability		Criteria	Price	Safety	Durability	Row average		Consistency		No of comparision	3		n	RI
3	Price	1	4	6		Price	0.705882	0.75	0.6	0.685294118		3.10944206		Average consistency	3.054144		2	0
4	Safety	0.25	1	3		Safety	0.176471	0.1875	0.3	0.221323529		3.03986711		CI	0.027072		3	0.58
5	Durability	0.166667	0.333333	1		Durability	0.117647	0.0625	0.1	0.093382353		3.01312336		RI	0.58		4	0.9
6																	5	1.12
7	Total	1.416667	5.333333	10		Total	1	1	1	1		9.16243253		Consistency	0.046676		6	1.24
8																	7	1.32
9																	8	1.41
10																	9	1.45
11																	10	1.51
12																		

FIGURE 11.62
Function in Excel.

O15 fx =O12/O13

	A	B	C	D
9				
10	Price	A	B	C
11	A	1	4	7
12	B	0.25	1	3
13	C	0.142857	0.333333	1
14				
15	Total	1.392857	5.333333	11
16				
17				
18	Safety	A	B	C
19	A	1	0.25	0.125
20	B	4	1	0.2
21	C	8	5	1
22				
23	Total	13	6.25	1.325
24				
25				
26	Durability	A	B	C
27	A	1	4	1
28	B	0.25	1	0.333333
29	C	1	3	1
30				
31	Total	2.25	8	2.333333
32				

	F	G	H	I	J
10	Price	A	B	C	Row average
11	A	0.717949	0.75	0.636364	0.701437451
12	B	0.179487	0.1875	0.272727	0.213238151
13	C	0.102564	0.0625	0.090909	0.085324398
15	Total	1	1	1	1
18	Safety	A	B	C	Row average
19	A	0.076923	0.04	0.09434	0.0704209
20	B	0.307692	0.16	0.150943	0.206211901
21	C	0.615385	0.8	0.754717	0.723367199
23	Total	1	1	1	1
26	Durability	A	B	C	Row average
27	A	0.444444	0.5	0.428571	0.457671958
28	B	0.111111	0.125	0.142857	0.126322751
29	C	0.444444	0.375	0.428571	0.416005291
31	Total	1	1	1	1

	L	N	O
10	Consistency	No of comparision	3
11	3.06750208	Average consistency	3.032576
12	3.02277386	CI	0.016288
13	3.00745318	RI	0.58
15	9.09772911	Consistency	0.028083
18	Consistency	No of comparision	3
19	3.01607585	Average consistency	3.095939
20	3.06756757	CI	0.047969
21	3.20417335	RI	0.58
23	9.28781677	Consistency	0.082706
26	Consistency	No of comparision	3
27	3.01300578	Average consistency	3.009208
28	3.0034904	CI	0.004604
29	3.01112878	RI	0.58
31	9.02762496	Consistency	0.007938

FIGURE 11.63
Function for Price, Safety and Durability in Excel.

3. Now take row average highlighted values obtained in the previous two steps and make the following model.

   The overall ranking of the three cars is obtained by inserting the following functions.

   Function insert in cell C40 = SUMPRODUCT($B$36:$B$38,C36:C38) and copy this function up to cell E40 (Figure 11.64).

C40 fx =SUMPRODUCT($B$36:$B$38,C36:C38)

	A	B	C	D	E
34					
35	Criteria	Weight	A	B	C
36	Price	0.685294	0.701437	0.213238	0.0853244
37	Safety	0.221324	0.070421	0.206212	0.7233672
38	Durability	0.093382	0.457672	0.126323	0.41600529
39					
40	Car model score		0.539015	0.203567	0.25741804
41		Rank	1	3	2
42					

FIGURE 11.64
Solution for Car Ranking in Excel.

Example based on dynamic programming using Excel

EXAMPLE 1:

Three types of fruit cartons-mango cartons, grapes cartons, and strawberry cartons are to be loaded into a tri-wheeler. The weight of each mango carton is 100 Kg and the owner of the truck will be paid Rs.20 per mango carton for transport. The weight of each carton of grapes is 200 kg and the owner of the cart will be paid Rs.50 for each carton of grapes. The weight of each strawberry carton is 400 kg and the owner of the truck is paid Rs. 80 for each pineapple carton. The truck can carry a maximum weight of 500 kg. Make the above model in Excel and determine how many cartons of mangoes, grapes, and strawberries should the owner of the tri-wheeler be loaded to make the maximum profit? *(BT level 6 and 3)*

Solution:

1. First, write the problem in Excel and make the model, as shown in Figure 11.65.

2. Now in the Data tab click Solver. The solver box appears as follows (Figure 11.66).

Set the target cell for the objective function value – i.e., $C$19.

Check the Equal to Max.

For the changing variable cell, select the solution values of the variables – i.e., $E$16:$G$16.

For the subject to the constraints, click on the Add option and add the following constraints:

$C$21 = 500,

$E$16:$G$16 = integer.

Click on Make unconstrained variable non-negative.

Select a solution method as Simplex LP.

	A	B	C	D	E	F	G
1							
2		Fruit	Weight of carton (Kg)	Charge/carton (Rs.)			
3		Mangos	100	20			
4		Grapes	200	50			
5		Strawberry	400	80			
6							
7							
8							
9					Weight of mangos carton	Weight of grapes carton	Weight of strawberry carton
10					100	200	400
11							
12					Charge/carton for mangos	Charge/carton for grapes	Charge/carton for strawberry
13					20	50	80
14							
15					x1 (Mangos carton)	x2 (Grapes carton)	x3 (Strawberry carton)
16					1	2	0
17							

FIGURE 11.65
Function in Excel.

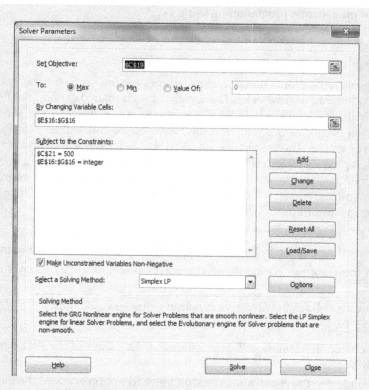

FIGURE 11.66
Solver Parameters in Excel.

3. Now click the Solve button.

4. After selecting the solve button, the solver results appear in a window; the default option has a keep solver solution. Click on the Answer in the Reports section on the right-hand side. Click the OK button to get the results.

5. Finally, the Excel Solver gives the solution values for variables and for the objective function. (Figure 11.67)

Function insert in cell C19 = SUMPRODUCT(E13:G13,E16:G16)

Function insert in cell C21 = SUMPRODUCT(E10:G10,E16:G16)

	C19		f_x	=SUMPRODUCT(E13:G13,E16:G16)
	A	B	C	D
18				
19		Max profit	120	
20				
21		st	500	<= 500
22				

FIGURE 11.67
Solution in Excel.

Multiple Choice Questions

1. Which one of the following is not correct?
   Linear programming problems must have an
   a) Objective we aim at maximizing or minimizing
   b) Variable decision which we need to decide
   c) Constraints to be specified
   d) Unrestricted decision variable

2. A feasible LPP solution
   a) should fulfil all constraints at the same time
   b) does not fulfil all the constraints, just some of them
   c) must be a feasible region corner
   d) all of the above

3. Simplex method of linear programming uses
   a) all the feasible region points
   b) intermediate points in the unfeasible region
   c) just the corners of the feasible region
   d) only interiors in the feasible region

4. The variable used to balance the two sides of equations is graded as less than or equal to constraint equations
   a) Solving variable
   b) slack variable
   c) condition variable
   d) positive variable

5. Right-hand constant in ith constraint in primal must be equal to the objective coefficient for
   a) ith dual variable
   b) jth primal variable
   c) ith primal variable
   d) jth dual variable

6. Which of the following is the correct statement?
   All constraints in the standard form of a linear programming problem are
   a) less than or equal to
   b) greater or equal to.
   c) equations.
   d) some constraints are less than equal to and some greater than equal to.

7. A variable of no physical meaning but used to achieve an initial basic feasible solution to the LPP is called
   a) basic variable
   b) non-basic variable
   c) artificial variable
   d) basis

8. If in an LPP the solution of a variable can be made infinitely large without violating the constraints, the solution is known as
   a) unbounded
   b) infeasible
   c) alternative
   d) none of these

9. Every LPP associated with another LPP is called
   a) primal
   b) non-linear programming
   c) dual
   d) none of the above

10. A linear programming problem with mixed constraints can be solved using
    a) Big M method
    b) simplex method
    c) Vogel's approximation method
    d) none of these

11. If a transportation matrix contains m sources and n destinations, the total number of basic variables in a basic feasible solution is
    a) m
    b) m+n+1
    c) m+n-1
    d) m+n

12. In a transportation problem, the least cost method would provide
    a) an optimum solution
    b) an initial feasible solution
    c) a Vogel's approximate solution
    d) none of these

13. The transportation problem is balanced if
    a) number of sources are equal to the number of destinations
    b) total supply equals total demands irrespective of the number of sources and destination
    c) total supply equals total demands and the number of sources are equal to the number of destinations
    d) none of these

14. An assignment problem of $n \times n$ has the following number of total decision variables
    a) n
    b) 2n
    c) n^2
    d) 2n − 1

15. Select the correct sentence for an assignment problem.
    a) An assignment problem can be solved only if the number of rows and columns are equal.
    b) The transportation method can be used to solve an assignment problem.
    c) Unbalanced assignment problem cannot be solved.
    d) None of these

16. The goal of network analysis is to
    a) minimize overall project duration
    b) minimize delays, interruptions, and conflicts in production
    c) maximize project overall time
    d) none of these

17. The analysis of PERT is based on
    a) optimistic time
    b) pessimistic time
    c) most likely time
    d) all the above

18. The shortest time possible for an activity under ideal circumstances is known as
    a) optimistic time
    b) pessimistic time
    c) most likely time
    d) expected time

19. For network analysis with dummy activity, which one of the following is not correct?
    a) Dummy activity is always zero in duration in the project network.
    b) It is desirable to have as many dummy activities as possible in a project network.
    c) A dummy activity from a given node extends all the activities that end at that node.
    d) A network should only employ the requisite dummy activities.

20. A critical activity in PERT analysis has
    a) zero float
    b) maximum float
    c) minimum cost
    d) maximum cost

21. The activity cost of the crash time is called
    a) cost slope
    b) normal time
    c) crash cost
    d) critical time

22. Select the incorrect statement.
    a) CPM is event oriented and PERT is activity oriented.
    b) CPM is a one-time estimate, while PERT is a three time estimate.
    c) If CPM floats are measured while in PERT, slack is calculated.
    d) CPM and PERT are used for project situations.

23. The earliest finish time of the activity should be found as
    a) EST + Activity duration
    b) EST – Activity duration
    c) LFT + Activity duration
    d) LFT – Activity duration

24. The difference between total and free float is
    a) free
    b) interference
    c) total
    d) independent

25. The critical path in the CPM network indicates the
    a) path to using the maximum resources
    b) path to using the least resources
    c) the path where one activity's delay extends the project completion duration
    d) Path that is automatically monitored

26. Dijkstra's algorithm solves
    a) all pairs of shortest paths type problem
    b) network flow type problem
    c) one-source shortest path type problem
    d) none of these

27. Choose the correct statement for a sequencing problem.
    a) The selection of a suitable order for a series of jobs on finite facilities
    b) All jobs shall be completed on the basis of a first come first served
    c) A service facility may process more than one job at a time
    d) None of the above

28. The general assumption that is not correct in solving a sequencing problem is that
    a) the time spent in switching from one machine to another by different jobs is negligible
    b) a job on a machine will be done uninterrupted to the point of completion
    c) at a given time, a machine can process more than one job
    d) the processing time on different machines is independent of the order of the various jobs on them

29. Johnson's rule is applicable to the
    a) sequencing problem
    b) assignment problem
    c) transportation problem
    d) none of these

30. The group replacement strategy is ideal for specific low-cost items that are likely to
    a) fail suddenly
    b) fail fully and suddenly
    c) fail over a period
    d) be gradual and retrogressive

31. The replacement problem is not concerned about the
    a) items that deteriorate graphically
    b) items that fail suddenly
    c) determination of optimum replacement interval
    d) maintenance of an item to work out profitability

32. Replace an item is done when
    a) the average up-to-date cost is equal to the current maintenance cost.
    b) the average up-to-date cost is higher than the current maintenance cost.
    c) the average up-to-date cost is lower than the current maintenance cost.
    d) next year's running cost is more than the average of the nth year's costs

33. Which one of the following is not correct?
    a) All items are replaced under group replacement policy irrespective of whether the item has failed or not.
    b) An item is replaced immediately after its failure under the individual replacement policy
    c) The optimal group replacement interval shall be determined at the point at which the sum of the group replacement by the unit of time and the costs of the individual replacement are the minimum.
    d) Individual and group replacement are carried out in group replacement.

34. When the sum of one player's gains is equal to the sum of losses for another player in a game, this is known as the
    a) two-person game
    b) zero-sum game
    c) non-zero-sum game
    d) none of these

35. Each player should follow the same strategy, regardless of the strategy of the other player, in which of the following games?
    a)   Pure strategy
    b)   Mixed strategy
    c)   Both (a) and (b) above
    d)   None of these

36. A payoff matrix's saddle point is always the
    a)   matrix's largest number
    b)   the column with the smallest number and the row with the smallest number
    c)   matrix's smallest number
    d)   the largest number in the column and the smallest number in the row

37. In the context of a mixed strategy, each player should optimize the
    a)   lower game value
    b)   min. losses
    c)   max loss
    d)   expected gain

38. Select the wrong statement in the context of game theory.
    a)   The game value may be positive, negative or zero.
    b)   The game value can only be established if the game has a saddle point.
    c)   The saddle point in the payoff matrix is the equilibrium point.
    d)   A game is strictly determinable and fair if the point value of the saddle is zero.

39. A two-person game is said to be zero-sum, if
    a)   one player's gain does not match another player's loss
    b)   one player's gain matched by a loss to the other exactly
    c)   the two players must have the same number of strategies
    d)   payoff matrix diagonal entries are zero

40. A common assumption concerning the players in a game is that
    a)   none of the players knows the matrix of payoff.
    b)   players have different payoff matrix information.
    c)   only one player has a rational strategy.
    d)   players' specific identity is irrelevant to game play.

41. Which is not an assumption in popular queuing mathematical models?
    a)   Arrivals come from an infinite population or very large population.
    b)   Arrivals are first-in and first-out treatment.
    c)   The average rate of arrival is faster than the average rate of service.
    d)   Customers don't balk or renege.

42. The case of the customer entering a line but leaving before serving is called
    a)   balking
    b)   reneging
    c)   infinite length
    d)   none of these

43. The calling population is expected to be infinite if
     a)   the system's capacity is infinite.
     b)   the rate of service is faster than the rate of arrival
     c)   arrivals are mutually exclusive
     d)   all customers arrive immediately

44. For a single server infinite population with Poisson exponential queuing model, which of the following is not correct?
     a)   The system has a single service facility.
     b)   The arrivals take place in Poisson fashion.
     c)   The service rate is based on the exponential distribution.
     d)   The population of the source is small and finite in size.

45. For most simple queuing models, the population size is assumed
     a)   finite
     b)   infinite
     c)   constant
     d)   none of these

46. In context with the integer programming problem, select the correct statement.
     a)   A pure integer programming problem is one where all decision variables are either zero or unity.
     b)   A mixed integer programming problem requires mixed constraints.
     c)   Integer programming problem is a linear programming problem with an integer-restricted decision variable.
     d)   A linear programming problem with only one integer variable is not an integer programming problem.

47. In the cutting plane algorithm, each cut involves the introduction of the
     a)   constraints on equality
     b)   less than or equal to constraint
     c)   greater than or equal to constraints
     d)   an artificial variable

48. Which of the following isn't a goal programming problem requirement?
     a)   Prioritizing goals
     b)   One objective function
     c)   Linear constraints
     d)   Linear objective function

49. Dynamic programming addresses
     a)   problems of multi-stage decision making
     b)   problems of single-stage decision making
     c)   problems of time-independent decision making
     d)   problems that set the various decision variables to optimize the objective function

50. When taking a risk-based decision, which of the following is a valid decision-making criterion?
     a) Maximax
     b) Maximin
     c) Minimax regret
     d) Minimize expected opportunity loss

Answers to Multiple Choice Questions

1.   d)
2.   a)
3.   c)
4.   b)
5.   a)
6.   c)
7.   c)
8.   a)
9.   c)
10. a)
11. c)
12. b)
14. c)
15. a)
16. a)
17. d)
18. a)
19. b)
20. a)
21. c)
22. a)
23. a)
24. b)
25. c)
26. c)
27. a)
28. c)
29. a)
30. c)
31. d)
32. a)
33. d)
34. c)
35. a)

36. d)
37. d)
38. b)
39. b)
40. d)
41. c)
42. b)
43. c)
44. d)
45. b)
46. c)
47. b)
48. b)
49. a)
50. d)

Bibliography

Bunday, B. D. 1996. *An Introduction to Queueing Theory*. Wiley, Hoboken, NJ.

Chatterjee, K., and W. F. Samuelson. 2001. *Game Theory and Business Applications*. Springer, New York.

Dantzig, G. B., and M. N. Thapa. 1997. *Linear Programming 1: Introduction*. Springer, New York.

Gregory, G. 1988. *Decision Analysis*. Plenum, New York.

Hall, R. W. 1997. *Queueing Methods: For Services and Manufacturing*. Prentice Hall, Upper Saddle River, NJ.

Hillier, F. S., and M. S. Hillier. 2008. *Introduction to Management Science: A Modeling and Case Studies Approach with Spreadsheets*. McGraw-Hill, New York.

Horner, P. 2002. History in the Making. *OR/MS Today* 29(5): 30–39.

Kanti, S. 1980. *Operations Research*. Sultan Chand & Sons, New Delhi, India.

Michael C., C. Camille, and R. Ghaith. 2018. *Operation Research A Practical Introduction*. CRC Press, Boca Raton, FL.

Natarajan, A. M., P. Balasubramanie, and A. Tamilarasi. 2014. *Operations Research*. Pearson, London.

Richard, A. G., and H. David. 2019. *Game Theory: A Modelling Approach*. CRC Press, Boca Raton, FL.

Srinath, L. S. 1973. *PERT and CPM: Principles and Applications*. East-West Press Private Limited, New Delhi, India.

Steven, O. K., and C. L. Hoong. 2016. *Business Analytics for Decision Making*. CRC Press, Boca Raton, FL.

Taha, H A. 2017. *Operations Research*. Pearson, London.

Vanderbei, R. J. 2008. *Linear Programming: Foundations and Extensions*. Springer, New York.

Webb, J. N. 2007. *Game Theory: Decisions, Interaction and Evolution*. Springer, New York.

Wiest, J., and F. Levy. 1977. *Management Guide to PERT-CPM*. Prentice Hall, Upper Saddle River, NJ.

Winston, W. L., and S. C. Albright. 2001. *Practical Management Science*. Duxbury Press, Pacific Grove, CA.

Index

Page numbers in *italic* indicate figures. Page numbers in **bold** indicate tables.

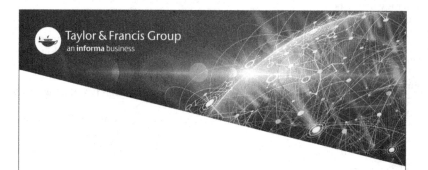

Printed in the United States
By Bookmasters